Essential Physics in the World Around Us

Essential Physics in the World Around Us

JERRY B. MARION
University of Maryland

JOHN WILEY & SONS
New York/Santa Barbara/London/Sydney/Toronto

Cover design: Eileen Thaxton
Cover photo: Monument Valley Arizona,
Pete Turner/The Image Bank

Copyright © 1977, by John Wiley & Sons, Inc.

All rights reserved. Published simultaneously in Canada.

No part of this book may be reproduced by any means, nor transmitted, nor translated into a machine language without the written permission of the publisher.

Library of Congress Cataloging in Publication Data

Marion, Jerry B
    Essential physics in the world around us.

    1. Physics.   I. Title.
QC23.M358   530   76-46545
ISBN 0-471-56905-4

Printed in the United States of America

10 9 8 7 6 5 4 3 2 1

Preface

The familiar, everyday world around us abounds in examples of physical ideas and principles at work. Even the most casual observers can easily make a long list of questions about phenomena and devices they see regularly. What goes on inside a refrigerator to produce cooling? Why does an electric bulb emit white light whereas a neon sign produces red light? How is a rocket launched on a mission to the Moon? Questions like these raise points about the fundamental behavior of physical systems. Moreover, answers to these questions can be given without complicated mathematics. This is the spirit in which physics is approached in this book. My purpose is to provide a nonmathematical survey of some of the basic concepts of physics by closely relating these ideas to everyday things.

This book is not a compendium of facts. Instead, I have elected to concentrate on a relatively small number of important physical ideas. Each chapter poses a question about a familiar phenomenon or device. The answers are developed by introducing and explaining in simple terms the background physics while keeping before the reader the notion that we are building step by step the answer to a specific question. Thus, each chapter has a definite theme. Although a good deal of physics can be introduced in this way, there is always a central focus for the discussion.

The approach used here necessarily bypasses many interesting and important physical phenomena. But there is no attempt here to be comprehensive or exhaustive. Rather, I believe that it is more important to provide nonscience students, toward whom this book is directed, with an in-depth look at a few of the ways that physics appears in the world around them rather than to deluge them with an endless series of facts. The emphasis here is on the forest, not on the trees.

This book has been designed as a textbook for a one-semester course that has no mathematical prerequisites except for the usual one year of required high-school mathematics. (Alternatively, the book is easily adapted for one- or two-quarter courses, as well.) Because the emphasis is conceptual instead of mathematical, the use of mathematics has been maintained at a minimum level. The *Questions and Exercises* sections at the end of each chapter, for example, are almost entirely nonnumerical. For those few mathematical exercises that are included, the answers are always given. Students who wish to pursue a topic further will find at the end of each chapter a section entitled *Additional Details for Further Study*. These sections do not include material that is required for the following chapters and are therefore entirely optional.

Jerry B. Marion

Contents

1. *How Far? How Long? How Much?*
   Our System of Measurement — 1
2. *How Do Rockets Work?*
   Momentum and Force — 21
3. *Planets, Moons, and Spacecraft—Why Do They Stay in Orbit?*
   Gravitation and Planetary Motion — 51
4. *What Is the Energy Crisis?*
   Energy, Power, and Natural Resources — 79
5. *How Do Electric Motors Work?*
   Electricity and Magnetism — 111
6. *How Do Refrigerators Work?*
   Heat and Thermodynamics — 149
7. *What Is a Sound Wave?*
   Waves, Sound, and Noise — 181
8. *How Do Cameras Work?*
   Light and Optical Instruments — 217
9. *Time Travel—Is It Really Possible?*
   Relativity Theory — 259
10. *How Do Lasers Work?*
    Atoms and Atomic Radiations — 295
11. *How Is Matter Put Together?*
    Molecules and the Properties of Matter — 333
12. *What Will Radiation Do FOR You and TO You?*
    Radiation and Its Effects — 367
13. *How Do Nuclear Reactors Work?*
    Nuclei and Nuclear Power — 403

Index — 441

Essential Physics in the World Around Us

How far? How long? How much?

1 OUR SYSTEM OF MEASUREMENT

We see in the world around us an astounding variety of natural processes and man-made things at work. Natural events do not happen by sheer accident, nor are our technical gadgets designed by guesswork. We like to say that all of these processes and things are governed by the laws of nature. By this we mean that through many years—indeed, many centuries—of observation, experiment, and reasoning, we have discovered a number of rules that seem always to be followed when particular classes of events take place or when certain kinds of operations are carried out. We search for these rules or laws not only because they permit us to understand what *has* happened around us, but also because they allow us to predict what *will* happen in particular circumstances. From the study of gravitation, for example, we have learned why an apple falls downward from a tree and why the Earth follows its characteristic path around the Sun. The development of these ideas also permits us to predict with precision the occurrence of solar eclipses and to plan the launch of a communications satellite so that it has the greatest usefulness to us. We seek to discover the laws of nature not only to satisfy our curiosity about the way the world works, but also to provide for the greater benefit of mankind.

We live in a technological society. Essentially every household contains a refrigerator, a television set, an automobile, a transistorized radio, and many other electrical and mechanical gadgets. The electricity we use is produced in huge and complex generating plants; much of our commerce depends on electronic computers; and our national security is ensured (we hope) by highly sophisticated radar systems, laser devices, and nuclear weapons. Each of these things is an outgrowth of some fundamental discovery about the way nature behaves. The design of a refrigerator depends on a knowledge of the behavior of gases under changing conditions of pressure and temperature. The study of the way electrons move in solid materials resulted in the invention of the transistor. And nuclear power plants could be constructed only after thorough studies had been made of the nuclear fission process.

In this book we will discuss some of the natural laws that govern the world in which we live. We will be concerned particularly with basic physical concepts and the way in which they relate to everyday things around us. By connecting every physical idea with some process or device that we have seen or used or heard about, we will gain a better appreciation of how natural phenomena influence our lives.

The emphasis in this book is on *ideas*, not *mathematics*. However, we cannot completely divorce physics from numbers, even in this elementary survey. Otherwise, how could we ever describe in a quantitative way a physical object or physical event? We must, from time to time, be able to discuss the *size* of a thing, or how *much* it weighs, or how *long* a process takes. Actually, we are accustomed to relating numbers to physical ideas in many situations. You drive 8.4 miles from home to class at an average speed of 28 miles per hour in an automobile that has a 90-horsepower engine. You buy potatoes in 5-pound bags, milk in half-gallon cartons, and aspirin in 100-milligram tablets. These are all familiar units of measure. When we state that a speed is 28 miles per hour, the number "28" is the *amount* or *value* of the speed and "miles per hour" is the *unit*. All physical quantities—quantities such as length, force, and energy—have units, and these units must be specified before we can report a meaningful result. It makes no sense to state that a certain distance is "60" unless we also state whether the unit is feet or meters or miles.

Fortunately, in science we use a restricted set of units. The unit (or *dimension*) of every physical quantity is made up of combinations consisting, at most, of the units for length, mass, and time. A simple example of the combination of units is in the quantity *speed* (which we will discuss later in this chapter). We are accustomed to measuring speed in units of miles per hour or feet per second. That is, the units for speed consist of a *length* unit divided by a *time* unit.

The fundamental units of measure in science are those of *length, mass,* and *time*. These are familiar concepts, but because they are so basic to the description of physical events and phenomena, we will briefly describe the units of measure for each of these quantities. In so doing, we will list several numbers. Do not be dismayed at this; these numbers are collected together here primarily as a matter of convenience. You will see more numbers in the next section than in any chapter that follows!

Length, Mass, and Time

Most Americans are accustomed to thinking about distance and length in terms of inches, feet, yards, and miles. And we are familiar with ounces, pounds, and tons for the measurement of mass. These are all units in the *British system* of measurement. Today, the scientific community universally uses the *metric system* of measurement. Indeed, even for everyday matters, most of the world (with the primary exception of the United States) uses metric measure. In this country we are gradually making a change away from the outdated and cumbersome British system, but several decades will probably be required before we will be using the metric system exclusively.

The basic unit of length in the metric system is the *meter* (m). Various multiples of the meter are also commonly used. For example,

$$1000 \text{ m} = 1 \text{ kilometer (km)}$$
$$\tfrac{1}{100} = 1 \text{ centimeter (cm)}$$
$$\tfrac{1}{1000} \text{ m} = 1 \text{ millimeter (mm)}$$

The correspondence between the metric length units and those in the British system is as follows (see also Fig. 1-1):

1 inch (in.) = 2.54 cm = 25.4 mm
1 yard (yd) = 3 feet (ft) = 0.9144 m
1 mile (mi) = 5280 ft = 1.609 km

Approximately, we have

1 in. = 2.5 cm   1 cm = 0.4 in.
1 yd = 0.9 m   1 m = 1.1 yd
1 mi = 1.6 km   1 km = 0.6 mi

While adjusting to thinking in metric terms, it will be helpful to use these approximate figures. For example, if you see a road sign that states the distance to a certain city is 100 km (such signs are beginning to appear here), then you know that the distance in miles is approximately 0.6 (or $\tfrac{3}{5}$) of this figure, or 60 miles. (For better accuracy, use a factor of 0.62 or $\tfrac{5}{8}$ to convert from kilometers to miles.)

The metric unit of mass is the *kilogram* (kg), which is about twice as large as the *pound* (lb). In fact,

1 lb = 0.454 kg

or, approximately,

1 kg = 2.2 lb

as illustrated in Fig. 1-2. Smaller and larger metric units of mass are the *gram* (g), the *milligram* (mg), and the *metric ton* (or *tonne*):

Figure 1-1 *Correspondences between British and metric length units.*

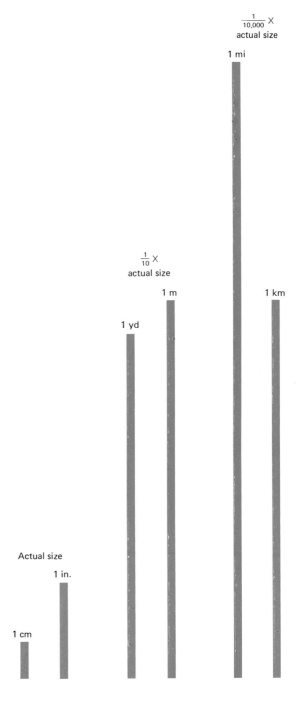

Our System of Measurement

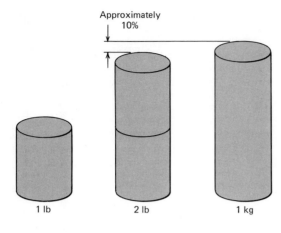

Figure 1-2 *A kilogram is approximately 10 percent greater than 2 pounds: 1 kg = 2.2 lb.*

$\frac{1}{1000}$ kg = 1 g
$\frac{1}{1\,000\,000}$ kg = $\frac{1}{1000}$ g = 1 mg
1000 kg = 1 tonne

If you have examined the labels on various products lately, you will have noticed that metric units of mass are being used on more and more items. Grocery products are frequently labeled with the mass in pounds and in kilograms. And pharmaceutical products are now almost always measured in milligrams.

Time is measured in the same units in both the British and the metric systems. The basic unit of time is the *second* (s). Several other time units are also used in everyday and scientific matters:

1 minute (min) = 60 s
1 hour (h) = 3600 s
1 day = 86 400 s
1 year (y) = 31 600 000 s

The numbers collected in this section have been included for your convenience in adjusting your thinking from British to metric units. How many of these numbers must you remember? Actually, only a few. First of all, the various metric units for length and mass differ only by simple multiples of 10, and these multiples are indicated by *prefixes;* for example,

centi- (c) means $\frac{1}{100}$: 1 centimeter (cm) = $\frac{1}{100}$ m
milli- (m) means $\frac{1}{1000}$: 1 millimeter (mm) = $\frac{1}{1000}$ m
kilo- (k) means 1000: 1 kilogram (kg) = 1000 g

Two other prefixes often used are *micro-* (μ, 1 *millionth*) and *mega-* (M, 1 *million*).

The basic number that allows the conversion between British and metric units of length is

1 in. = 2.54 cm

All other length relationships can be obtained from this number together with the familiar expressions, 1 ft = 12 in., 1 mi = 5280 ft, and so forth. In this way, you can obtain, for example, the connection between miles and kilometers. However, it is probably easier to remember this relationship than to compute it each time it is needed:

1 mi = 1.6 km

Finally, the connection between the units for mass is

1 kg = 2.2 lb

These are the three key relationships that will permit you to shift between the British and metric systems of units. (Notice that these expressions are all stated in such a way that each numerical factor is approximately 2.)

Mass and Weight

We have indicated that kilograms and pounds are the units of measure for *mass*. You are probably accustomed to giving your *weight* as a certain number of pounds. Are mass and weight therefore the same? No, they are not. *Mass* and *weight* actually refer to two different physical ideas. *Mass* is a measure of the amount of matter in an object. We could specify the mass of an iron bar in terms of the number of iron atoms in the bar. This would not be practical, however, because the number of atoms in a bar is enormous and there is no way by which they can be counted accurately. But the number of atoms in the bar does not change if we move the bar from place to place or change its size or shape. The number of atoms in an object—that is, the *mass* of the object—is a property of that object and does not depend on what we might do to the object. We say that *mass is an intrinsic property of an object*.

The *weight* of an object is a measure of how strongly gravity pulls on the object. The gravitational attraction of the Earth for an object is nearly the same at any position on or near the Earth's surface. Because most of us (astronauts are exceptions) spend our lives close to the surface of the Earth, we never experience any substantial change in the pull of gravity. The weight of an object therefore seems to be the same whatever its location. Away from the surface of the Earth, however, the pull of gravity and, hence, the weight of an object *does* change. When an astronaut is on the surface of the Moon, he experiences a much smaller effect due to the Moon's gravity than he does due to the Earth's gravity on the Earth's surface. Thus, an astronaut on the Moon has a smaller weight than he does on Earth (about one-sixth as great). The weight of an object is a variable quantity, whereas mass is not.

In scientific terms we say that mass is *mass* but that weight is *force*. Indeed, mass and weight, being different physical concepts, have different physical units. Mass and weight are definitely *not* the same thing. (For more details, see the discussion on page 76.)

Speed

Many of the quantities that are useful in science (and often in everyday matters as well) are measured in units that are combinations of the basic units, meters, kilograms, and seconds. The quantity of this type that is easiest to understand is probably *speed*. We are all familiar with the measurement of speed in units of miles per hour (otherwise written as mi/h or mph). The speedometer in an automobile is an instrument that automatically gives the vehicle's speed in these units (or in kilometers per hour, km/h, in European cars not intended for export to the United States). The standard automotive speed unit is not the only possible unit for specifying speed. We could, in fact, use *any* length unit divided by *any* time unit. Some possible speed units are feet per minute (ft/min), centimeters per hour (cm/h), or furlongs per fortnight. In the metric system we usually measure speed in meters per second (m/s), kilometers per second (km/s), or kilometers per hour (km/h).

How do we determine the speed in a particular situation? If we make a 40-km trip in 1 h, we say that our speed has been 40 km/h. Or, if the same trip required 4 h, the speed was 10 km/h. That is, we specify speed by dividing the distance traveled by the time required for the trip:

$$\text{speed} = \frac{\text{distance}}{\text{time}}$$

and the units for speed are automatically those of length divided by time.

To make the computation of speed according to this procedure, we need only the total distance traveled and the total time required. But we know that in any automobile trip from one place to another we rarely make the entire journey with the speedometer reading a steady value. Most trips involve starts, stops, periods of slow driving, and periods of fast driving. What do we mean by the *speed* for such a trip? If we divide the *total* distance traveled by the *total* time required, we obtain a value that we call the *average speed* for the trip. The average speed applies to the trip *as a whole*. In a particular case we may have a certain average speed for the entire trip and yet no appreciable part of the distance will have been covered at that speed. For example, suppose that you undertake a trip of 150 km. You travel the first 30 km in 1 h at a steady speed of 30 km/h. Then, you are able to increase your speed and you travel the remaining 120 km in 2 h at a steady speed of 60 km/h. Now, if you consider the trip as a whole, the total distance traveled was 150 km and the total elapsed time was 3 h. Therefore, the average speed was 50 km/h, a value different from that for either segment of the trip (see Fig. 1-3). The *average speed* is that speed, which if maintained constant

Figure 1-3 *The average speed for an entire trip is not necessarily the speed during any segment of the trip. Here, there are two segments with speeds of 30 km/h and 60 km/h; the average speed for the trip as a whole is 50 km/h.*

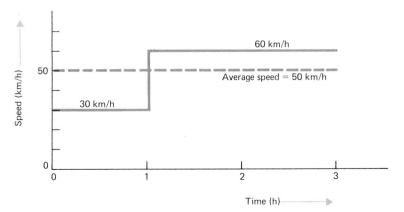

during the trip, would allow completion of the trip in the same total time as for the actual trip.

To state the average speed for a trip gives useful information, but it must be remembered that the average speed applies only for the time or distance interval that is specified and not necessarily to smaller intervals. This is always the way with average values: Very few members of the group for which an average is calculated (or even none at all) will have the average value. For example, the class average for an examination may be 76.8, but probably no member of the class will have received this score.

The average speed for a long trip tells us very little about how the trip was actually made. How do we obtain more information? One way would be to divide the trip into a number of segments and to give the average speed for each segment. The larger the number of segments, the greater the amount of information that is given about the trip. We can imagine that the intervals during which the average speeds are measured are made extremely small. We could have intervals of 1 s, or $\frac{1}{10}$ s, or $\frac{1}{1000}$ s, or even smaller. In fact, we could make the intervals so small that we can specify the speed essentially at one *instant* of time. We call such a value the *instantaneous speed*. The value of the speed that we read from a speedometer is just the instantaneous speed of the automobile at that instant. Usually, when we refer to *the* speed of an object, we mean the instantaneous speed.

Acceleration

When you "step on the gas" in your automobile, you cause the speed to increase. Or, when you step on the brake, you cause the speed to decrease. In both cases, the essential feature of the motion is that the speed *changes*. Whenever the speed of an object changes, we say that the object has undergone *acceleration*. Indeed, the gas pedal in an automobile is appropriately called the *accelerator*. (The brake pedal could be called a "decelerator" or "negative accelerator.") Acceleration is familiar in other situations as well. If you drop an object from a height, it is easy to see that the object gains speed as it falls. The greater the distance of fall, the greater will be the final speed. An automobile is accelerated by the push of the engine; a falling object is accelerated by the downward pull of the Earth's gravity.

How do we measure acceleration? Just as we measure speed as the change in *position* per unit of time, we measure acceleration as the change in *speed* per unit of time:

$$\text{acceleration} = \frac{\text{change in speed}}{\text{time}}$$

Also in analogy with the case of speed, we can define *average* acceleration and *instantaneous* acceleration.

What are the units of acceleration? These are less familiar than the units for speed, but they are easy to find from the definition of acceleration. The definition of speed is distance divided by time and the units for speed are a length unit divided by a time unit. Similarly, the definition of acceleration is speed change divided by time, so the units for acceleration must be a speed unit divided by a time unit. If the speed of an object increases from 2 m/s to 10 m/s, the speed change is 8 m/s. If this change took place during a time interval of 2 s, the acceleration would be 8 m/s per 2 s, or 4 meters per second per second. That is, during each second of acceleration the speed increased by 4 m/s. We usually write this kind of result using a shorthand notation:

4 meters per second per second = 4 m/s/s   or   4 m/s^2

Do not be confused by this kind of notation. We would read 4 m/s^2 as "4 meters per second squared." There is, of course, no physical significance to a unit of 1 s^2 (1 square second) as there is for a unit of 1 m^2 (1 square meter). But in the combination m/s^2, we mean "meters per second per second."

If we were to drop an object from a tall building and measure the object's instantaneous speed at the end of each second of fall, we would find a steady increase in speed. At the end of 1 s the speed would be 10 m/s; at the end of 2 s the speed would be 20 m/s; at the end of 3 s the speed would be 30 m/s; and so forth. During each second of fall the speed increases by 10 m/s. That is, the acceleration is 10 m/s per second, or 10 m/s^2. (If we performed such an experiment we would actually find a value of 9.8 m/s^2.)

In discussing speed and acceleration we have found that the units used to measure a physical quantity can always be obtained directly from the definition of the quantity in terms of

12   *Our System of Measurement*

other physical quantities. It is unnecessary to define new units for every physical quantity that we use (although we frequently do so for convenience), because the units for any quantity can always be expressed in terms of the units for length, mass, and time in some combination.

Vectors

Suppose that you have an object initially at a certain position *A*. You then move the object by 6 m to a new position *B*. If you state that the object has been moved 6 m from position *A* you have not provided sufficient information to locate the object in its new position. In addition to the *distance* through which the object has been moved you must also give the *direction* of movement in order to locate the object precisely in its new position. (See also Fig. 1-4.)

Any physical quantity that requires *direction* as well as *size* (or *magnitude*) for a complete specification is called a *vector*. As we have just seen, *net movement* or *displacement* is a

Figure 1-4 *If you start at El Paso, Texas, and take a trip of 1900 km, you could end in Portland, Oregon, or Tallahassee, Florida. Only if you travel 1900 km* northeast *will you end in Chicago. To specify completely a movement, you must give* direction *as well as* distance. *That is,* net movement *(or* displacement*) is a* vector *quantity.*

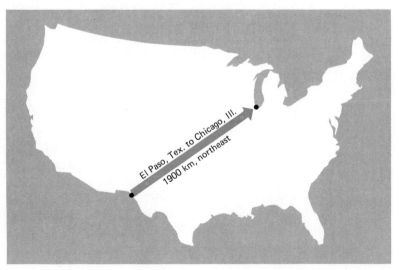

vector quantity. Not all physical quantities are vectors; quantities such as time, mass, and temperature do not require direction for their specification.

If you state that you are moving with a speed of 50 km/h, you have not completely described your motion. In the case of motion, also, direction is important. We usually describe motion in terms of a *velocity* vector—for example, 50 km/h *southeast* or 35 m/s *upward*. Notice that we have made a distinction here between *speed* and *velocity*. We use the term *velocity* when we wish to imply the importance of direction; that is, velocity is a vector quantity. We use the term *speed* when direction is not important. The speed of an object is the magnitude of the object's velocity vector.

Vectors provide us with a convenient way of describing many situations. Suppose that you have a boat that is capable of a speed of 10 mi/h. Suppose that you run this boat in a river in which the current flows at 4 mi/h. How fast will your boat move relative to the river bank? The answer, of course, depends on the direction of motion of the boat with respect to the direction of the river current. If the boat moves *downstream*, the velocity of the boat *adds* to the velocity of the current and the boat moves at 14 mi/h with respect to the land. We can express this statement in a pictorial way by means of vectors, as in Fig. 1-5a. (Vectors are usually denoted by boldface characters.) We add the vector $\mathbf{v}_{boat}$ to the vector $\mathbf{v}_{river}$ to obtain the resultant vector $\mathbf{v}_{net}$ which has a magnitude of 14 mi/h and a direction the same as that of $\mathbf{v}_{boat}$ and $\mathbf{v}_{river}$. Notice

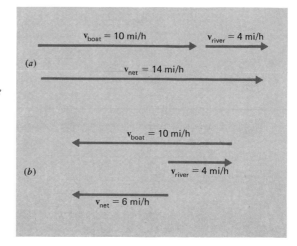

Figure 1-5 *The addition of two vectors.* (a) *A case in which both vectors have the same direction. The magnitude of the resultant vector is the* sum *of the magnitudes of the individual vectors.* (b) *A case in which the vectors have opposite directions. The magnitude of the resultant vector is the* difference *of the magnitudes of the individual vectors.*

that we show the magnitude of a vector by the length of the arrow that represents the vector. (The magnitude of a vector **v** is written as v; thus, the magnitude of **v** = 10 mi/h, *east* is the *speed*, v = 10 mi/h.) If we add two vectors that have the same direction, the magnitude of the sum is just the sum of the two individual magnitudes.

If you reverse the direction of the boat and run *upstream,* we know that the velocity relative to the river bank will be less than in the previous situation. Figure 1-5b shows how we analyze this case using vectors. The vector $\mathbf{v}_{river}$ is still directed toward the right, but now $\mathbf{v}_{boat}$ points to the *left* (that is, upstream). The diagram indicates the way in which we add $\mathbf{v}_{boat}$ and $\mathbf{v}_{river}$ to obtain $\mathbf{v}_{net}$, which now has a magnitude of 6 mi/h and a direction upstream. (The more complicated case in which $\mathbf{v}_{boat}$ is directed *across* the river is discussed in the section on *Additional Details* at the end of this chapter.)

Questions and Exercises

1. Suppose that you wish to describe to a friend how to go from one place in the city to another. Could you do this without using numbers? (Remember, to say that you go a block north and turn right *does* use a number: "a block" = "1 block.") Try a simple case.

2. Without using a dictionary, try to devise a definition for the concept of *time*. (The definition should be of the *concept* of time, not the *unit* of time.) After you have a satisfactory definition, compare with that in a dictionary.

3. What are some of the advantages of the metric system of measure? Does the British system have any advantageous characteristics?

4. Use a dictionary or other reference source and see how many different nonmetric units of length you can find. (Some examples are *rods, cubits,* and *furlongs*.) The size of your list should convince you of the great simplicity of the metric system!

5. A common unit of length in machine shop practice is the "mil" or $\frac{1}{1000}$ of an inch. How many mils equal one millimeter (1 mm)? [Ans. approximately 40.]

6. In 1976 the price of gold was about $130 per ounce. (Gold is

measured in *troy* ounces; there are 12 troy ounces in a pound.) What was the price of gold per kilogram? [Ans. about $3400.]

7. In adventure or western movies you often see the heroes (or villains) making off with "millions" in gold. Use the information in the previous exercise and estimate the value of gold in a box that two men could lift and carry conveniently. [Ans. about $200,000 at 1976 prices—a lot less during the days of the Old West.]

8. Express 1 megasecond (1 Ms) in more conventional units. [Ans. 11.6 days.]

9. During a trip along an interstate highway, you notice the mile markers at various times:

   9:00 A.M.   Mile 60
   9:30 A.M.   Mile 85
   10:00 A.M. Mile 100
   11:00 A.M. Mile 150

   What was your average speed (a) during the first half hour, (b) during the first hour, (c) during the entire 2-h trip? [Ans. 50 mi/h, 40 mi/h, 45 mi/h.]

10. A certain boat can travel at a speed of 15 km/h in still water. This boat moves upstream in a river whose current is 4 km/h. At some time during the trip a passenger walks toward the stern of the boat at a velocity of 3 km/h. What is the velocity of the passenger relative to the river bank? Draw a vector diagram to illustrate this case. [Ans. 8 km/h.]

Additional Details for Further Study

Powers-of-10 Notation

In describing natural phenomena we often encounter numbers that are very large or very small. For example, the distance from the Earth to the Sun is approximately 150 000 000 000 m (150 billion meters) and the diameter of a hydrogen atom is about 0.000 000 0001 m (1 ten-millionth of a meter). Dealing with so many zeroes in a number is awkward and sometimes leads to errors in transcribing or in calculating. In order to convey the same information in a simpler way, we use a notational device involving *powers of 10*.

If we multiply the number 10 by itself several times, we obtain, for example,

$10 \times 10 \times 10 = 1000$
$10 \times 10 \times 10 \times 10 \times 10 = 100\,000$

In the first case, we have *three* factors of 10 and we say that 1000 is equal to the *third power* of 10. In the second case, we have *five* factors of 10 and we say that 100 000 is equal to the *fifth power* of 10. Thus, we can write

$1000 = 10^3$
$100\,000 = 10^5$

The superscript attached to the number 10 is the *power* to which 10 is raised and corresponds to the number of tens that are multiplied together. Notice that this power is equal to the number of zeroes that follow the *one* in the ordinary way of writing the number.

Any number can be expressed in powers-of-10 notation. For example, consider the number 137 000. We can write this number as

$137\,000. = 1.37 \times 100\,000 = 1.37 \times 10^5$

Notice that in going from 137 000. to 1.37, we move the decimal *five* places to the left, and *five* is the power of 10 in the final expression.

For numbers that are less than 1, we use *negative* powers of 10 according to the following rule:

$0.1 = \dfrac{1}{10} = 10^{-1}$

$0.001 = \dfrac{1}{1000} = \dfrac{1}{10^3} = 10^{-3}$

That is, we express the number as a fraction with the denominator equal to a power of 10. (Notice that $10^1 = 10$.) Then, we indicate that this power of 10 is in the denominator by attaching a *negative* sign to the power in the final expression.

Again, any decimal number can be written in terms of a negative power of 10; for example,

$0.000\,027 = \dfrac{2.7}{100\,000} = \dfrac{2.7}{10^5} = 2.7 \times 10^{-5}$

Notice that in going from 0.000 027 to 2.7, we move the decimal *five* places to the right, and *negative five* is the power of 10 in the final expression.

Multiplying or dividing large and small numbers is made easy by using the powers-of-10 notation:

$$40\,000 \times 2\,000\,000 = (4 \times 10^4) \times (2 \times 10^6)$$
$$= (4 \times 2) \times (10^4 \times 10^6)$$
$$= 8 \times 10^{10}$$

Notice that the product $10^4 \times 10^6$ is equal to 10^{10}; that is, the powers of 10 are *added* to obtain the resultant power of ten: $4 + 6 = 10$. Also,

$$\frac{40\,000\,000}{2000} = \frac{4 \times 10^7}{2 \times 10^3} = \frac{4}{2} \times (10^7 \times 10^{-3})$$
$$= 2 \times 10^4$$

Notice that the final power of 10 is again obtained by adding the individual powers of ten: $7 + (-3) = 4$.

$$\frac{0.000\,008}{0.004} = \frac{8 \times 10^{-6}}{4 \times 10^{-3}} = \frac{8}{4} \times (10^{-6} \times 10^3)$$
$$= 2 \times 10^{-3}$$

Notice that $1/10^{-3} = 10^3$ and that $10^{-4} = 1/10^4$; that is, in moving a power of 10 from the denominator to the numerator (or vice versa), the *sign* of the power is changed.

The Addition of Vectors

In the last section of this chapter we discussed the addition of vectors that have the same or opposite directions. How do we combine vectors that are directed at other angles with respect to one another?

Suppose that you hike from point A to point B by going 3 km east and then 4 km north. How would you describe your net movement by means of vectors? The diagram below shows the way to do this. First, we lay out a vector that has a length of 3 units (representing 3 km) and directed toward the east from point A. Next, we place a vector with a length of 4 units (north) with its foot at the tip of the first vector. The sum of these two vectors is constructed by connecting the foot of the first vector to the tip of the second vector and represents the net movement or *displacement* during the hike. The diagram

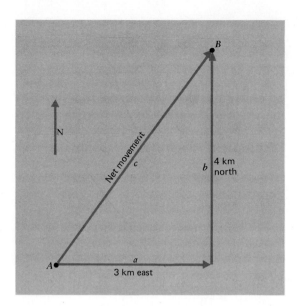

below shows how we apply this rule to the addition of *three* vectors. In this case, the sum of the vectors **A, B,** and **C** equals the vector **D**; that is,

A + B + C = D

In the special case that the two vectors being added are at *right angles* (as in the first diagram above), we can easily calculate the magnitude of the resultant vector. For a *right triangle,* the Pythagorean theorem of geometry states that the sum

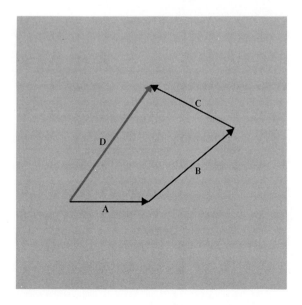

of the squares of the lengths of the two sides is equal to the square of the length of the hypotenuse. In the diagram the two sides have lengths a and b, and the hypotenuse has length c. Therefore,

$$c^2 = a^2 + b^2$$

or

$$c = \sqrt{a^2 + b^2}$$

Inserting the values given in the diagram, we find

$$c = \sqrt{(3 \text{ km})^2 + (4 \text{ km})^2}$$
$$= \sqrt{9 + 16} = \sqrt{25} \text{ km}$$
$$= 5 \text{ km}$$

Thus, the straight-line distance from A to B is 5 km.

Additional Exercises

1. Write the following as powers of 10:
   (a) 36 000 000   (b) 0.000 000 013
   [Ans. 3.6×10^7; 1.3×10^{-8}.]

2. Express *three million* multiplied by *four hundred thousand* as a power of ten. [Ans. 1.2×10^{12}.]

3. Calculate the following:
   (a) $\dfrac{300\ 000 \times 6000}{2\ 000\ 000}$

   (b) $\dfrac{0.003 \times 400\ 000}{6 \times 10^{-6}}$

   [Ans. 9×10^2; 2×10^8.]

4. Refer to Fig. 1-5a. Suppose that the helmsman of the boat attempts to steer the boat directly *across* the river. Show the actual direction of the boat by means of a vector diagram.

5. For the situation in Exercise 4, calculate the speed of the boat with respect to the land. [Ans. $\sqrt{116} = 10.8$ mi/h.]

6. A pilot wishes to fly his aircraft (which has a speed of 200 km/h) directly north when there is a crosswind of 50 km/h from the west. Show in a vector diagram how he must orient his plane in order to accomplish this.

7. In Exercise 6, what will be the *ground speed* (the speed relative to the ground) of the aircraft? [Ans. 193.6 km/h.]

How do rockets work?

2 MOMENTUM AND FORCE

We have all seen, in photographs or on television, the impressive sight of a NASA rocket launch. Huge clouds of vapor cover the launch pad and surround the rocket as firey gases spew from the nozzles and slowly lift the rocket into the air. Gradually, the rocket gains speed until finally it is placed into orbit around the Earth or is propelled away from the Earth on a mission into space.

There is another type of rocket motion that everyone has seen. Suppose that you blow up a toy balloon and then release it without tying closed the filling tube. The balloon will fly away on a short and erratic flight. In both of these cases, the vehicle—rocket or balloon—is propelled in some way by the ejection of gases from the rear of the vehicle (see Fig. 2-1). Exactly how do the gases leaving a rocket propel it forward? Do the gases push on the air molecules, thereby giving a forward thrust to the rocket? If this is so, how do we explain the

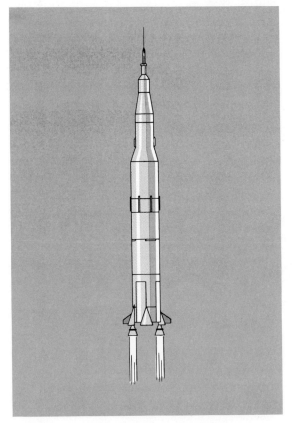

Figure 2-1 *How do the gases ejected from the rear of a rocket propel it forward?*

fact that rockets can operate in the vacuum of space? What is the true nature of a rocket push?

In this chapter we will answer these questions as we examine the problem of how things move.

The Concepts of Force and Momentum

In order to discuss the motion of rockets and other objects, we need to introduce two fundamental physical ideas, namely, *force* and *momentum*. Some of our everyday experiences will guide us in thinking about these quantities.

We all know that an object can be set into motion in a particular direction by the application of a push or a pull in that direction. Also, we know that an object in motion can be slowed down or stopped by the application of a push or a pull in the direction opposite to the direction of motion. A push or a pull constitutes a *force*. In fact, we can define force as any influence that can act to *change* the state of motion of an object.

When a rocket is set into motion, clearly, some force acts on the rocket. Let us think about this force in the following way. Suppose that a shooter fires a rifle. As the bullet speeds through the barrel, the shooter receives a push or recoil kick from the butt of the rifle (Fig. 2-2). The firing of a bullet and the accompanying recoil of the rifle is very similar to the action of a rocket. The ejection of the gases from a rocket provides the push that propels the rocket forward. In the same way, the ejection of the bullet from the rifle is responsible for the push that the shooter experiences as the recoil. We can go a step further in our reasoning. If the shooter uses a greater powder charge to fire an identical bullet with a higher speed, the recoil is also greater. Or, if he increases the mass of the bullet and maintains the same bullet speed, the recoil again is greater than in the first case. We must conclude that the magnitude of the recoil experienced by the shooter depends on two factors, the mass and the speed of the fired bullet. We should be able to apply the same reasoning to the case of rocket motion: The forward push experienced by a rocket should depend on the mass and the speed of the ejected gases. As we will see, this is exactly correct.

Let us make more quantitative the connection between recoil or push and the combination of mass and speed. Look at the simplified situation shown in Fig. 2-3. Here we have two

24   Momentum and Force

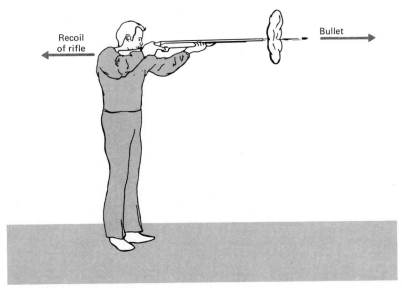

Figure 2-2 *When the shooter fires the rifle, he receives a "kick" from the recoil of the rifle. Can you see the similarity between this situation and that shown in Fig. 2-1?*

blocks with equal masses. Between the balls is a coiled spring (but the spring is not attached to either block). Suppose that the spring is compressed and tied in this condition with a thread (Fig. 2-3a). Next, we gently snip or burn the thread (so that our system is not disturbed in the process). The spring uncoils and the two blocks are pushed aside. If we measure

Figure 2-3 (a) *Two blocks with equal masses have between them a coiled spring.* (b) *When the spring uncoils, each block moves away from the original position with the same speed. (We use here the symbol V which stands for velocity.)*

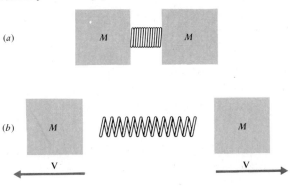

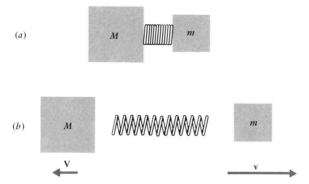

Figure 2-4 (a) *Two blocks with unequal masses have between them a coiled spring.* (b) *When the spring uncoils, the smaller block moves away from the original position with a speed that is greater than the speed of the larger block.*

the speeds of the blocks, we find that they are the same. This is easy to understand: There is nothing to distinguish one block from the other, so they react to the uncoiling spring in exactly the same way. Each block experiences a push only as long as the spring is in contact with both blocks. (Can you see why?) Because the blocks are identical, when the push of the spring is completed, each block has the same speed.

In Fig. 2-4 we have a slightly different case. Here we have substituted a smaller block on the right-hand side. What happens now when we release the spring and allow it to push on the two blocks? Remembering the high speed of a fired bullet compared to the much slower recoil motion of the rifle, we know that the smaller block will move away with a greater speed than will the larger block. We can measure the masses and the speeds of the blocks, and by using different blocks and different springs we can study a wide range of cases. If we carry out such an experiment, we discover an interesting and important result. No matter how we change the system, we always find that the product of mass and speed for one block of a pair is exactly equal to the product of mass and speed for the other block:

(mass) × (speed) for left-hand block =
 (mass) × (speed) for right-hand block

Because this result appears to hold for a wide variety of different cases, the product (mass) × (speed) must have some special significance. We call this product *momentum:*

momentum = (mass) × (speed or velocity)

Isaac Newton (1642–1727) was the first scientist to appreciate the importance of momentum. (Newton used the term "quantity of motion.") He incorporated momentum into his famous theory of motion, published in 1687.

How can we use the idea of momentum in discussing the results for the recoiling blocks? First, we note that before the spring is released, there is no motion and each block has zero momentum (Figs. 2-3a and 2-4a). After release of the spring, each block possesses a certain amount of momentum. But let us take a view of the system *as a whole* (that is, *both* blocks), not merely a single block. We know that the description of velocity requires the specification of *direction* as well as magnitude (Chapter 1). Multiplying velocity by mass produces a quantity (namely, *momentum*) that also has direction. (We say that momentum is a *vector* quantity.) Thus, in Fig. 2-4b we should state that the right-hand block has a momentum mv *directed to the right* and that the left-hand block has a momentum MV *directed to the left*.

From the results of our measurements, we know that mv and MV have the same magnitude. What happens, then, if we write down the total momentum of the pair of blocks by adding mv (to the right) and MV (to the left)? The result of adding two quantities (vectors) with *equal magnitudes* and *opposite directions* is *zero* (Chapter 1). Therefore, the total momentum of the recoiling blocks is zero. But the original momentum of the blocks was also zero. That is, the act of releasing the spring and thereby causing the blocks to move away from one another has not changed the total momentum of the pair.

We can reinforce this idea by thinking about some everyday experiences. Suppose that two automobiles that have equal masses move directly toward one another with the same speeds, as shown in Fig. 2-5a. In this diagram, notice that the velocity vectors have the same length, indicating that the *speeds* are the same. The *directions,* however, are *opposite,* so one vector is labeled **v** and the other is labeled $-\mathbf{v}$.

What happens when the cars collide? Apart from the extensive damage that will result, the important point is that the cars come to rest, as in Fig. 2-5b. (We consider the case in which the cars lock together and do not rebound.) We analyze this situation by noting that the two moving cars have equal momenta which are oppositely directed. Adding these two momenta gives zero for the total momentum of the cars before

The Concepts of Force and Momentum

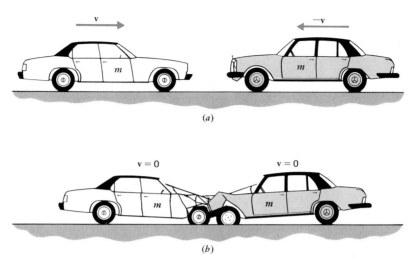

Figure 2-5 (a) *Two automobiles with equal masses approach one another with equal speeds.* (b) *Upon collision the cars lock together and come to rest. The total momentum of the pair of cars does not change during the process.*

collision. The collision does not change the total momentum of the pair so that the final momentum is also zero and the cars come to rest. Notice that this example is just the reverse of the case shown in Fig. 2-3. In both situations the original total momentum is zero and the final total momentum is also zero.

Next, suppose that we replace one of the cars with a large truck, as in Fig. 2-6. Even though the initial speeds are equal, the momenta are not. It is clear that in this case the vehicles will not come to rest after collision. Instead, the truck-car combination (again locked together) will move in the direction of the truck's original motion but with a reduced speed.

In the car-car collision (Fig. 2-5), the final condition is one of rest. But in the truck-car collision (Fig. 2-6), the vehicles remain in motion after contact. In what way are these two cases similar? In the first case, the total momentum before collision is zero and it is zero after collision; the total momentum does not change. In the second case, the total momentum after collision is not zero, but a measurement would show that it is the same as the total momentum before collision. In this case, too, the total momentum does not change. Before they come into contact, the truck and the car move with the same speed. But because the mass of the truck is larger than that of the car, the truck's momentum is greater than the car's mo-

28 *Momentum and Force*

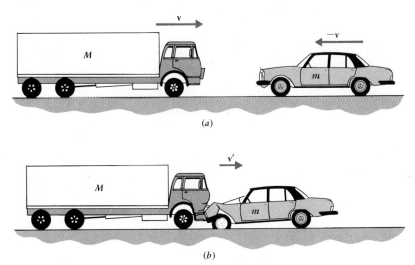

Figure 2-6 (a) *A truck is substituted for the left-hand car in Fig. 2-5.* (b) *After collision the truck-car combination moves to the right with reduced speed. The total momentum of the pair does not change during the process.*

mentum. The sum of these two momenta (which are oppositely directed) is a certain momentum in the direction of the truck's motion. We can represent the addition of the two momenta by the diagram in Fig. 2-7. The total momentum of the combination is the quantity that is not changed by the collision. After collision the product of the truck-car mass and their common speed is the final momentum, and this momentum is equal to the sum of the individual momenta before collision.

We have here an important result concerning the motion of objects of all kinds. The total momentum of an object or group of objects remains the same as long as the object or group is not acted upon by an outside agency. In our examples, the momenta of the individual blocks and cars *did* change because they were acted upon by other blocks and cars. But these blocks and cars were parts of the systems we were studying. Thus, the total momentum in each case remained constant. That is,

Total momentum is conserved.

Using symbols for mass and velocity, we can express the

Rocket Propulsion

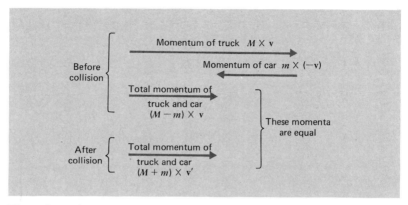

Figure 2-7 *The sum of the truck's momentum and the car's momentum (Fig. 2-6) yields the total momentum of the combination. This total momentum is in the direction of the truck's motion and is the same before the collision as after the collision.*

momentum conservation law as

$$(m \times v)_{\text{before}} = (m \times v)_{\text{after}}$$

If the system consists of several parts or objects, we must sum ($m \times v$) for each member of the system.

In describing physical processes, we have learned that it is much easier to analyze situations if we can identify some quantity that remains unchanged throughout. Total momentum is such a quantity, and the statement, *total momentum is conserved,* is one of the important *conservation laws* of physics. (Other conserved quantities, as we will see, are mass, energy, and electrical charge.)

Rocket Propulsion

Let us now apply the ideas of momentum and momentum conservation to the problem of rocket propulsion. First, we must dispense with a popular misconception. It is sometimes said by the uninformed that a rocket is propelled forward by the push of the exhaust gases on the atmosphere. In 1919, Robert H. Goddard, the principal pioneer of American rocketry, published a paper entitled "A Method of Reaching Extremely High Altitudes." In this paper, Goddard made one of the first practical proposals for rocket flight above the atmosphere.

30 *Momentum and Force*

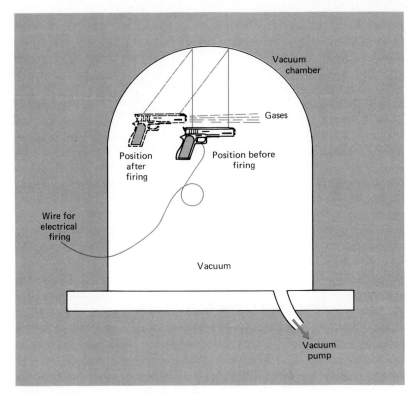

Figure 2-8 *Robert H. Goddard's demonstration to convince the skeptics at* The New York Times *that a rocket will operate in the vacuum of space. Goddard suspended a pistol in a vacuum chamber and electrically fired a blank cartridge; the ejected gases caused the pistol to swing backward.*

However, *The New York Times* sneered editorially that rockets would not work in space beyond the Earth's atmosphere. Goddard countered by demonstrating to *The Times* in a convincing way that rocket propulsion has nothing to do with the presence or absence of an atmosphere.

The apparatus used by Goddard for his demonstration is shown schematically in Fig. 2-8. A pistol containing a blank cartridge (that is, a cartridge with powder but no bullet) was suspended with strings inside a chamber from which the air could be exhausted by means of a vacuum pump. A wire leading into the chamber provided a way to fire the pistol electrically. After pumping the air from the chamber to simulate the vacuum of space, Goddard fired the pistol. As the gases were ejected from the barrel, the pistol recoiled on the suspension strings, in the manner indicated in Fig. 2-8. Because there was no air for the gases to push against, *The Times* reporters were suitably convinced that rocket propulsion was indeed feasible in the vacuum of space.

When a rocket is fired, the exhaust gases do not push against the air. Instead, the gases push against the *rocket,* and the rocket pushes against the gases. This is exactly the situation in Fig. 2-3: each block pushes on the other block (through the spring) and each is set into motion. Before firing, the momentum of a rocket and its propellant material is zero. After firing, the total momentum of the system (rocket plus gases) must still be zero. The rearward momentum of the gases must be balanced by the forward momentum of the rocket. If the rocket is fired from the surface of the Earth, then during the initial portion of the flight, the resisting effect of the air must be overcome; that is, air friction impedes the movement of the rocket. But when the rocket escapes from the atmosphere, air resistance is no longer a factor. Therefore, instead of not working at all in vacuum, a rocket will actually work *better* when there is no air to slow its motion!

How does a rocket actually work? Consider the simplified system shown in Fig. 2-9. Here we have a rocket with a mass M which is being propelled forward by the ejection of small pellets from the rear. If a single pellet with mass m is ejected with a velocity $\mathbf{v}$, the pellet carries a momentum $m\mathbf{v}$ to the left in Fig. 2-9. Momentum conservation tells us that the rocket must acquire an equal momentum to the right. Because the rocket mass M is much greater than the pellet mass m, the velocity change experienced by the rocket is much smaller than the pellet velocity $\mathbf{v}$. Each successive pellet that is ejected will give an equal velocity increment to the rocket. If

Figure 2-9 *A rocket (not a practical one!) is propelled forward by ejecting small pellets from its rear.*

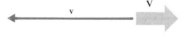
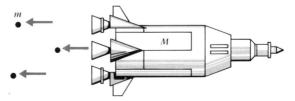

we arrange for the pellets to be ejected at closely spaced intervals of time, there would be a smooth increase of the rocket's velocity.

As the rocket moves forward, we measure its velocity with respect to the Earth or some other convenient point. But what about the velocity of the ejected pellets? From the standpoint of momentum conservation, it is always the velocity of the pellets with respect to the rocket that is important. Therefore, even though the rocket is moving with an increasing velocity with respect to the Earth, the pellet velocity remains the same with respect to the rocket (assuming that each pellet is fired in exactly the same way).

Clearly, if we wish the forward velocity of the rocket to be large, then we must make the momentum carried by each pellet, mv, as large as possible. There is no particular advantage in making m large: We could always achieve the same results by simply firing more pellets at the same velocity. Therefore, in order to increase the rearward momentum, we must concentrate on increasing the *velocity*. Now, the pellets must be set into motion in some way. For example, the pellets could be rifle bullets fired by the expanding gases from chemical explosions in the cartridge cases. But if rapidly moving gases must be produced to propel the bullets, then we really do not need the bullets at all! (Remember Goddard's demonstration for *The Times*.) Instead of ejecting bullets, we should eliminate the bullets and carry more chemical fuel that can be ignited to produce high-speed gases. Only the total mass of material ejected per second and the velocity of ejection are important in determining how rapidly a rocket will gain forward speed; the mass of the individual ejected particles does not matter.

For this reason, all present-day rockets operate by expelling high-speed gases—not solid particles—that result from the combustion of chemical fuels. Various types of fuels, both liquid and solid, have been used. The earliest rockets, made by the Chinese more than 700 years ago, were propelled by the burning of a type of gunpowder. Robert Goddard's first rockets used black powder, and later models employed smokeless powder. In 1926, Goddard launched his first liquid-fueled rocket; it reached a height of 41 ft while traveling 184 ft "down range." By 1937, one of Goddard's rockets had reached an altitude of 8500 ft. The first truly successful modern rocket was the German V-2, which became operational and was used as a bombardment weapon during World War II. The V-2 was powered by liquid oxygen and alcohol. The most effective liquid-fueled rocket engines so far developed use liquid oxygen and liquid hydrogen. These rockets have been used extensively in our program of space exploration.

Liquefied gases, such as liquid oxygen and liquid hydrogen, require special low-temperature (or *cryogenic*) techniques for production, storage, and handling. The procedure for filling a rocket's tanks with cryogenic liquids is long and involved. Therefore, essentially all military rockets (for example, ICBMs, intercontinental ballistic missiles, as well as those launched from submarines) employ solid fuels. Although the best solid fuel is only about two-thirds as effective as a liquid oxygen-liquid hydrogen mixture, the advantages in long-term storage more than compensate for the decreased effectiveness.

Rocket Flight

Figure 2-10 shows a simplified schematic diagram of a liquid-fueled rocket. The internal tanks contain the fuel. Actually,

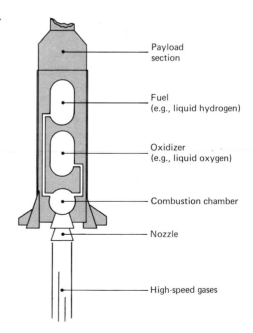

Figure 2-10 *Schematic diagram of the arrangement of the essential parts of a liquid-fueled rocket.*

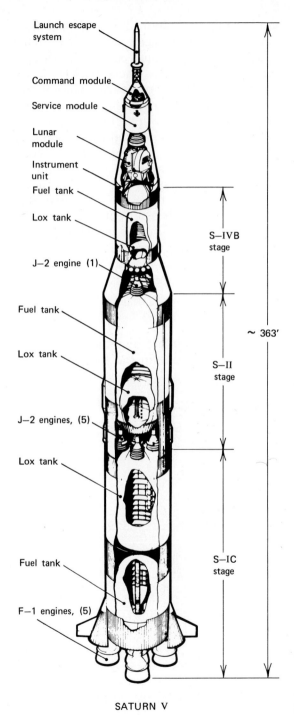

Figure 2-11 *Cutaway drawing of a Saturn V rocket. The fuel used in the first stage is kerosene; in the upper stages, the fuel is liquid hydrogen. The oxidizer is liquid oxygen (LOX). Courtesy of NASA.*

there are two separate components to the fuel supply—the *fuel* proper and the *oxidizer*. All high-performance rockets carry their own supply of oxygen or an oxidizing agent. Even while moving through the atmosphere it is too inefficient to use the oxygen in the air to support the combustion of the fuel. (Internal combustion engines, such as those used in automobiles and aircraft, *do* rely on atmospheric oxygen.) And when moving through the vacuum of space, an internal oxygen supply is essential. Each fuel component is fed by special cryogenic pumps into the combustion chamber where ignition takes place. The combustion chamber and the nozzle are shaped in a particular way to provide the most efficient use of the fuel. Of course, an operational rocket is much more complex than the simplified diagram shown in Fig. 2-10. Figure 2-11 is a cutaway drawing of a Saturn V rocket, long the mainstay of the U. S. space program. The interior of this rocket is a maze of tubes, pumps, and electrical controls, but the basic operation is still that shown in Fig. 2-10.

Notice in Figs. 2-10 and 2-11 that the fuel storage and combustion regions of the rocket require considerably more space (and mass) than the payload section. This is generally true of all types of rockets—the fuel-to-payload mass ratio is always high. We can explain this in the following way. A particular fuel-oxidizer combination will produce gases that leave the rocket with a certain velocity. (A typical value is 9000 ft/s or 10 000 km/h.) Let us measure the final velocity attained by the rocket (the *burnout* velocity) in terms of the gas velocity. That is, we form the ratio

$$\text{velocity ratio} = \frac{\text{final velocity of rocket}}{\text{velocity of exhaust gases}}$$

Further, let us measure the mass of the fuel supply in terms of the mass of the rest of the rocket (frame, payload, tanks, pumps, and so forth). That is, we form the ratio

$$\text{mass ratio} = \frac{\text{mass of fuel (and oxidizer)}}{\text{mass of rocket}}$$

These two ratios are related in the manner shown in Table 2-1. Here we see that if a final rocket velocity of a few times the gas velocity is desired, then the mass of the required fuel supply is far greater than the mass of the rocket. The minimum velocity that a rocket must have to escape the Earth's gravitational pull and proceed into space is approximately

Momentum and Force

Table 2-1

Velocity Ratio	Mass Ratio
1	1.7
2	6.4
3	19.1
4	53.6
5	147.4
10	22 025.5

40 000 km/h. This is four times the exhaust velocity of the rocket gases, so that the mass of the fuel supply must be more than 50 times the mass of the rocket. Remember that the rocket mass includes the mass of the frame and tanks as well as that of the payload. The Apollo vehicles each had a mass of approximately 3 000 000 kg when launched on their missions to the Moon, and yet the mass actually carried to the Moon was about 50 000 kg. The capsule in which the astronauts returned to Earth had a mass of only 5000 kg.

In order to reduce the mass that is carried to high velocities, the usual procedure is to *stage* the rocket launch. As the fuel is consumed, there is no need to carry along the empty tanks and the portion of the frame that houses the tanks. Most of the very large rockets are of the three-stage design, as shown in

Figure 2-12 *Instead of carrying the entire rocket on a space mission, the empty fuel containers are dropped and fall back to Earth. The final stage powers the payload into orbit or into space.*

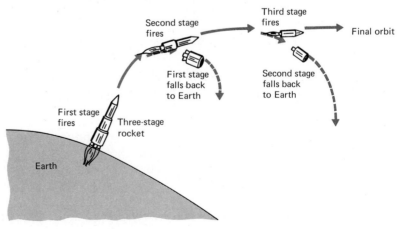

Table 2-2 Important Events in the History of Space Travel

1902	First paper on rocket travel published by K. E. Tsiolkovskii, a Russian school teacher.

Robert H. Goddard

1926	First liquid-fueled rocket launched by Robert H. Goddard at Auburn, Massachusetts; a height of 41 ft was reached.
1944	First truly successful liquid-fueled rocket, the German V-2, used to bombard England; the 14-ton missiles reached heights of about 50 mi.
1949	First two-stage rocket fired from White Sands, New Mexico; the booster stage was a modified V-2 and the second stage was a WAC-Corporal rocket. A height of 244 mi was reached by the upper stage.
1957	First launching of an artificial satellite, the Russian Sputnik I, into orbit around the Earth.
1958	First U.S. satellite, Explorer I, placed into orbit; the Earth's radiation belts were first detected on this flight.
1959	First rocket, the Russian Lunik I, to reach the Moon.

Yuri Gargarin

1961	First space flight by Man; the Russian Cosmonaut Yuri Gargarin completed one orbit around the Earth in 1 h and 48 min in Vostok I.
1961	First American in space; Astronaut Alan Shepard made a suborbital flight in Freedom 7.
1961	First space flight to Venus by Mariner II.
1962	First communications satellite, Telstar I, placed into orbit.
1965	First space-docking operation performed by Astronauts Stafford and Schirra in Gemini 6.
1968	First manned flight to the Moon; Astronauts Borman, Lovell, and Anders, in Apollo 8, orbit but do not land on the Moon.

 Eagle

1969	First landing on the Moon; Astronauts Neil Armstrong and Edwin Aldrin land in Eagle, the lunar module of Apollo 11.
1972	Completion of the series of Apollo missions to the Moon. Launch of Pioneer 10 to Jupiter and beyond.
1973–1974	Skylab missions; scientific experiments and observations carried out in Earth orbit.
1975	Joint U.S.-Soviet mission in Earth orbit; Apollo-Soyuz linkup.
1975–1976	Viking mission to Mars; search for Martian life.

Fig. 2-12. The largest portion of the fuel is used in the initial part of the flight. Therefore, at an altitude of about 65 km, when the velocity is about 8500 km/h, the first-stage fuel section is detached and falls back to Earth. The remaining sections continue to gain speed, powered by the second-stage rocket. At an altitude of about 150 km, when the velocity is nearly 25 000 km/h, the second stage is also dropped, and the third stage powers the remainder of the rocket into orbit or on a mission into space. Additional firings of the final stage will be required if the mission involves a landing on the Moon and a return to Earth. In every phase of the flight, the sections of the vehicle that are no longer useful are dropped in order to minimize the mass that the remaining fuel must propel forward.

By staging a rocket launch, advantage can be taken of the fact that the second- and third-stage motors begin firing after the vehicle has already been boosted to a high velocity. Consequently, the payload section can be driven to considerably greater velocities than Table 2-1 would indicate on the basis of the mass ratio at lift-off.

How Far into Space?

The star nearest the Sun is Alpha Centauri. Light from Alpha Centauri, traveling with a speed of 3×10^8 m/s, requires 4.3

years to reach the Earth. We say that Alpha Centauri lies at a distance of 4.3 *light years,* which corresponds to 4×10^{16} m or 4×10^{13} km, 40 trillion kilometers. Can we ever hope to travel by rocket to our nearest neighbor in space? Let us assume that our rocket can be driven to a velocity of 100 000 km/h. (This is not an unreasonable velocity for a chemical-fuel, three-stage rocket; in fact, planetary probes have achieved such velocities.) In a year's time at this speed, the rocket will move a distance of approximately 8×10^8 km. Therefore, to travel the 4×10^{13} km to Alpha Centauri would require 50 000 years! It is clear that chemical-fuel rocket technology is completely inadequate to provide us with a practical means of traveling even to the nearer stars.

Some improvement could be provided by alternative propulsion systems now being developed or considered. For example, one system envisages the use of a nuclear reactor to heat hydrogen gas to extremely high temperatures; the gas atoms could then be ejected with high velocities to propel the spacecraft forward. Such a *nuclear propulsion system* involves an exceptionally large mass (for the reactor and the shielding that must be provided to protect the crew from the nuclear radiations) and would require a large booster rocket for launching. Nevertheless, nuclear propulsion systems do appear feasible for certain types of missions. But no designs (or proposed designs) even approach the requirements for interstellar travel.

Other ideas for propulsion systems include the ejection of high-speed electrically charged atoms (*ions*) or light beams from special rockets. A light-propelled rocket would be the ultimate design because no material particle (according to relativity theory) can ever achieve a speed that is equal to or greater than the speed of light. A light beam carries momentum, and so projecting such a beam from the rear of a rocket would propel it forward. However, even with the most intense light beams that we know how to produce (or can imagine producing), the rate at which the rocket would gain speed is extremely slow. For example, suppose that we could convert into a light beam the entire output of our largest nuclear power reactor. If we used this system to propel a 100 000-kg spaceship, it would build up speed at a rate of about 35 000 km/h for each year of operation. Such a reactor would require far more than 100 000 kg of fuel in order to achieve a speed that is even a small fraction of the speed of light. We must conclude that our present technology provides us with no conceivable method of propeling a spaceship to a speed that begins to be comparable with the speed of light.

Although we have no idea of a way to propel a rocket to a speed that is a substantial fraction of the speed of light, if we could do so we could think seriously about voyages to other stars. According to relativity theory (Chapter 9), a clock aboard a fast-moving rocket will run more slowly than will an Earth clock. However, this effect is important only if the speed of the rocket is close to the speed of light. To an astronaut aboard a spaceship moving with a speed equal to 99.5 percent of the speed of light, ten Earth years would seem to be only one year. That is, a voyage that would require 400 years, according to Earth time, could be accomplished in only 40 years according to the astronaut. As exciting as this prospect may be, we have absolutely no conception of a method to propel a spaceship to such an enormous speed.

There is little likelihood that a space vehicle will ever be constructed that will carry people on a deep space voyage. At least, that is the view with our present understanding of physical principles and with our best guesses as to the way in which technology will develop. However, manned missions within the solar system, even to the outer planets, are by no means impossible. The series of Apollo missions to the Moon is now complete, and there will probably be no additional manned flights beyond the immediate vicinity of the Earth for at least 20 years. But plans to send astronauts on an expedition to Mars have been developed, and the mission may actually be carried out during the next century, perhaps using some sort of nuclear propulsion system.

We will return to the discussion of planetary space missions in the next chapter when we study the effects of gravitation in the solar system.

More About Force

When a force (a push or a pull) is applied to an object and the object then speeds up or slows down, the result is a *change* in the motion of the object. But, as we saw in Chapter 1, a change in the speed of an object means that the object undergoes an *acceleration*. Thus, force and acceleration are intimately connected.

More About Force   41

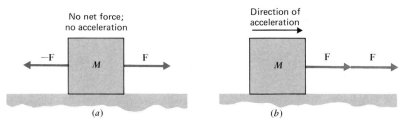

Figure 2-13 (a) *Two equal and oppositely directed forces are applied to an object; the net or total force is zero and there is no acceleration.* (b) *If the two forces are applied in the same direction, the forces combine to produce a net force toward the right and the acceleration is in this direction.*

It is the *net* force on an object that causes acceleration. Remember, force has direction and therefore is a *vector* quantity. In order to find the net force acting on an object, we must vectorially add all of the individual forces that act on the object. Figure 2-13 illustrates this point. In each part of Fig. 2-13 we have two forces of equal magnitude applied to an object. In Fig. 2-13a the forces act in opposite directions. The net force is the sum of the two applied forces and is therefore zero ($-\mathbf{F} + \mathbf{F} = 0$). Consequently, the object undergoes no acceleration: Zero net force means zero acceleration. In Fig. 2-13b both forces are directed toward the right and produce a net force in this direction. As a result, the object accelerates toward the right.

Suppose that we apply a force to an object and set it into motion. We then remove the force and watch the subsequent motion. If the object is a hockey puck sliding over smooth ice, we see that the puck slides with nearly constant velocity for a considerable distance. After we remove the force from the puck, there is no other force acting (except for a small frictional force). With zero applied force, there can be no acceleration of the puck. That is, the velocity cannot change and therefore remains constant. If, instead of pushing the puck over smooth ice, we set it to sliding over a rough concrete surface, we would see the puck quickly lose speed and come to rest. In this case there *is* a force—the force of friction—still acting on the puck after the applied force is removed. It is this frictional force that causes the velocity to decrease.

In the vacuum of outer space, there is no friction to slow down a moving space vehicle. If such a vehicle is moving in deep space with a certain velocity (with the rocket engines turned off), it will continue to move indefinitely with the same velocity. Therefore, once a rocket has been accelerated to its cruise velocity, it will "coast" through space with this same velocity until some force is exerted to cause the velocity to change.

The Force on a Rocket

The acceleration of an object (*any* object) is always the result of an applied force. Similarly, the application of a net force to an object will always result in an acceleration. Net forces and accelerations are inseparable: One never occurs without the other. When we discussed rocket motion, however, we did so entirely in terms of momentum considerations. The ejection of high-speed gases from a rocket causes the rocket to move forward, to accelerate. Therefore, some force must be acting on the rocket. What is the origin of this force?

In order to answer this question, let us look in detail at what happens in the combustion chamber. If the rocket fuel is liquid hydrogen plus liquid oxygen, the combustion reaction is

$$2H_2 + O_2 \rightarrow 2H_2O$$

That is, two molecules of hydrogen (H_2) combine with one molecule of oxygen (O_2) to produce two molecules of water (H_2O). In this combustion reaction, the product molecules (the water molecules) have much higher speeds than the original molecules. (This is always the case in a combustion reaction.) We assume for simplicity that the H_2 and O_2 molecules are at rest when they interact. When the two water molecules leave the interaction site, they move away with equal speeds in opposite directions (conservation of momentum!). This is the same situation as that illustrated in Fig. 2-3. Figure 2-14 shows the motion of a pair of water molecules in the combustion chamber of a rocket. In Fig. 2-14a, the molecule moving toward the right strikes and is absorbed by the chamber wall. The impact of a molecule with a mass m and a velocity v means that a momentum mv is imparted to the rocket, thereby increasing the rocket's velocity. We can see that a momentum mv must be gained by the rocket because the molecule moving to the left carries a momentum of exactly this magnitude in the direction opposite to that of the rocket's motion. In this case,

The Force on a Rocket   43

(a)

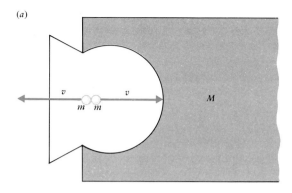

(b)

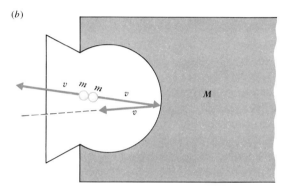

Figure 2-14 *Imparting momentum to a rocket. Two high-speed water molecules are produced by the combustion of hydrogen and oxygen. The molecules move away from the combustion site with equal and opposite velocities. In* (a) *the molecule moving to the right strikes and is absorbed by the wall of the combustion chamber, thereby imparting a momentum* mv *to the rocket. In* (b) *the molecule is reflected from the wall and leaves the chamber with a velocity* v, *thereby imparting a momentum* 2 mv *to the rocket.*

the rocket simply absorbs the momentum of the molecule moving to the right and this momentum is exactly equal in magnitude to that of the ejected molecule. The total momentum is conserved throughout the process.

Figure 2-14b shows a slightly different situation. In this case, the molecule originally moving toward the right strikes and is reflected by the chamber wall. This molecule then moves toward the left and leaves the nozzle with the same velocity and in the same direction as the other molecule. How much momentum has been imparted to the rocket in this case? We can answer this question by again looking at the momenta of the molecules. Each molecule carries a momentum mv to the left; therefore, the momentum of the rocket (to the right) is increased by an amount $2mv$.

We see that the force exerted on a rocket is simply the push of the rapidly moving molecules. Each individual push involves a change in the momentum of a molecule and a corresponding change in the momentum of the rocket.

If the molecule moving toward the right is absorbed by the chamber wall (Fig. 2-14a), the momentum transfer to the rocket is only half as large as that which results from reflection (Fig. 2-14b). In the combustion chamber of an operating rocket, the walls are so hot that any molecule striking a wall will be immediately boiled away. Therefore, the absorption case (Fig. 2-14a) never really occurs, and all of the molecules eventually leave the chamber thereby imparting the maximum amount of momentum to the rocket.

Notice that if we wish to make a complete analysis of the rocket's motion, we must take account of the fact that each molecule leaving the combustion chamber reduces the mass of the rocket. Such an analysis is necessary to compute the velocity values listed in Table 2-1.

The burning of fuel in a rocket combustion chamber produces high-speed molecules that exert a push on the rocket as they strike the chamber walls. In a real case, the molecules move in a complicated way within the chamber. But we do not have to examine each individual molecule and take account of the way in which it interacts with the chamber wall. Instead, the law of momentum conservation tells us that we need look only at the amount and the speed of the gases ejected from the chamber. The momentum carried by these gases must exactly equal the momentum imparted to the rocket.

Questions and Exercises

1. If you stand on the stern of a small boat (which is at rest) and dive into the water, what happens to the boat? Why?
2. When an astronaut carries out an EVA (extravehicular activity or *spacewalk*), how does he propel himself about?
3. A rifle fires a bullet vertically into the air while the rifle butt is on the ground. We observe no recoil of the rifle. Why? Is total momentum conserved? Explain.
4. Imagine a boxcar that can roll without friction along a set of rails, as shown in the diagram. A man braces himself against one end of the boxcar (which is at rest) and fires a rifle toward the opposite end. A moment later the bullet strikes and embeds itself in a wooden block that is attached to the boxcar. Describe what happens (a) while the bullet is in flight and (b) after the bullet comes to rest in the block.

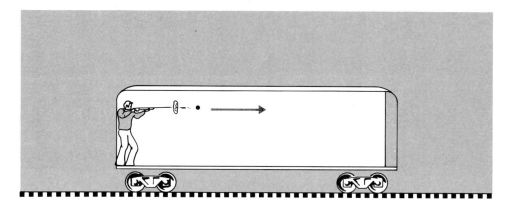

5. What do you think about the idea that UFOs (unidentified flying objects) are visitors from other planetary systems?

6. Suppose that a bullet is fired at a certain velocity into a stationary block of wood. In order to discuss the motion of the block we could use Newton's law of motion, which tells us that the force exerted on the block by the bullet is equal to the mass of the block multiplied by its acceleration. But the force exerted by the bullet as it comes to rest in the block is very complicated. Even so, we can easily compute the final speed of the block. How?

7. A moving automobile is brought to a stop by applying the brakes. What happened to the automobile's momentum? Is total momentum really conserved in this process? Explain.

8. A block rests on a floor and is against a brick wall, as shown in the diagram. If you push on the block, thereby applying to it a

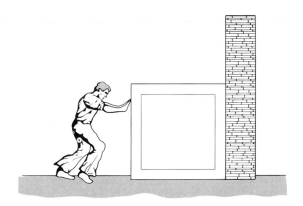

force, the block does not move. Why? What is the net force on the block? Make a sketch of all the forces acting on the block.

9. Two identical hockey pucks slide over smooth ice at the same speed. The directions of motion of the two pucks are perpendicular (at right angles). If the pucks collide and stick together, what will be the resultant motion of the pair?

10. In Fig. 2-14a the molecule moving toward the left has no direct effect on the motion of the rocket. The entire push is exerted by the molecule moving toward the right. Would it be possible to double the push by arranging for both molecules to move initially toward the right and to be reflected from the chamber wall? Explain.

Additional Details for Further Study

Momentum

When writing momentum equations, we usually give to momentum the symbol **p:**

p = $m \times$ **v**

Momentum is a *vector* quantity. The magnitude of **p** is equal to the product $m \times$ **v**, and the direction of **p** is the same as the direction of the velocity **v**.

For the situation in Fig. 2-6, we would write

Before collision $\begin{cases} \mathbf{p}(\text{truck}) = M \times \mathbf{v} \\ \mathbf{p}(\text{car}) = -(m \times \mathbf{v}) \end{cases}$

After collision $\{\mathbf{p}(\text{combination}) = (M + m) \times \mathbf{v}'$

Notice that the initial momentum of the car is given a negative sign because the car is moving in the direction opposite to that of the truck. The negative sign means that the direction of the vector **p**(car) is opposite to that of the vector **p**(truck). (Compare Fig. 2-7.) Then, momentum conservation is expressed as

p(before) = **p**(after)

or

$(M \times \mathbf{v}) - (m \times \mathbf{v}) = (M + m) \times \mathbf{v}'$

In the event that the initial velocities of the truck and the car are different (they are the same in Fig. 2-6), we would use different symbols for the velocities and would write

$$(M \times V) - (m \times v) = (M + m) \times v'$$

For example, suppose that a 4000-kg truck moving with a velocity of 10 m/s collides head-on (as in Fig. 2-6) with a 2000-kg car moving with a velocity of 8 m/s. What is the final velocity of the combination? Solving the equation above for v', we find

$$v' = \frac{(M \times V) - (m \times v)}{M + m}$$

We know that

M = 4000 kg
V = 10 m/s
m = 200 kg
v = 8 m/s

Then,

$$v' = \frac{(4000 \times 10) - (2000 \times 8)}{4000 + 2000}$$

$$= \frac{40\,000 - 16\,000}{6000}$$

$$= \frac{24\,000}{6000}$$

$$= 4 \text{ m/s}$$

Newton's Law of Motion

We have learned that force is connected with acceleration. Moreover, we have found that an object experiencing no net force will move with constant velocity: Zero force means zero acceleration. We also know from experience that a force applied to an object with small mass will produce a greater effect (a greater acceleration) than will the same force applied to an object with a large mass. Isaac Newton found that he could summarize all of our knowledge about the way forces act on bodies by the simple statement,

$$\begin{pmatrix} \text{net force applied} \\ \text{to an object} \end{pmatrix} = \begin{pmatrix} \text{mass of} \\ \text{the object} \end{pmatrix} \times \begin{pmatrix} \text{acceleration of} \\ \text{the object} \end{pmatrix}$$

This is Newton's famous law of motion. In symbols we write

$\mathbf{F} = m \times \mathbf{a}$

which means that a net force **F** acting on a mass m produces an acceleration **a**. If the mass is 1 kg and if the acceleration is 1 m/s², the force is 1 kg-m/s². To this force we give the special name, 1 *newton* (N). That is,

1 N = 1 kg-m/s²

Force is a *vector* quantity. The magnitude of **F** is equal to the product $m \times a$, and the direction of **F** is the same as the direction of the acceleration **a**.

What will be the acceleration of a 4-kg block that is pushed across a floor with a 10-N force if the retarding frictional force is 2 N? The frictional force is in the direction *opposite* to that of the applied force. Therefore, the net force is

$F(\text{net}) = 10 \text{ N} - 2 \text{ N} = 8 \text{ N}$

Then,

$$a = \frac{F(\text{net})}{m} = \frac{8 \text{ N}}{4 \text{ kg}} = 2 \text{ N/kg} = 2 \text{ m/s}^2$$

Additional Exercises

1. A 20 000-kg railway car moves along a track with a velocity of 6 m/s and strikes a 40 000-kg car that is at rest. The two cars couple together and move away. With what velocity does the combination move? [Ans. 2 m/s.]

2. In the previous exercise, suppose that the 40 000-kg car is originally moving with a velocity of 3 m/s in the same direction as the 20 000-kg car. What will be the final velocity of the pair of cars after they couple together? [Ans. 4 m/s.]

3. A ball of putty with a mass M is thrown horizontally with a velocity of 12 m/s toward a block with a mass $5M$ that is at rest on a table. The putty sticks to the block and the combination moves off together. What is the velocity of the combination? [Ans. 2 m/s.]

4. A horizontal force of 6 N is applied to a block that has a mass of 4 kg. The block slides across a floor with a constant velocity of 3 m/s. What is the frictional force acting on the block? [Ans. 6 N.]

5. A 4-kg block is accelerated from a velocity of 6 m/s to a velocity of 10 m/s during a time of 2 s. What magnitude of force was acting on the block? [Ans. 8 N.]

Planets, Moons, and Spacecraft—Why do they stay in orbit?

3

GRAVITATION AND PLANETARY MOTION

Throughout most of history, people considered the Earth to be the center of the solar system and, indeed, the center of the Universe. The Sun, the planets, and the stars were believed to move on gigantic spheres with the Earth at the center. Early in the sixteenth century, the Polish astronomer Nicolas Copernicus (1473–1543) proposed a simplified scheme for explaining planetary motions in which the Sun was placed at the center of the system of orbiting planets. We now recognize that the Copernican Sun-centered (or *heliocentric*) theory is the correct description of the solar system.

About a hundred years after the heliocentric theory was proposed, Johannes Kepler (1571–1630) improved upon Copernicus' scheme by showing that the orbits of planets around the Sun are *ellipses* (instead of *circles* as had been supposed). Although Kepler had provided an accurate method for describing planetary motions, there was no real understanding of *why* the planets move the way they do. It remained for Isaac Newton to develop the theory of planetary motion in terms of the universal law of gravitation.

In this chapter we will learn how gravity governs the motion of planets, moons, artificial satellites, and spacecraft, and we will see how we must contend with gravitational forces when we send rocket-powered vehicles on missions into space.

Motion Near the Surface of the Earth

Everyone is familiar with the effect of gravity: When you drop an object it falls downward, toward the Earth. But we can say more than this. If you observe a falling object, you will see that it gains speed as it falls, moving a greater distance during each second of fall than during the preceding second (Fig. 3-1). The higher the point from which an object is dropped, the greater will be its impact velocity at the surface of the Earth. By making measurements we can determine that a falling object gains velocity at a rate of approximately 10 m/s for every second of fall. That is, an object falling freely near the Earth's surface experiences an acceleration of 10 m/s^2. (A more precise value is 9.8 m/s^2, but we will use the approximate value of 10 m/s^2 for simplicity.) Because this acceleration due to gravity is the same (or very nearly so) anywhere

Motion Near the Surface of the Earth   53

Figure 3-1 *An object falling freely near the surface of the Earth gains speed as it falls. For each second of fall, the speed increases by approximately 10 m/s. That is, the acceleration due to gravity is* g = *10 m/s².*

near the surface of the Earth, we give this important number a special symbol and write $g = 10$ m/s². (In British units, $g = 32$ ft/s².)

A falling object is the simplest example of the effect of gravity. But we can also study the motion of an object that is thrown instead of dropped. Suppose that an object is thrown in a horizontal direction from some point near the Earth's surface. We know that the object will move along a curved path and will strike the ground some distance away from the position at which it was released (Fig. 3-2). The higher the point from which the object is thrown and the greater the horizontal velocity it is given, the farther the object will move before striking the ground.

Any object that is thrown by hand will not move a great distance before it strikes the ground. In any such case, therefore, we do not need to consider the fact that the Earth is actually round. However, we can imagine an object that is propelled with a very high initial velocity so that it moves a very great distance before striking the ground. (We ignore the fact that air resistance will drastically affect the distance that the object

54   *Gravitation and Planetary Motion*

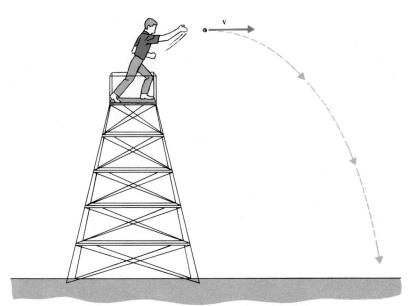

Figure 3-2 *An object that is thrown horizontally from a point near the Earth's surface will move along a curved path and will strike the ground some distance away.*

will travel; that is, we consider the motion in a world that has no atmosphere.) If the distance of travel is sufficiently great, the curvature of the Earth will be important. Figure 3-3 shows the paths of several objects that are projected with different horizontal velocities from the top of a tower. The object with the least initial velocity will strike the Earth at *A*. By increasing the velocity, the projectile can be made to travel to *B* or *C*. If the initial velocity is increased still further, the projectile will travel an even greater distance around the Earth. Some initial velocity can be found so that the projectile will actually move completely around the Earth, returning to its original position. In this case, the projectile's curving path of fall just matches the curvature of the Earth so that the projectile remains always at the same height above the surface. After leaving the mechanism that drives it to its initial horizontal velocity, the projectile falls freely, in the same way that the object in Fig. 3-2 falls freely. That is, the projectile is in free fall *around* the Earth. Because the projectile returns to its starting point with a velocity equal to its initial velocity, the projectile will begin again on its circular path. The projectile is in orbit around the Earth.

Force and Acceleration in Free Fall

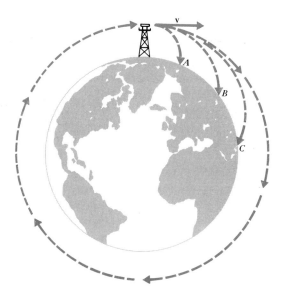

Figure 3-3 *A projectile is fired horizontally from a tower with an initial velocity* **v.** *If* **v** *is small, the projectile will strike the Earth at* A *or* B *or* C. *But if the velocity is sufficiently great, the projectile will travel all the way around the Earth and will return to its starting point. This projectile therefore orbits the Earth.*

Force and Acceleration in Free Fall

In Chapter 2 we learned the connection between force and acceleration. For an object dropped near the surface of the Earth, we can readily identify both of these quantities: The force is due to gravity and the acceleration is g. But what about an object that is falling freely *around* the Earth? In this case there is no change in the height of the object above the Earth's surface. Is such an object accelerating?

In order to answer this question, let us look more closely at the meaning of *acceleration*. We have previously defined acceleration in terms of a *change in velocity*. If the velocity of an automobile changes from 40 km/h to 60 km/h, some push has been required to increase the velocity and the automobile has clearly been accelerated. But we must also remember that velocity has *direction* as well as *magnitude*. A change in the direction of the velocity of an object involves acceleration, even if there is no change in the magnitude. That is, a change in either the magnitude or the direction of the velocity of an object means that acceleration has occurred (Fig. 3-4).

Acceleration implies the action of a force. If you are in an automobile that is coasting with constant velocity along a level road and if you wish to come to a stop, you know that a braking force is necessary to slow the automobile to a stop. Figure 3-5 shows how a force (and therefore an acceleration)

56   *Gravitation and Planetary Motion*

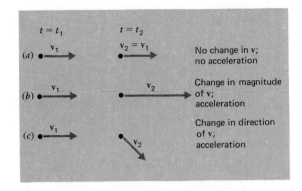

Figure 3-4 *A change in either the magnitude or the direction of the velocity involves acceleration. In each of the cases illustrated here, the object has velocity* v_1 *at the time* t = t_1; *at the later time* t = t_2, *the velocity is* v_2. *Cases* (b) *and* (c) *involve acceleration.*

is also involved in the change of *direction* of a moving object. In this situation, the motion is originally to the right with a velocity v_1. At point *A* the force F_1 acts on the object for a short time. For example, the object might be struck a sharp blow with a hammer. The direction of F_1 is perpendicular to the direction of motion of the object and, as a result, the object is deflected and moves toward point *B*. At *B* the object experiences another perpendicular force and then moves toward *C*. Each application of the force causes the object to change the direction of its motion, but the magnitude of the velocity remains the same.

The direction of acceleration is always the same as the direction of the applied force. In Fig. 3-4b, the acceleration (and therefore the force) is along the direction of motion. But

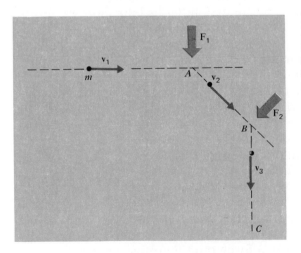

Figure 3-5 *When the forces* F_1 *and* F_2 *act at right angles to the motion, the object is accelerated and moves in a new direction, but the magnitude of the velocity does not change.*

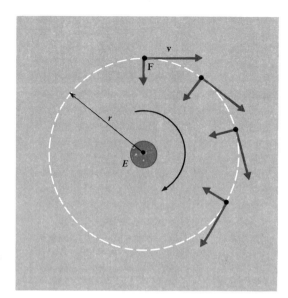

Figure 3-6 *A force (constant in magnitude) that always acts perpendicular to the direction of motion of an object causes the object to move in a circular path. The force* **F** *always points to the center of rotation E. The motion of the object corresponds to that of a satellite in orbit around the Earth.*

in Fig. 3-5, the acceleration is perpendicular to the direction of motion.

Figure 3-5 shows what happens when a force acts momentarily at right angles to the direction of motion of an object. What will be the result if the force acts *continuously* at right angles to the direction of motion? In terms of the situation in Fig. 3-5 this means that the individual pushes occur closer and closer together until they merge, and that the straight sections of the motion become shorter and shorter until finally the motion takes place in a *curved* path. Figure 3-6 shows the result: The object follows a *circular* path. The force is always directed perpendicular to the velocity at any instant and therefore always points toward the center of the circular path. This type of motion corresponds to that of a satellite in orbit around the Earth. The force of gravity on the object is always directed toward the Earth, which lies at the center of the satellite orbit.

By examining the way in which the velocity changes for an object moving in a circular path, it is not difficult to show that the acceleration of the object is equal to the square of its velocity divided by the radius of the orbit: $a = v^2/r$. This acceleration is directed toward the center of the orbit (because the force is directed toward the center) and is therefore called the center-seeking or *centripetal* acceleration.

In Fig. 3-3 we have an object that is freely falling around the Earth in a circular path. We now see that this motion is the result of a constant centripetal acceleration due to the force of gravity acting always toward the center of the Earth.

Newton and the Falling Moon

All of the discussion above was understood by Newton. Indeed, in one of his books Newton included a sketch very similar to Fig. 3-3 to explain how an object could fall freely around the Earth. Newton also knew that the Moon moves around the Earth in a (nearly) circular path. He reasoned that the Moon must be in free fall around the Earth, just like the object in Fig. 3-3. Newton wondered whether an apple that drops to the ground from a tree is subject to the same force that causes the Moon to move in a circular orbit. If so, the accelerations of the two objects should be related. In this way, Newton began his investigation of the *gravitational force*.

First, Newton computed the centripetal acceleration of the Moon, using the expression $a = v^2/r$. He knew the distance r to the Moon and he found the velocity v by dividing the circumference of the orbit $2\pi r$ (the distance traveled in one lunar month) by the time required, 27.3 days. He obtained $a = 0.0027$ m/s². (See the section on *Additional Details* at the end of this chapter.) This value is far smaller than the acceleration due to gravity near the Earth's surface, $g = 10$ m/s². What is the reason for the large difference?

Could it be that the Moon is so massive that the acceleration is thereby diminished? Newton understood that this could not be the explanation. He knew that if two objects with different masses are dropped at the same instant from the same height, they will strike the ground together (Fig. 3-7). (This is the experiment that has been attributed to Galileo, but he probably did not actually drop a cannon ball and a musket ball from the Tower of Pisa, as the story goes.) The fact that the two masses fall at the same rate means that they experience the same acceleration. Thus, the acceleration due to gravity does not depend on mass. Some other explanation must be found for the very small acceleration of the Moon.

Newton found the answer by drawing upon his knowledge of the properties of *light*. Consider a light source that radiates uniformly in all directions. Imagine that we surround this

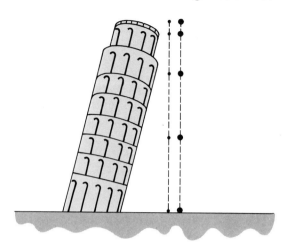

Figure 3-7 *According to the legend, Galileo dropped a cannon ball and a musket ball from the Tower of Pisa to demonstrate that all objects fall at the same rate regardless of mass. Air resistance actually has an effect in such an experiment, causing a small difference in the rates of fall. If the experiment were performed in a vacuum, however, all objects—whether a cannon ball or a feather—would fall at the same rate.*

source with a series of transparent spheres, the first with a radius of 1 m, the next with a radius of 2 m, then 3 m, and so forth. The source is at the center of each sphere, and the light radiated by the source passes entirely through each sphere in turn. It is clear that as the light spreads out from the source, it is distributed over a larger and larger area as it passes through the spheres with increasing size. Let us concentrate on the portion of the light that passes through a small area of the first sphere. In Fig. 3-8 this area is the shaded square at a distance of 1 m from the source *O*. As we can see from the diagram, the light that passes through this area at 1 m is distributed over *four* similar squares on the 2-m sphere and over *nine* similar squares on the 3-m sphere. We say that the *intensity* of the

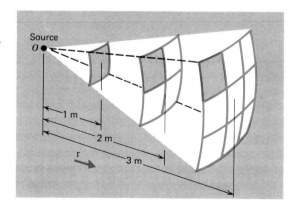

Figure 3-8 *Light that radiates uniformly from the source O decreases in intensity inversely as the square of the distance r from the source. (Each shaded square has the same area.)*

light (that is, the amount of light per unit of surface area) decreases with increasing distance from the source. Compared with the intensity at $r = 1$ m, the intensity at $r = 2$ m is one-fourth and at $r = 3$ m it is one-ninth. Because the square of 2 is 4 and the square of 3 is 9, we can summarize the situation with the statement: *The intensity of light decreases inversely with the square of the distance from the source.* That is,

intensity *is proportional to* $\dfrac{1}{(\text{distance})^2}$

or, using symbols,

$$I \propto \frac{1}{r^2}$$

Newton reasoned that the gravitational attraction that an object (such as the Earth) exerts on another object (such as the Moon) must be similar to the behavior of light intensity. The effect of gravity, he argued, must spread out uniformly in space in the same way that light does. This "effect of gravity" is the acceleration that gravity produces in freely falling objects. Because the Moon is much farther away from the Earth than an object near the Earth's surface, the gravitational acceleration experienced by the Moon should be correspondingly smaller. The amount of decrease should be inversely proportional to the square of the distance.

But what distances should be used for the Moon and for an object near the Earth's surface? Should the distance be measured from the surface of the Earth? Certainly, this cannot be the correct procedure. We can easily verify that an object experiences an acceleration of 10 m/s² whether dropped from a height of 1 m or 100 m above the Earth's surface. Newton reasoned (and was later able to show mathematically) that because the *entire* Earth exerts a gravitational attraction on an object, the distance should be measured from the *center* of the Earth. Newton knew that the distance to the Moon is approximately 60 times the radius of the Earth. That is, the Moon is approximately 60 times farther away from the center of the Earth than is an object near the surface (Fig. 3-9). Therefore, the gravitational acceleration of the Moon in its orbit should be $(60)^2$ or 3600 times smaller than $g = 10$ m/s². Indeed, if we divide 10 m/s² by 3600, we find an acceleration of 0.0027 m/s². This value is equal to the Moon's acceleration found from a knowledge of the Moon's orbit. Newton had succeeded in demonstrating that the effect of gravity decreases as $1/r^2$.

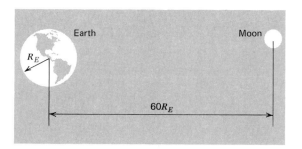

Figure 3-9 *The Moon is approximately 60 times farther away from the center of the Earth than is an object near the Earth's surface. The factor $(60)^2 = 3600$ relates the acceleration g of an object falling near the surface of the Earth to the acceleration of the Moon in its orbit.*

Newton went on to complete his analysis by noting that the *gravitational force* acting on an object must be equal to the *mass* of the object multiplied by its *acceleration*. (This is Newton's law of motion, $F = m \times a$, discussed in Chapter 2.) Furthermore, if the mass of the Earth were increased, it would exert a greater gravitational effect on the Moon. Therefore, the gravitational force exerted by the Earth on the Moon must be proportional to the product of the two masses divided by the square of the distance separating their centers:

$$\text{gravitational force} \propto \frac{\text{product of masses}}{(\text{distance})^2}$$

or, using symbols,

$$F_G \propto \frac{m \times M}{r^2}$$

This is the statement of *Newton's law of gravitation*.

The great importance of Newton's analysis is that the gravitational force law is not limited to the case of the Earth and the Moon. Newton applied his law to the motions of planets and comets with equal success. Indeed, as far as we know today, Newton's gravitational force law applies to *any* pair of objects—it is truly a *universal* law.

One of the first great successes of Newton's law of gravitation was in the discovery of the planet Neptune. The planets Mercury, Venus, Mars, Jupiter, and Saturn were all known to ancient peoples because they are bright and easy to observe with the unaided eye. The first discovery of a planet by telescope was the detection in 1781 of Uranus, the planet that follows Saturn in order of distance from the Sun. Study of the motion of Uranus over many years revealed that there was no orbit that could satisfactorily describe its motion, even after account was taken of the gravitational influences of the massive planets Saturn and Jupiter. Because of the discrepancy between theory and observation, some scientists thought that Newton's law of gravitation might contain a defect. Two young astronomers, John Couch Adams (1819–1892) of England and Urbain Leverrier (1811–1877) of France, independently hit upon the idea that Uranus was not following its prescribed orbit because it was being disturbed by the gravitational attraction of another planet, farther from the Sun and as yet undiscovered. Both men proceeded to calculate, on the basis of Newton's law, where this new planet must be in order to account for the observed motion of Uranus. When the results were finally transmitted to the astronomers at the Berlin Observatory in 1846, less than an hour was required to locate the planet we now call Neptune. This discovery remains a monument to the power of calculational techniques applied to Newton's law of universal gravitation!

Elliptical Orbits

In our discussions thus far we have assumed that a satellite moves around its parent body in a circular orbit. The Moon's orbit around the Earth is nearly circular but not exactly so. And the orbits of the various planets around the Sun likewise are close to but not exactly circular. Kepler was the first to realize that the observed positions of the planets in the sky are not consistent with the assumption of circular orbits. He succeeded in demonstrating that the paths followed by the planets around the Sun are *ellipses,* not circles (Fig. 3-10). Kepler announced this discovery in 1609.

An ellipse can be thought of as a flattened circle. The ellipse shown in Fig. 3-10 deviates much more from a circle than any planetary orbit. In the case of the Earth's orbit, the long dimension of the ellipse (called the *major diameter*) is only 3 percent greater than the short dimension (called the *minor diameter*).

Figure 3-10 indicates that there are two points, called the *foci,* that are important in the geometry of an ellipse, whereas there is only one point (the *center*) that is important for the circle. We can see the significance of the foci in the following way. Choose two points, F and F' (Fig. 3-11), and attach one end of a string to F and the other end to F'. Next, take the point of a pencil and pull aside the string until it is taut, as in

Elliptical Orbits   63

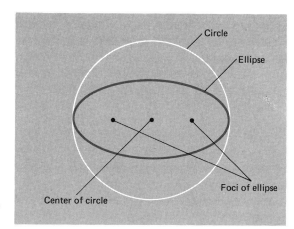

Figure 3-10 *An ellipse compared with a circle. The major diameter of the ellipse is the same as the diameter of the circle. The significance of the foci of the ellipse is explained in Fig. 3-11.*

Fig. 3-11. Finally, move the pencil point along the path that maintains the tautness of the string. The path that is drawn by the pencil point is an ellipse. Because the length of the string does not change as the path is drawn, we can say that an ellipse is defined by all of the points (such as point P in Fig. 3-11) for which the sum of the distances FP and $F'P$ remains constant. By changing the length of the string and the distance between the foci, ellipses with all possible shapes, from very thin to almost circular, can be drawn. When the two foci coin-

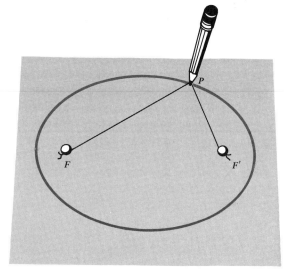

Figure 3-11 *To draw an ellipse, attach the ends of a string to the points* F *and* F'. *Then, draw the path that maintains the string taut.*

cide, the path is exactly circular. Thus, a circle is actually a special kind of ellipse.

The foci of an ellipse have a physical as well as a geometrical importance. If a small object orbits around a large object because of the gravitational attraction between the pair, the large object will be located at one of the foci of the ellipse followed by the small object. (The other focal point does not have any physical significance in this case.) Thus, if we construct the elliptical orbit followed by the Earth, we find the Sun located at one of the foci.

Although the planets follow paths around the Sun that are not very different from circles, the orbits of *comets* are highly elliptical. The most famous comet is *Halley's comet*, named for the English astronomer Edmund Halley (1656–1742), who first realized that several reports of comets over the years actually referred to a single comet that reappears at intervals of approximately 76 years. The most recent approach of Halley's comet to the vicinity of the Sun occurred in 1910, and the next visit will be in 1986. After swinging around the Sun in its elongated elliptical orbit (Fig. 3-12), Halley's comet recedes to the outer part of the solar system, reaching a maximum distance from the Sun outside the orbit of Neptune. A comet becomes visible when near the Sun because only then

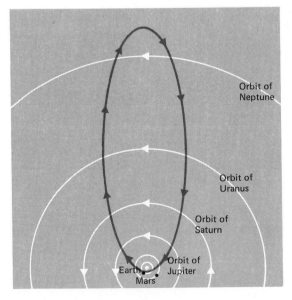

Figure 3-12 *Halley's comet moves in a highly elliptical orbit that reaches beyond the orbit of Neptune. Halley's comet reappears near the Sun every 76 years (next in 1986).*

Spacecraft Orbits

do the Sun's rays warm the cometary material, producing a flowing trail of vapor and dust.

Spacecraft Orbits

Many of the research satellites that have been placed into orbit around the Earth are designed to study the Earth's space environment over a range of distances from the Earth. In order to make measurements at various distances, these satellites must be placed into elliptical orbits. Let us look more closely at these orbits and how space vehicles are placed into them.

We know from the discussion of Fig. 3-3 that if an object has too small a velocity near the Earth, it will not achieve an orbit and will fall back to the Earth. (This is the situation for cases A, B, and C in Fig. 3-3.) Generally, the closer the orbit is to the Earth, the greater must be the velocity to maintain the object in the orbit. For example, the orbital velocity of the Earth around the Sun is approximately 30 km/s, whereas the orbital velocity of Venus (which is nearer to the Sun) is 35 km/s and that of Jupiter (which is farther from the Sun) is 13 km/s. This variation of orbital velocity with distance is true not only for circular orbits of different sizes but also for the different positions around an elliptical orbit. Figure 3-13 shows this effect for a spacecraft in an elongated elliptical orbit around the Earth. The point at which the spacecraft approaches closest to the Earth is labeled *P* (called the *perigee* of the orbit), and the orbit point farthest from the Earth is labeled *A* (called the *apogee* of the orbit). The different segments of the orbit that are denoted by the marks correspond

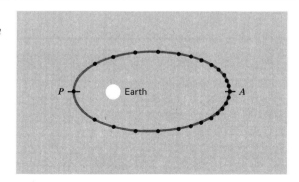

Figure 3-13 *The marks indicate equal time intervals in the motion of the satellite around the Earth. The orbital velocity is greatest at perigee* (P) *and least at apogee* (A).

to equal *time* intervals around the orbit. The spacecraft moves a much greater distance in one time interval near P than it does near A. That is, the orbital velocity is greatest at perigee and least at apogee. The spacecraft moves toward P with increasing speed, sweeps quickly through its closest approach to the Earth, and then slows down as it moves toward the most remote part of the orbit.

How can a spacecraft be placed into a particular circular or elliptical orbit? The procedure for a circular orbit is to fire the first two stages of the rocket (Chapter 2) in order to bring the vehicle to a height of several hundred kilometers. At the desired height, the spacecraft is maneuvered into a position parallel to the Earth's surface below and the third rocket stage is fired until the appropriate velocity is reached. For a circular orbit at a height of 200 km above the Earth's surface, the orbital velocity must be 7.8 km/s (28 100 km/h), and the orbit period will be 90 min.

Usually, this direct placement of an artificial satellite into a circular orbit is restricted to orbits with heights of a few hundred kilometers. If elliptical orbits or circular orbits with greater heights are desired, two or three steps are followed. The first step involves establishing a circular orbit at a height of 200 or 300 km in the manner just described. This orbit becomes a *parking orbit* from which further maneuvers are carried out to achieve a different orbit. Figure 3-14 shows this parking orbit. If an elliptical orbit is required, the rocket engine is fired again at point P. The velocity of the spacecraft is now too great for a circular orbit through P, so it loops farther out into space and follows an elliptical orbit that has its perigee at P.

This elliptical orbit may also not be the desired final orbit. In such a case it becomes another parking orbit (or *transfer ellipse*) for additional maneuvers. If the rocket engine is fired again at the apogee A of the orbit (Fig. 3-14), the orbital velocity can be increased until it is equal to that necessary for a circular orbit through A. This large-diameter circular orbit has been achieved in three steps: (1) original boost into a small-diameter parking orbit, (2) firing of rocket engine at P to produce an elliptical transfer orbit, and (3) firing of the engine again to achieve the desired final circular orbit. The reason this procedure is followed is that less fuel (and therefore less rocket bulk) is required compared with placing the satellite directly into the final orbit in a single step.

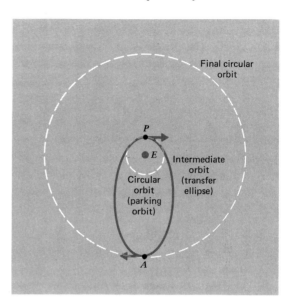

Figure 3-14 *In order to place a space vehicle into an Earth orbit with a large diameter, the vehicle is first boosted into a circular orbit at a height of 200 or 300 km. After remaining in this* parking orbit *for some period of time, the rocket engine is fired at* P *and the orbit becomes an elliptical orbit with perigee at* P. *This intermediate orbit (or* transfer ellipse) *is used to increase the altitude to point A where the engine is fired again to produce the final circular orbit.*

If a space vehicle is placed into a circular orbit at a height of 35 900 km above the Earth's surface (or 42 200 km from the center of the Earth), the orbital velocity will be 3.07 km/s. (Refer to the section on *Additional Details* at the end of this chapter.) At this speed, exactly 24 h will be required for the satellite to circle once around the Earth. But during this same 24-h period, the Earth turns once on its axis. If the satellite orbit lies in the plane that defines the Earth's equator, the Earth and the satellite will rotate together at the same rate around the same axis (Fig. 3-15). Thus, the satellite will appear to stand motionless above a spot on the equator. Such a stationary satellite is called a *synchronous satellite*. (To move in *synchronism* means to move at the same rate.)

Synchronous satellites are being used in increasing numbers in the international communications network. The importance of these devices lies in the fact that high-frequency radio signals (such as those used for television and other forms of communication) travel in straight lines and therefore do not propagate around the curvature of the Earth. A synchronous satellite, however, is in straight-line contact with nearly one-half of the Earth's surface. A radio signal can be transmitted from one station to another far around the Earth by using a synchronous satellite as a relay station (Fig. 3-16).

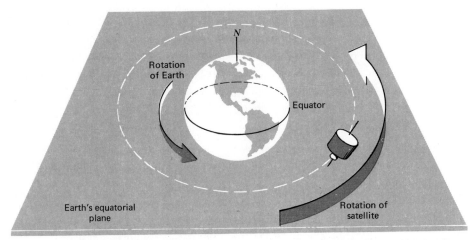

Figure 3-15 *A satellite that moves around the Earth in an orbit that lies in the Earth's equatorial plane and has a radius of 42 200 km will rotate at the same rate as the Earth. This* synchronous satellite *will appear to remain motionless above a spot on the equator.*

In addition to the synchronous communications satellites, there are in orbit around the Earth many nonsynchronous satellites which are used for a variety of purposes. Some of these are equipped to study the Earth's magnetic field and radiation belts. Some have special telescopes and radiation detectors for observing the Sun and stars. And others are instrumented to yield information about the Earth itself. Among these are

Figure 3-16 *A satellite placed in a synchronous orbit can be used to relay radio signals between points on the Earth that would not otherwise be able to communicate because of the straight-line propagation of high-frequency radio signals.*

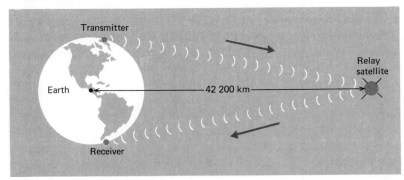

the Earth Resources Technology Satellites (ERTS, now renamed Landsat I and II), the Applications Technology Satellites (ATS), and various meteorological satellites. The instruments aboard these satellites have provided a wealth of information about geography, land use, agriculture, and weather. We are only now beginning to reap the full benefits of these remarkable devices.

Planetary Missions

Suppose that we wish to launch a spacecraft on a trip to another planet. How do we use our knowledge of gravitation and planetary orbits to plan such a mission? We cannot simply fire a rocket at the position in the sky where we observe the planet. We must take account of the motions of the Earth and the planet around the Sun and the gravitational effects that influence the motion of the spacecraft.

First, consider what will happen if a rocket is fired away from the Earth with a certain initial velocity. If the initial velocity is low, we know that the rocket will fall back to Earth (Fig. 3-17). As the velocity is increased, the rocket will rise to greater heights. Finally, if the velocity exceeds 11.3 km/s, the rocket will break the gravitational tie that binds it to the Earth. That is, with an initial velocity of exactly 11.3 km/s,

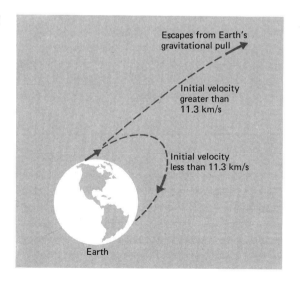

Figure 3-17 *If a rocket is launched from the surface of the Earth with an initial velocity of 11.3 km/s or greater, it will escape from the gravitational pull of the Earth.*

70 *Gravitation and Planetary Motion*

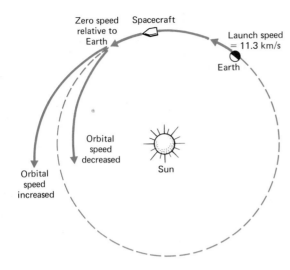

Figure 3-18 *A spacecraft in an Earth orbit can be launched toward another planet by firing the rocket motor to increase or decrease the orbital speed. The Earth's orbital speed is 30 km/s.*

the gravitational pull of the Earth will slow the object as it moves away, but the direction of motion will never be reversed, and the object will never return to the Earth. At a great distance from the Earth, the speed of the object relative to the Earth will approach zero. (If the launch speed exceeds 11.3 km/s, the object will always move away from the Earth.)

Suppose that we launch a spacecraft with a velocity relative to the Earth of 11.3 km/s and in a direction corresponding to the Earth's orbital motion around the Sun (Fig. 3-18). At some large distance along the orbit, the spacecraft will slow to a speed relative to the Earth that is essentially zero. In this condition, the spacecraft moves in the same solar orbit as does the Earth and with the same orbital speed (30 km/s). The spacecraft is essentially in a circular parking orbit *around the Sun*. If we wish to change this orbit and send the spacecraft toward another planet, we must fire the rocket motor and either increase or decrease the orbital velocity. If the orbital velocity is increased, the spacecraft will spiral outward; if the orbital velocity is decreased, the spacecraft will spiral inward. By timing the moment and duration of the rocket firing, the spacecraft could be made to approach either Mars (the next planet away from the Sun) or Venus (the next planet toward the Sun).

Actually, when a spacecraft is launched on a mission to another planet, the vehicle is not first placed in an Earth orbit, as indicated in Fig. 3-18. Instead, the spacecraft is placed in a high-altitude parking orbit around the Earth, and then the

Planetary Missions   71

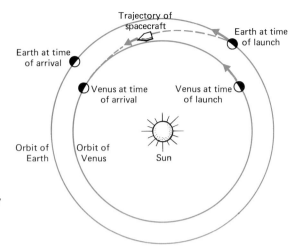

Figure 3-19 *Diagram of a 100-day space flight from Earth to Venus. The Earth-Venus distance at the time of arrival is 48 million km.*

engine is fired to increase the speed above the escape speed, sending the spacecraft along a transfer orbit to intercept the planet.

Figure 3-19 shows the trajectory for a mission to Venus. Notice that at the time of launch, Venus is located *behind* the Earth in its orbit. As the mission progresses, Venus overtakes the Earth because of its greater orbital speed. The time of travel of the spacecraft in the mission depicted here is 100 days. When the spacecraft finally arrives at Venus, the Earth is near its closest approach to the planet (approximately 48 million kilometers). In any interplanetary mission, the time of launch and the initial speed must be carefully coordinated in order that the spacecraft actually arrive at its desired destination. If it is determined, after the launch, that the trajectory is not accurate, a mid-course correction can be made by a brief firing of the rocket motor. Upon arrival at the planet, the motor may be fired again to place the spacecraft in an orbit around the planet. Thus far, four planets (Venus, Mars, Jupiter, and Mercury) have been visited by spacecraft. Several vehicles have been maneuvered into orbit around Mars and Venus.

Questions and Exercises

1. A spacecraft is placed into a circular orbit around the Moon at a height of 5 km above the surface. Such an orbit is practical

around the Moon (which has no atmosphere) but not around the Earth. Explain why.

2. You are driving along a certain road in an automobile and the speedometer reads a steady 60 km/h. (It's a European car.) A companion mentions that you are actually accelerating. Can he be right? Explain.

3. Two students are standing on a platform that rotates at a constant speed around an axis. Student A stands at a point 2 m from the rotation axis and student B stands at a point 4 m from the axis. What is B's velocity compared with A's? What is B's acceleration compared with A's?

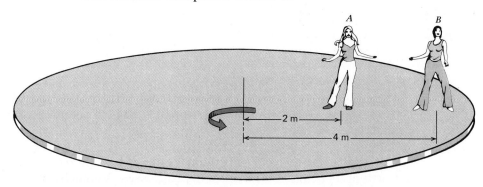

4. The gravitational force on an object is proportional to its mass. Why, then, do two objects with different masses fall at the same rate even though the force on the more massive object is greater than the force on the less massive object?

5. The symbol g stands for the acceleration due to gravity. Near the surface of the earth, g has the value 10 m/s^2; near the surface of the Moon, g has the value 1.6 m/s^2. Is there a point between the Earth and the Moon where $g = 0$? Explain.

6. When an ICBM (intercontinental ballistic missile) is fired, the rocket engine operates for a short period and then the missile coasts to its distination. What is the shape of the path followed by an ICBM?

7. What do you suppose the terms *perihelion* and *aphelion* mean? (*Helios* = Sun.)

8. The orbital speed of a spacecraft in a high-altitude circular orbit around the Earth is smaller than that of a spacecraft in a low-altitude orbit. Why is it not possible to increase the orbit altitude by simply firing the rocket engine to decrease the speed of the spacecraft?

9. A spacecraft is in a circular parking orbit. The spacecraft is maneuvered until the nose is pointing in the direction *opposite* to its orbital motion. The rocket engine is then fired briefly. Sketch the new orbit and compare with the original orbit. Why would the astronauts aboard the spacecraft ever execute such a maneuver?

10. An astronomer claims to have discovered a new planet between the orbits of Jupiter and Saturn that has an orbital period of 240 days. What do you think of his claim and why?

Additional Details for Further Study

Some Useful Data

Radius of the Earth = 6.38×10^6 m = 6380 km
Radius of the Moon = 1.74×10^6 m = 1740 km
Earth-Moon distance = 3.84×10^8 m = 384 000 km
Mass of Sun = 2.0×10^{30} kg
Mass of Earth = 6.0×10^{24} kg
Mass of Moon = 7.4×10^{22} kg
Gravitational constant, $G = 6.67 \times 10^{-11}$ N-m²/kg²

Centripetal Acceleration of the Moon

The centripetal acceleration of an object moving with a speed v in a circular orbit with a radius r is

$$a = \frac{v^2}{r}$$

The speed v is equal to the circumference of the orbit, $2\pi r$, divided by the time required for one complete revolution, the period τ:

$$v = \frac{2\pi r}{\tau}$$

Therefore,

$$a = \frac{(2\pi r/\tau)^2}{r} = \frac{4\pi^2 r}{\tau^2}$$

For the moon,
$r = 3.84 \times 10^8$ m
$\tau = 27.3$ days $= 2.36 \times 10^6$ s

From which we obtain

$$a = \frac{4\pi^2 \times (3.84 \times 10^8 \text{ m})}{(2.36 \times 10^6 \text{ s})^2} = 0.0027 \text{ m/s}^2$$

The Gravitational Force Law

The gravitational force between two objects with masses m and M whose centers are separated by a distance r is

$$F_G = G \frac{m \times M}{r^2}$$

where the constant G is called the *gravitational force constant* and has the value

$$G = 6.67 \times 10^{-11} \text{ N-m}^2/\text{kg}^2$$

That is, if the masses are stated in *kilograms* and the distance is stated in *meters*, and if G is set numerically equal to 6.67×10^{-11}, the force will be given in *newtons*. For example, the gravitational force between the Earth and the Moon can be calculated by using

mass of Moon $= m = 7.4 \times 10^{22}$ kg
mass of Earth $= M = 6.0 \times 10^{24}$ kg
Earth-Moon distance $= r = 3.84 \times 10^8$ m

Then,

$$F_G = (6.67 \times 10^{-11}) \frac{(7.4 \times 10^{22}) \times (6.0 \times 10^{24})}{(3.84 \times 10^8)^2}$$
$$= 2.0 \times 10^{20} \text{ N}$$

As a check on this result, remember that the force exerted on the Moon must also be given by Newton's law, $F = ma$. We know the mass of the Moon and its acceleration (0.0027 m/s^2). Therefore,

$$F = m \times a = (7.4 \times 10^{22} \text{ kg}) \times (0.0027 \text{ m/s}^2)$$
$$= 2.0 \times 10^{20} \text{ N}$$

which agrees with the previous calculation.

Synchronous Orbits

A satellite in a circular orbit around the Earth experiences a centripetal acceleration $a = v^2/r$. The force on the satellite is equal to its mass multiplied by this acceleration:

$$F = m \times a = m \times \frac{v^2}{r}$$

According to the gravitational force law, the force on the satellite can also be expressed as

$$F = G \frac{m \times M}{r^2}$$

where M is the mass of the Earth. Equating these two expressions for the force, we have

$$G \frac{m \times M}{r^2} = m \times \frac{v^2}{r}$$

The satellite mass m cancels in this equation. This means that the mass of a satellite is irrelevant in determining the orbit characteristics. Solving for v^2,

$$v^2 = G \frac{M}{r}$$

and taking the square root,

$$v = \sqrt{\frac{GM}{r}}$$

For a distance r (measured from the center of the Earth) equal to 42 200 km = 4.22×10^7 m,

$$v = \sqrt{\frac{(6.67 \times 10^{-11} \text{ N-m}^2/\text{kg}^2) \times (6.0 \times 10^{24} \text{ kg})}{4.22 \times 10^7 \text{ m}}}$$
$$= 3.07 \times 10^3 \text{ m/s} \quad \text{(or 3.07 km/s)}$$

At this velocity, the period of the motion is

$$\text{period} = \frac{\text{circumference of orbit}}{\text{velocity}}$$

Thus,

$$\tau = \frac{2\pi r}{v}$$

$$= \frac{2\pi \times (4.22 \times 10^7 \text{ m})}{3.07 \times 10^3 \text{ m/s}}$$

$$= 86\,400 \text{ s}$$

which is exactly equal to 1 *day*. (Number of seconds in 1 day: 24 h × 60 min/h × 60 s/min = 86 400 s/day.) That is, a satellite in a circular orbit at a distance of 42 200 km moves in synchronism with the Earth's rotation. If a satellite is placed in such an orbit directly above the Equator, it will appear to remain stationary. Satellites placed in synchronous orbits are used as "fixed" relay stations in the international communications network. News events can now be televised around the world "via satellite."

Weight

We sometimes hear the terms *mass* and *weight* used interchangeably. But these two terms actually refer to different physical ideas. The *mass* of an object is a measure of the amount of matter in the object, and the mass does not change with changes in position of the object. A 1-kg object will have a mass of 1 kg whether it is on the surface of the Earth, on the Moon, or in space. The *weight* of an object, on the other hand, is the gravitational force acting on the object. Because it is a force, weight can be specified by using Newton's force equation, $F = ma$. We identify the acceleration in this equation as the acceleration due to gravity g, and we identify the force as the weight w. Therefore, the weight of an object with a mass m and subject to a gravitational acceleration g is

$$w = mg$$

Notice that weight is measured in the same units as force: namely, in *newtons*. Thus, the weight of an 80-kg man is

$$w = (80 \text{ kg}) \times (10 \text{ m/s}^2) = 800 \text{ N}$$

Because weight depends on the acceleration due to gravity, the weight of a particular object will be different at places where the gravitational acceleration is different. A 1-kg mass on the surface of the Earth has a weight of 10 N. But this same mass, when on the surface of the Moon (where the acceleration due to gravity is only 1.6 m/s²) will have a weight of

Weight

(1 kg) × (1.6 m/s²) = 1.6 N. An "average" man whose mass is 75 kg (165 lb) has a weight of 750 N.

Additional Exercises

1. How does the acceleration of an automobile compare with the acceleration of gravity g? Suppose that the automobile can accelerate from rest to a speed of 100 km/h in 14 s. [Ans. $a = 2$ m/s².]

2. The acceleration due to gravity of an object falling freely near the Earth's surface is 10 m/s². What will be the acceleration at a height of 6380 km above the Earth's surface? [Ans. 2.5 m/s².]

3. A 2-kg ball swings in a horizontal circle at the end of a 1.5-m rope. The speed of the ball is 6 m/s. What force must be exerted on the rope to maintain the ball in its circular path? [Ans. 48 N.]

4. The gravitational force on an object near the Earth's surface is equal to its mass multiplied by the gravitational acceleration ($F = ma$). The gravitational acceleration is g. The force is also given by Newton's law of gravitation, $F = GmM/r^2$, where M is the mass of the Earth and r is the Earth's radius. Equate these two expressions for the force and solve for the gravitational acceleration g. Use the known values of G, M, and r, and compute the value of g. [Ans. $g = GM/r^2 = 9.8$ m/s².]

5. A spacecraft is in a circular parking orbit at an altitude of 200 km above the Earth. What is the orbital speed of the spacecraft, and what is the period of the orbit? [Ans. 7.8 km/s; 1.5 h.]

6. A spacecraft is sent to the Moon and there is placed into a circular orbit at a height of 5 km above the Moon's surface. What is the orbital speed of the spacecraft and what is the period of the orbit? [1.7 km/s; 1.8 h.]

7. What is the weight of an 80-kg astronaut who is at a height above the Earth's surface equal to the radius of the Earth? [Ans. 200 N.]

What is the energy crisis?

4

ENERGY, POWER, AND NATURAL RESOURCES

Energy is a term that we are very much accustomed to hearing and using. We read that our energy supplies are dwindling and that we are faced with an "energy crisis." When we pay our fuel bills (or *energy* bills)—for gasoline, heating oil, and natural gas—we are made painfully aware that energy costs are rising. And we are reminded again of this fact when we receive the charges for the electricity that we use.

Are we actually in danger of running out of useful energy? Are we faced with the prospect of darkened cities, a lack of transportation facilities, and no heat for our homes? Actually, the world's energy sources are plentiful. Reserves of coal are sufficient to last for several centuries; we receive vast amounts of energy from the Sun; there is a huge and almost untapped reservoir of heat within the Earth; and the supplies of nuclear fuels are almost inexhaustible. In view of these facts, why should we be concerned about an "energy crisis"?

In this chapter we will first discuss the ideas of *work, energy,* and *power.* Then we will look at the situation with regard to energy sources and supplies in order to determine the extent of the "energy crisis."

Work and Energy

We all have some intuitive feeling for what is meant by the terms *work* and *energy*. We know, for example, that we must buy gasoline to supply the energy that does the work of running our automobiles. And we know that an electric motor can do work for us if it is supplied with energy from electric power lines.

"Work" is a common term and we frequently use it in ordinary conversation. We might say, for example, that a particular job requires a great amount of *work*. If you lift a number of heavy boxes and place them on a high shelf, you feel tired after the job is completed—you know that you have done *work*. This is exactly right. Gravity pulls the boxes downward and when you lift the boxes, you are doing work against the gravitational force.

In its physical meaning, *work* always involves overcoming some opposing force. Suppose that instead of lifting one of the boxes, you push it across a rough floor. In this case, you are

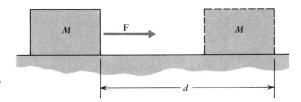

Figure 4-1 *The work done by the force **F** is* F × d.

not working against the gravitational force because the box is at the same height throughout the movement. Instead, you are now working against the frictional force that exists between the moving box and the floor.

How do we measure work? The amount of work done in any situation depends on how much force was exerted and on how far the object moved. Increasing either the applied force or the distance through which the object is moved increases the amount of work done. That is, the work done is proportional to both the applied force and the distance through which the force acts (Fig. 4-1). We can state that

work = (force) × (distance)

or, using symbols,

$W = F \times d$

In the case of lifting a box and placing it on a shelf, the force F that we must work against is the gravitational force. We know from Newton's law, $F = m \times a$, that the gravitational force is equal to the mass of the object multiplied by the acceleration due to gravity; that is, $F = m \times g$. Therefore, the work done in lifting a box with a mass m through a height h is (Fig. 4-2)

work = (gravitational force) × (height)

or

$W = mgh$

The greater the mass of the box or the greater the height through which it is lifted, the greater the amount of work that is done. This agrees exactly with our experience.

Driving a stake into the ground requires work. With each blow, the stake moves a certain distance against the resisting force of the ground. The amount of work done is equal to the product of the resisting force and the distance moved. One

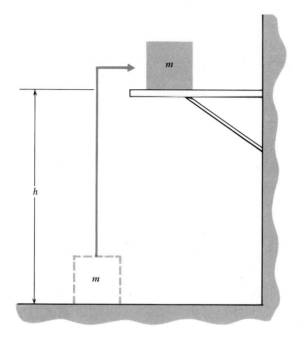

Figure 4-2 *The amount of work done in lifting the box from the floor to the shelf is* mgh.

way of driving a stake is to raise a heavy block to some height and allow it to fall on the stake (Fig. 4-3). An amount of work *mgh* is expended in raising the block to the height *h*. When the block is released, it will fall, strike the stake, and drive it into the ground. The block is raised at some time, and at a later time the falling block delivers the work done on it to the stake. That is, the capacity to do work is stored in the block in its raised position. We say that the raised block possesses *potential energy*. This energy is stored as long as the block remains in its raised position and can be recovered by releasing the block, allowing it to fall and do work by driving the stake into the ground. In fact, one way to define *energy* is

Energy is the capacity to do work.

Whenever work is done against some opposing force and this work is recoverable, the object or the system possesses potential energy. For example, in Fig. 4-4a we have a relaxed spring and a mass at rest; this system has no potential energy. In Fig. 4-4b, however, we see the result of pushing on the block to compress the spring. If the block is now released, the spring will push on the block and set it into motion. The

Work and Energy   83

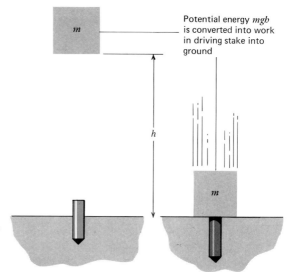

Figure 4-3 *In the raised position, the block has a potential energy equal to the work done* (mgh) *in raising the block to the height* h. *This potential energy can be recovered as useful work by allowing the block to fall and drive the stake into the ground.*

moving block can then do work on some other object (such as driving a nail into a wall).

There is an important point here that we must emphasize. Potential energy does not itself do work; potential energy represents a stored *capacity* to do work. Notice what happens in the case of the raised block and in the case of the compressed

Figure 4-4 *If work is done to compress the spring, potential energy is stored in the spring. By releasing the spring, this potential energy can be recovered.*

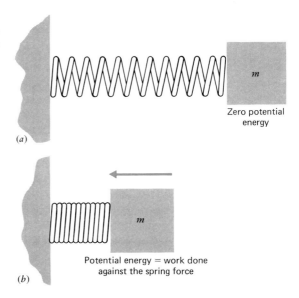

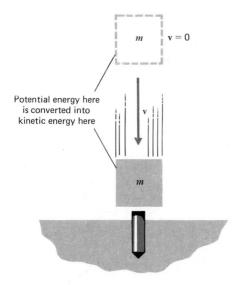

Figure 4-5 *In the raised position, the block possesses potential energy which is converted into kinetic energy during the fall. This kinetic energy is, in turn, converted into the work done in driving the stake into the ground.*

spring. Before either block can do useful work, it must be set into motion. It is the *moving* block in each case that does work by driving the stake into the ground or the nail into the wall. Look again at Fig. 4-3. By the time the falling block nears the ground, it has lost most of its original potential energy. But the block is moving with a velocity **v**. Even though it has almost no potential energy in this position, the block is still capable of driving the stake into the ground. That is, the block possesses energy (the capacity to do work) because of its *motion*. We call this energy the motional or *kinetic* energy. The potential energy of the block in its raised position is converted into kinetic energy during the fall (Fig. 4-5). At the highest position, when the block is at rest, the energy of the block is entirely potential energy. During the fall, the energy is partly potential and partly kinetic. At ground level, the energy is entirely kinetic energy.

Conservation of Energy

In Chapter 2 we discussed momentum and pointed out that the conservation of momentum is one of the important principles that assist us in describing the way Nature behaves. Another quantity that is always conserved is *energy*. Refer again to the example of the falling block that drives a stake into the ground (Fig. 4-3). An amount of work *mgh* was required to lift the block from ground level to the height *h*. In the raised position, the block possessed a potential energy *mgh*. During the fall, the block's energy was partly potential and partly kinetic, but the *total* energy was always *mgh*. Finally, the block did an amount of work *mgh* in driving the stake into the ground. In every phase of the process the total amount of energy was the same; energy was converted from one form into another, but no energy was lost. *Energy was conserved.* The same is true for *every* process that takes place in Nature.

The true importance of the principle of energy conservation cannot be fully understood or appreciated unless it is realized that energy appears in many forms. If we add up all of the energy in its various forms that an isolated system possesses before an event or process takes place and then do the same afterward, we always find an exact balance. We can make this calculation only if we know all of the ways in which energy can appear. If we did not realize the existence of potential energy, for example, we would discover many situations in which there is an apparent increase or decrease in energy.

The Various Forms of Energy

We know that a moving object has the capacity to do work on some other object or system—this we call *kinetic energy*. We can also identify kinetic energy in other kinds of situations. For example, in heated objects we refer to *thermal* energy. The atoms and molecules in every piece of matter—solid, liquid, or gas—are in a state of continual motion. This random, agitated motion represents an *internal* kinetic energy or thermal energy. An object possesses this type of kinetic energy even though the object as a whole may not be in motion. An increase in the internal energy of an object can be brought about by doing work on the object. We could do this work by supplying heat from a flame or by mechanical action (for example, by hitting a block of metal with a hammer). In either case, the atoms and molecules are caused to move more rapidly and we sense this change in thermal energy as an increase in the *temperature* of the object. (We will have more to say about heat and temperature in Chapter 6.)

The transmission of *sound* from one point to another takes place when the sound source (for example, a vibrating speaker cone) sets into motion the air molecules in its immediate vicinity. These molecules collide with other nearby molecules and further molecular collisions cause the propagation of the sound to other points. Thus, sound is due to molecular motions and constitutes another form of kinetic energy.

When we lift an object we do work against the gravitational force and the raised object has the capacity to do work on another object or system—this stored energy we call the *gravitational potential energy*. There are other forms of potential energy because there are other forces in addition to the gravitational force against which work can be done and from which energy can be recovered. In Fig. 4-4b we have a case in which energy is stored in a compressed spring. What is the nature of this potential energy? To answer this question, we must real-

Figure 4-6 *Three situations in which work is done by or against an electrical force.* (a) *In moving the positive charge from* A *to* B, *work must be done by an outside agency against the repulsive electrical force between the two positive charges. The result is an increase in the electrical potential energy associated with the two charges.* (b) *If the charge at* A *is released, it will move away from the other charge because of the repulsive electrical force. At* B *the charge will be moving with a certain velocity and some of the original electrical potential energy will have been converted into kinetic energy.* (c) *A negative charge moves from* A *to* B. *This represents the movement that an atomic electron might experience in a chemical reaction. Energy is* released *in this case. Can you see why? (The negative charge is repelled by the other negative charge and is attracted by the positive charge.)*

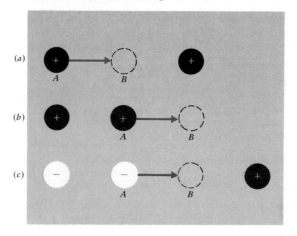

ize that all matter is composed of atoms and molecules that are held together by *electrical forces,* the forces that act between electrically charged objects. When a spring is compressed (or stretched), work is done against the electrical forces between the positively and negatively charged particles that make up the material (that is, between electrons and nuclei). Thus, the energy stored in a spring is *electrical potential energy.*

There are many other situations in which the stored energy is electrical potential energy. Any process in which atoms are displaced or in which atoms are rearranged in molecules involves work by or against electrical forces. Three situations involving electrical potential energy are shown in Fig. 4-6.

When gasoline burns or when dynamite explodes, the potential energy stored in the substance is converted into heat or motional energy. When we burn the fuel *methane,* CH_4 (the primary component of *natural gas*), the reaction is

$$CH_4 + 2O_2 \longrightarrow CO_2 + 2H_2O$$

That is, methane combines with oxygen to produce carbon dioxide (CO_2) and water (H_2O).

The burning of methane releases energy. Where does this energy come from? We can represent the reaction in the following schematic way:

$$\underset{CH_4}{H-\underset{\underset{H}{|}}{\overset{\overset{H}{|}}{C}}-H} \; + \; \underset{2O_2}{\begin{matrix} O=O \\ O=O \end{matrix}} \longrightarrow \underset{CO_2}{O\!\!=\!\!C\!\!=\!\!O} \; + \; \underset{2H_2O}{\begin{matrix} H \\ O{<}_H \\ H \\ O{<}_H \end{matrix}}$$

$$CH_4 \quad + \quad 2O_2 \quad \longrightarrow \quad CO_2 + 2H_2O$$

where each short line connecting two element symbols represents a pair of electrons that bind the two atoms together. In order for the reaction to proceed, several atomic bonds must be broken and new ones formed. There is a certain amount of electrical potential energy in every molecule due to the arrangement of the electrons around the positively charged nuclei. Some arrangements of electrons and nuclei have more potential energy than others. There is *more* electrical potential energy in the combination $CH_4 + 2O_2$ than when the same atoms are in the arrangement $CO_2 + 2H_2O$. Thus, the burning of methane to produce carbon dioxide and water *releases* energy. All forms of *chemical energy* are basically *electrical* in character.

The rearrangement of atoms to form different molecules involves energy changes. The chemical processes that take place when fuels are burned *release* energy. Similarly, the rearrangement of neutrons and protons to form different nuclei also involves energy changes. For example, when a uranium nucleus undergoes *fission* and splits apart into two nuclei of smaller mass, energy is released. In this case, the stored energy is *nuclear energy*. The force that holds nuclei together—the *nuclear force*—is much stronger than the electrical force that holds molecules together. Therefore, the chemical rearrangements involving the atoms in molecules release only small amounts of energy compared to the nuclear rearrangements involving protons and neutrons in nuclei. The fission of all of the nuclei in 1 kg of uranium releases about one million times more energy than the burning of 1 kg of methane!

Energy occurs in many forms and can be converted from one form into another. We have already seen, for example, how gravitational potential energy can be converted into kinetic energy and then into thermal energy. We make use of energy conversion processes every day. When we turn on an electric switch, we make use of electrical energy that has been provided for us by the power company from the burning of coal or oil (or from the fission of uranium in a nuclear reactor). If the switch turns on an electric light bulb, electrical energy is converted into radiant energy (light) and thermal energy (heat). Our bodies convert the chemical energy (actually, *electrical* energy) in foodstuffs into mechanical energy whenever we move a muscle. And we use the stored chemical energy in batteries to operate radios, flashlights, and the electrical starters in our automobiles.

The processes by which energy is converted from one form into another are extremely important in the energy distribution network. The electrical energy required to operate a motor, for example, can be traced to its ultimate source in the Sun. Figure 4-7 shows in a schematic way the various steps that are involved. The Sun's energy is derived from nuclear reactions that take place deep in the Sun's interior. These reactions release energy that emerges from the Sun as light and heat radiations. Radiant energy is responsible for evapo-

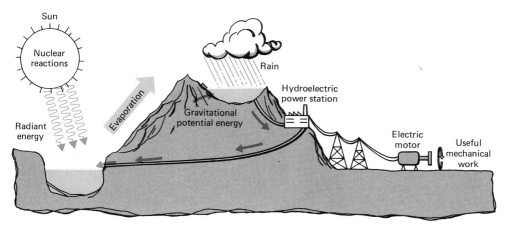

Figure 4-7 *Energy changes its form many times between the release of nuclear energy in the Sun's core and the final utilization of that energy in doing useful mechanical work.*

ration that raises water from the oceans to clouds and then this water, falling as rain, collects in streams, lakes, and rivers at high elevations. In flowing back to the sea, the water is used to produce electricity in hydroelectric power plants. Finally, this electrical energy is used by an electric motor to do useful mechanical work.

If the electrical energy in our example had been produced in a coal-burning power plant instead of a hydroelectric station, we would still identify the Sun as the starting point for the chain of energy conversions. Radiant energy from the Sun is converted into chemical energy in green plants. Coal that we now dig from the ground was formed by decaying plants millions of years ago. Thus, energy received from the Sun in ages past is stored as chemical energy in coal (and also in oil and natural gas). This energy is released by burning and is converted to thermal energy. This thermal energy, in turn, produces steam, which drives the electrical generating equipment.

We have seen how energy can be converted from one form into another. And we have emphasized the fact that energy is never lost, that energy is conserved. Why, then, is there any problem at all about sources of energy? It would seem that we could convert energy from one form into any other form that we require, never losing or running out of energy in the process. Unfortunately, it is not quite this simple. Every time that we convert energy from one form to another or use energy to do useful work, some part of the energy appears as heat. We may do this deliberately (as in the burning of a fuel), or it may occur even though unwanted (as in the production of heat by friction). Once this heat has been dissipated into the surroundings, it can no longer be recovered and used to do work. The heat generated by friction in an electric motor cannot be recovered and used to run the motor. The heat transmitted from your home to the outside air on a cold day cannot be recovered and used to warm the house again. In these cases we say that energy has been *lost*, but what we really mean is that energy has been converted into a form that is no longer useful to us.

Because of losses to heat, we can never extract as useful work the full amount of energy involved in any conversion process. For example, only about 20 to 25 percent of the chemical energy in gasoline appears as work when we use gasoline to run an automobile. And only about 35 to 40 percent of the chemical energy in the coal or oil burned in a power plant appears as electrical energy. To describe these situations, we say that the *efficiency* of an automobile engine is about 25 percent and that the *efficiency* of a power plant is about 40 percent.

Power

Suppose that you walk up a flight of stairs. In so doing, you have done work against the gravitational force by raising the mass of your body to a certain height. Suppose, next, that you *run* up the same flight of stairs. Again, you have done work, but this time you feel a different bodily sensation. You feel tired or out of breath after running up a flight of stairs but not after walking up the same stairs. In each case the same amount of work was done. Why, then, do you feel differently?

The difference between walking and running up a flight of stairs is in the *time* during which the work is done. The human body is accustomed to converting chemical energy into work at a certain *rate*. When this normal rate is exceeded (in running up the stairs), the body calls upon its reserves. The rapid expenditure of energy causes a greater feeling of tiredness than if the same amount of energy is used over a longer period of time.

The rate at which energy is used or work is done is called *power*. That is,

$$\text{power} = \frac{\text{energy used or work done}}{\text{time}}$$

The power produced in your body when you are running is greater than when you are walking. The higher the power level of bodily activity, the greater the feeling of exhaustion after a given amount of work is done.

We can apply the idea of *power* to any situation in which energy is expended or work is done. For example, the rate at which energy is used by an electric light bulb is expressed in terms of its power rating. A typical power value stamped on a light bulb is "100 W," which stands for 100 *watts*. What does this power mean?

To understand power values in everyday terms, consider the following situation. Suppose that you lift a mass m to a height h in a time t. At what rate have you done work? According to our discussion earlier in this chapter, the amount of work done in this case is mgh. Therefore, the power P for the lifting movement is

$$P = \frac{mgh}{t}$$

If we supply values of m, g, h, and t in metric units, the power value will be expressed in *watts* (W). Thus, if an object with a mass of 1 kilogram is raised to a height of 1 meter in 1 second,

$$P = \frac{(1 \text{ kg}) \times (10 \text{ m/s}^2) \times (1 \text{ m})}{1 \text{ s}} = 10 \text{ W}$$

and the power required is 10 watts. If the mass is increased to 100 kg, the power becomes 1000 W; we call this power 1 *kilowatt* (1 W):

1000 W = 1 kW

An even larger power unit is the *megawatt* (MW) or million watts:

1 000 000 W = 1000 kW = 1 MW

We almost always refer to power in terms of watts, kilowatts, or megawatts. For example, a light bulb may be rated at 60 W, an electric heater at 1.5 kW, and an electrical generating plant at 800 MW.

It is important to understand the difference between *energy* and *power*. To repeat, *power* is the *rate* at which *energy* is used. Power and energy are related in the same way that distance and speed are related. *Speed* is the *rate* at which *distance* is traveled. You would never confuse distance and speed; and you should never confuse energy and power.

Just as we write (power) = (energy) ÷ (time), we can also write (energy) = (power) × (time). If we wish to know how much energy is consumed by a 100-W light bulb that operates for 10 h, we have

energy = (power) × (time)
 = (100 W) × (10 h)
 = 1000 Wh or 1 kWh

Thus, the energy used is 1000 watt-hours or 1 kilowatt-hour (1 kWh). This unit, the kWh, is the unit most often used to measure the amount of electrical energy used. At today's prices, each kWh used costs about 3 cents. Thus, a 100-W light bulb could be used 10 h per day for a month at a cost of about 90 cents.

The use of the kWh is not confined to electrical energy; it can be used to specify the amount of energy of any type. For example, the amount of chemical energy in 1 gallon of gasoline is approximately 36 kWh. In this book we will always express energies in units of kWh (except that we will introduce another unit when we discuss energy in atoms and nuclei in later chapters).

The energy content (in kWh) of some common fuels is given in Table 4-1.

Table 4-1 Energy Content of Some Fuels

Fuel	Amount	Energy (kWh)
Coal	1 metric ton (1000 kg)	8600
Oil	1 barrel (42 gallons)	1700
Natural gas	1 cubic foot (ft^3)	0.29
Gasoline	1 gallon	36
Uranium (fission)	1 kg	23 000 000

How Much Energy Do We Use?

When we discuss the amount of energy used by the entire world or by the United States alone, the figures are so tremendous that it is easy to lose sight of their significance. Therefore, let us begin by looking at the energy balance sheet for a typical U.S. citizen. First, the average intake of food energy amounts to about 3.5 kWh per day for persons in the United States. (Do not be confused by the use of kWh to specify the amount of food energy. Remember, the kWh can be used for any type of energy. In more conventional terms we would say that the daily intake of food energy is about 3000 Calories; this is the same as 3.5 kWh.)

In addition to food energy, people use energy in many other forms: electrical energy for home and office lighting, chemical energy to operate automobiles, the energy represented in the products we buy, and so forth. In fact, food energy accounts for only a small fraction of the total energy used by any individual in a civilized country today. The average amount of energy used by a person in the United States per year is approximately 100 000 kWh, about 75 times more than the food energy alone! A typical U. S. citizen uses energy at an average rate of a little more than 10 kW.

If we multiply 100 000 kWh (or 10^5 kWh) by the U.S. population (220 million or 2.2×10^8), we arrive at a figure of 2.2×10^{13} kWh for the annual consumption of energy in all forms in the United States—this is 22 *trillion* kWh!

The United States is the world's largest user of energy. This country accounts for about one-third of the worldwide consumption of energy. That is,

Current U.S. energy consumption:
  approx. 2×10^{13} kWh/y
Current worldwide energy consumption:
  approx. 6×10^{13} kWh/y

Not only does the United States consume more energy than any other country, we also use more energy per person than any other nation in the world. The United States has only 6 percent of the world's population, and yet we account for one-third of the world's energy consumption. That is, the per capita use of energy in the United States is more than five times the worldwide average. A person in the United States consumes, on the average, about twice as much energy as a person in one of the highly developed countries of Western Europe. But compared to the poor nations of Asia, the per capita use of energy in this country is about 30 times greater.

The history of energy consumption has been one of continual increase. During the last 100 years, the U.S. population has grown by a factor of 5 and the per capita usage of energy has increased by a factor of 4. That is, in this country we now use about 20 times as much energy per year as we did 100 years ago (Fig. 4-8). In recent years, the rate of energy usage in the United States has increased by about 4 percent per year. However, the downturn in the economy together with a variety of conservation measures actually resulted in a "zero growth" situation during 1974 and 1975. By 1976, the usage of energy was again on the rise.

During the 1930s, the United States began a program to develop and expand its electrical generating capacity. (The Tennessee Valley Authority, or TVA, was established in 1933 to develop hydropower facilities in the Tennessee River basin.) Since that time, there has been a steady increase in the fraction of our total energy consumption represented by electrical energy. In 1974, about 10 percent of the energy used in the United States was electrical energy. By the year 2000, probably one-third of our energy consumption will be electrical. Our society is becoming more and more electrically oriented.

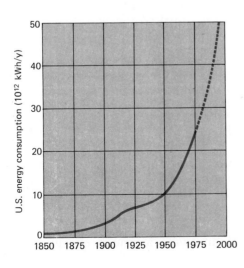

Figure 4-8 *Consumption of energy per year in the United States since 1850. The dashed line is one of the predictions for the future increase.*

Will these energy usage patterns continue into the future? If they do, we can look forward to a doubling of energy consumption in the United States every 18 years. Can we continue to tolerate this rapid growth of energy usage in the future? Where will we find the required sources of energy? And what will be the consequences to our environment and mode of life?

Sources of Energy—Chemical Fuels

Until about 150 years ago, the primary sources of energy were wood, water, and wind, plus, of course, the heating effect of the Sun's direct rays. We still make use of these sources, but only water power in the form of electricity generated by huge hydroelectric plants is now a significant factor in our overall energy supply. Most of the energy used today is the result of the burning of various chemical fuels. The breakdown is shown in Fig. 4-9 and the data are listed in Table 4-2. The projections for 1985 indicate that about 40 percent more chemical fuels will be consumed than in 1974 but that the relative proportions will be about the same. Also, the percentage of our energy that is produced in nuclear power plants may increase significantly in the future.

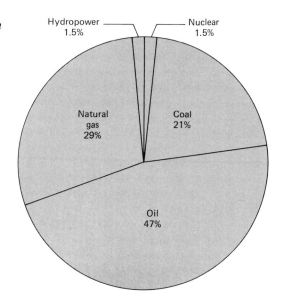

Figure 4-9 *Sources of energy in the United States in 1974.*

Table 4-2 Contributions of Various Sources to the Total Energy Consumption in the United States During 1974

Source	Amount	10^{12} kWh	Percentage
Coal	542×10^6 metric tons	4.6	21
Oil	6.0×10^9 barrels[a]	10.2	47
Natural gas	22×10^{12} ft^3 [a]	6.4	29
Hydropower		0.3	1.5
Nuclear		0.3	1.5
Other (wood, solar, geothermal)		small	small
		21.8	100

[a] These are not metric units, but they are the units in which these fuels are usually measured.

Among the primary sources of our energy, only hydropower constitutes a "renewable" resource; all of the other sources are "depletable." That is, there are only certain definite amounts of coal, oil, and natural gas that are available to us in the Earth's crust. When the supplies of these chemical fuels near exhaustion, we will be forced to turn to other sources of energy. What other sources are available? The answer is to be found in those sources that are now only minor contributors to our overall energy supply—namely, nuclear energy, solar energy, and geothermal energy. Fortunately, as we will see later in this chapter, the amounts of energy represented by these "minor" sources are truly enormous.

Energy Resources

How long will our supplies of chemical fuels last? First, let us look at the situation concerning our coal supplies. Coal-bearing layers in the Earth's crust usually extend over vast areas. Mining this coal by means of deep shafts can be carried out to depths of about 2000 m. Once a coal seam has been located, it can often be traced for hundreds of miles. Similarly, the near-surface layers of coal that are suitable for strip mining can easily be followed over a wide region. Consequently, estimates of our coal resources can be made with reasonably high reliability. According to the figures shown in Table 4-3, the United States has about 200 billion metric tons of coal that are economically recoverable at the present time.

Table 4-3 *Chemical Fuel Resources of the United States, Including Alaska and Offshore Areas (1974 estimates)*

Fuel	Resources		
	Identified and economically recoverable		Total
Coal	200×10^9 metric tons		3000×10^9 metric tons
Lifetime[a]	370 y		5500 y
	Measured	Inferred	Probable
Oil	49×10^9 barrels	$25-45 \times 10^9$ barrels	$200-400 \times 10^9$ barrels
Lifetime[a,b]	13 y	6–12 y	50–100 y
	Measured	Inferred	Probable
Natural gas	266×10^{12} ft^3	$130-250 \times 10^{12}$ ft^3	$1000-2000 \times 10^{12}$ ft^3
Lifetime[a]	12 y	6–11 y	45–90 y

[a] Assuming present rate of consumption (Table 4-2).
[b] Assuming that we continue to import 35 percent of our total oil consumption.

If this amount of coal were consumed at the 1974 rate (Table 4-2), the supply would last for more than 400 years. It is certain that our usage of coal will accelerate in the future, thereby hastening the depletion of our supplies. On the other hand, it also seems certain that as fuel costs increase and as improved mining methods are developed, a greater fraction of the total coal resources will become available to us. Therefore, even though we place a greater emphasis on coal as a fuel in the coming years, we have a comfortable supply for several centuries.

The reason that we must look forward to an increasing dependence on coal is that our oil and natural gas resources are much more limited than our coal supplies. Liquid crude oil and natural gas usually occur together in the Earth. In recent years, the yield of natural gas has been reasonably constant at about 6000 ft^3 per barrel of oil. Unfortunately, we do not have estimates of oil and natural gas supplies that are as reliable as those for coal. Locating deposits of oil and natural gas is not as simple as locating minable coal because petroleum deposits tend to be more localized than those of coal. Even in a proven field, not every well that is drilled will yield oil or gas. The search for petroleum deposits is therefore much less certain than locating sites for coal mines. As a result, our estimates of oil and natural gas reserves are subject to considerable uncertainties.

These uncertainties in petroleum supplies are reflected in the way in which the supply estimates are reported. Table 4-3 shows the three categories of resources. First, there are the *measured* resources; these figures are considered quite reliable. The next column gives the additional quantities *inferred* from test drilling and geologic projections. The least reliable are the *probable* figures, which are based on broad geologic evidence and theoretical considerations.

In Table 4-3 we can see one of the main reasons for the present concern about future supplies of energy. The domestic reserves of oil and natural gas are by no means plentiful. If we continue to use oil and natural gas at the 1974 rate (Table 4-2), we will have depleted the domestic *measured* and *inferred* resources in 20 to 25 years. However, there are two additional points that we must remember. First, the bulk of the large *probable* supplies of oil and natural gas are located beneath the ocean floors of the continental shelves that lie off the coasts of the United States. We have begun to exploit some of these supplies, particularly in the Gulf of Mexico where about 18 000 wells have been drilled; in addition, some wells have been drilled off the coast of southern California. A considerable number of offshore wells are producing oil from beneath the waters of the Middle East and Indonesia, and from the extensive deposits that underlie the North Sea. But no oil has yet begun to flow from the deposits beneath the Atlantic continental shelf. Although we know that oil exists in this region, extensive explorations have not been made and the true size of the resources in the area are not yet known. Drilling along the Atlantic coast has been delayed because of the very great concern for the damage to the coastal environment that could result if any large oil leaks or spills occurred. An additional factor is the high cost of drilling into the continental shelf: An offshore well costs between one and several million dollars, depending on the water depth and the well depth. Nevertheless, it seems clear that we will eventually begin a sizable campaign to tap the vast oil and natural gas deposits that lie off our coasts. This action will substantially prolong the lifetime of our domestic petroleum resources.

Second, it has been known for many years that there is an enormous amount of oil that is trapped in fine-grain rocks called *shale*. Extensive deposits of oil-bearing shale are located in Colorado, Utah, and Wyoming. There are probably more than 600 billion barrels of oil in shale that contains at least 25 gallons per metric ton. This represents at least as much oil as the known reserves of the entire Middle East. When the price of oil was low (prior to 1973), it was uneconomical to attempt to extract oil from shale. But now that oil prices have quadrupled compared to pre-1973 levels, the recovery of oil from shale becomes a real possibility. Before it becomes a reality, however, serious technical and environmental problems must be solved. The processing of shale oil requires large amounts of water. Can adequate supplies of water be provided in an area that has no great abundance of water? Moreover, huge tracts of land will have to be worked in order to uncover useful quantities of oil-bearing shale. How will this land be restored? These are only two of the important questions that must be answered before our huge shale-oil resources can be made part of the general supply. At the present time, only a token effort is being made to tap our shale-oil resources.

One additional point must be made concerning the petroleum resource figures in Table 4-3. The history of drilling for oil has been that as exploration proceeds and fields are opened, new discoveries are made which add to our reserves. This will not continue indefinitely, however, because the supplies are limited. But it seems most probable that more oil and natural gas exist in sites from which they can be recovered than we now suspect. Furthermore, some experts believe that the figures in Table 4-3 (which are from a U.S. Geological Survey report) are too conservative. One other group estimates that there are 200 billion barrels of recoverable oil in Alaska and under Alaskan waters. And another report puts the estimate of recoverable offshore oil at 800 billion barrels.

In addition to using our domestic supplies, we also import large quantities of oil into this country. In 1974, approximately 35 percent of the oil we used was imported. Although it is the announced policy of the United States to reduce substantially its dependence on foreign oil, we will probably continue to import oil for many years. (Oil reserves in the rest of the world, particularly in the Middle East, are about 10 times greater than the U.S. reserves.) We were made painfully aware of the fact that we rely heavily on foreign producers when our supplies from the Arab countries of Africa and Western Asia were cut off following the Middle East war in October 1973. Shortages of gasoline and heating oil quickly developed even though less than 6 percent of our oil was then imported from Arab countries. The embargo was lifted after a few months and imports returned to normal. But during 1973 and 1974, the oil exporting nations of the world, led by the Middle Eastern and African countries, forced the price of oil to a level about four times that of 1972. It seems unlikely that the price of oil, although it may undergo occasional decreases, will ever return to pre-1973 levels.

In summary, the situation with regard to chemical fuels is as follows. The worldwide supplies of oil and natural gas, and particularly the domestic supplies of the United States, are being rapidly depleted. Even by exploiting the Alaskan deposits and by extensive offshore drilling, we have at best about a hundred-year supply remaining. The development of shale oil will extend this outlook but not by a large factor. Furthermore, we must remember that we cannot consider burning all of our petroleum reserves as fuel because these materials represent the all-important feedstocks for the petrochemical industry, which produces a wide variety of materials including plastics, artificial rubber, fertilizer, and other chemicals.

The extensive deposits of coal that have been found in this country represent an energy resource that will remain useful for hundreds of years. Although these supplies will eventually be depleted, the severe problem that we face in the immediate future centers mainly around oil and natural gas.

Oil is the source of nearly half of the energy that we use in the United States (Table 4-2). As long as we continue to import a substantial amount of the oil that we consume, our supplies will be subject to the uncertainties of the international political situation. These facts, coupled with the high cost of oil, the increasing concern with environmental problems, the slow development of petroleum refining facilities, and our tardiness in facing the situation by pursuing alternative sources of energy, have produced our present "energy crisis."

The solution to our energy problems will require several actions:

1. We must take positive steps to curtail the unessential use of petroleum products. Indeed, such conservation measures are the *only* steps we can take to produce immediate results.

2. We must substitute coal for oil wherever possible, particularly as a fuel in the generation of electrical power. Also

we must pursue the development of methods to convert coal into gaseous and liquid fuels that can be substituted for oil and natural gas.

3. We must increase our efforts to bring alternative sources of energy into widespread usage. The only such source that we now know how to develop is nuclear energy. But solar energy and geothermal energy are abundant and attractive alternates, at least in the long-range picture.

Nuclear Energy

The first nuclear reactor to produce electrical power on a commercial basis began operating in 1957. By 1976, 60 nuclear power stations in the United States were producing 8 percent of the total amount of electrical power generated in this country. The fraction will be about 15 percent in 1980 and 30 percent in 1985. By the year 2000, it is possible that more electrical power will be generated in the United States by nuclear reactors than by all other methods combined.

Are nuclear reactors the answer to our energy problems? Can we replace our dwindling chemical fuel supplies by adding nuclear power stations? In part, we can. But the problem is much more complex than these questions imply. First, nuclear reactors produce *electrical* power, and although this contribution to the electrical network is important (and will become more important), it does not assist materially in solving our short-range problems in the nonelectrical sector. In the future, as our electrical requirements increase from the present 10 percent of total energy usage to a much higher fraction, nuclear energy will become substantially more important in our economy.

Second, to increase significantly the output of energy from any source—whether coal, oil, or nuclear reactors—requires long development and construction times. This is particularly true for nuclear power stations. Typically, six or eight years elapse between the time that a license to build a station is requested and the time that the station actually goes "on line." (Opening a new oil field or a new series of coal mines or a new petroleum refining complex requires three to five years. It is for this reason, as we have already mentioned, that conservation measures offer the only real hope for immediate relief to our energy shortage problems.)

Third, supplies of fuel for nuclear reactors—namely, uranium—are vast but they are not unlimited. Within 20 or 30 years we will have consumed almost all of our high-grade uranium ores. Then we will be forced to turn to lower-quality ores that are more expensive to mine and to process. The best prospect to forestall a shortage of nuclear fuel is through the development of *breeder reactors,* which produce electrical power and at the same time convert plentiful materials (for example, thorium) into additional nuclear fuels. Several breeder reactors are currently in operation (none in the United States), but they have not yet been developed to the point that they can assume a significant role in the production of electrical power. By 1990, however, breeder reactors will probably represent an important segment of our electrical generating capacity.

Finally, the most troublesome aspect of nuclear power is the problem of the radioactive waste materials that are produced in all nuclear reactors. There are two parts to this problem. If the cooling systems of an operating reactor were to fail, the reactor could suffer a *meltdown* and release radioactive materials to the surrounding countryside. Although extraordinary measures have been taken to prevent such an accident (and none has occurred in the United States), it is still a disquieting possibility. Even though a reactor operates completely safely, there still exists the problem of what to do with the radioactive wastes that build up in the fuel rods as uranium undergoes fission to release energy. Some of the radioactive materials decay away in relatively short periods of time, but others continue to emit radiations for hundreds or thousands of years. These long-lived radioactive substances must be stored in safe places and monitored closely to ensure that no leakages can endanger human life. This is a formidable task and one for which we have not yet found a permanent solution.

(We will discuss more fully nuclear reactor operations and the risks and benefits of nuclear power in Chapter 13.)

The type of nuclear reactor we have been discussing makes available energy by the process of *fission* in which a heavy nucleus such as uranium splits apart into two fragments and releases energy. Nuclear energy can also be released by combining two light nuclei into a single heavier nucleus. This is the process of *fusion.* There are two distinct advantages that fusion holds over fission as a source of energy. First, far less radioactivity is produced in the release of a certain amount of energy by fusion than when the same amount of energy is released by fission. Second, the supplies of fuel for fusion reactors—deuterium and lithium—are so plentiful that they are essentially inexhaustible. Unfortunately, we do not yet know how to harness nuclear fusion to produce electrical power. (Fusion energy has been tapped in an explosive way in the hydrogen bomb.) Much progress has been made toward understanding how a fusion reactor might work, but a pilot plant is still many years in the future and a network of fusion power stations will probably not exist before the year 2020 or 2050. Nevertheless, fusion offers the bright prospect of inexpensive, abundant energy for use by mankind in some future era.

Geothermal and Solar Energy

Heat from within the Earth and radiations from the Sun represent abundant energy resources that we have barely begun to exploit. At many places over the surface of the Earth, hot water and steam are released from the heated interior and appear as hot springs and geysers. Underground steam at The Geysers near San Francisco is used to produce as much electrical power as is generated by many coal-fired plants. Similar generating facilities run by natural steam are in operation in Italy, New Zealand, the Soviet Union, Japan, Iceland, Mexico, and Kenya.

Of even greater potential than underground steam and hot water is the heat energy stored in subsurface rocks. There is probably 10 times as much energy that could be recovered from heated rocks than is available from natural steam and hot water. This energy could be tapped by drilling holes into the rocks and pumping water through the heated region. At the surface the hot water or steam could be used to operate electrical generating equipment. One area near Marysville, Montana, only 3 km by 8 km, is believed to have the potential of supplying 10 percent of the U.S. electrical needs for 30 years. If the exploitation of geothermal energy were pursued vigorously, we could be producing, by the year 2000, almost as much geothermal electrical power as we now produce in this country by conventional means.

The source of energy most readily available to us is sunlight. If we could find a way to make efficient use of solar energy, we would have a perpetual supply of "free" energy. In the relatively cloudless desert regions of the southwestern United States, the solar energy received by an area of 5 km^2 during the course of a year is the same as the output of a large coal or nuclear generating plant. If 10 percent of the incident solar energy could be utilized, about 8 percent of the area of Arizona would produce an amount of electrical power equal to the total installed capacity of the United States in 1975 (470 000 MW).

Although we have put solar radiations to use in generating electricity on a limited scale (as, for example, in the use of solar cells aboard spacecraft), we have not yet devised an economical method for the widespread use of solar power. Estimates indicate that any large solar generating facility constructed with present technology would cost several times that for a conventional facility. Until there is a substantial reduction in the costs of the special materials required to convert solar radiation into electricity, or until other forms of energy become too expensive or too burdensome on the environment, we will be unable to take full advantage of "free" solar energy.

The large-scale production of electrical power from sunlight appears to be many years in the future. However, the heating (and cooling) of homes and businesses with solar radiations may become widespread within a much shorter time. Some of today's new construction includes equipment that will allow solar heating to replace a part (in some cases, 50 percent) of the conventional heating. Within the next decade, probably 10 percent of new homes will use some form of solar heating.

Energy in the Future

Our supplies of oil and natural gas will cease to be major sources of energy by the middle of the next century. But we have abundant supplies of coal, of nuclear fuels for fission reactors, and of geothermal energy. And the potential represented by nuclear fusion and solar power is essentially unlimited. Why, then, is there an "energy crisis" at all? The primary reason for our immediate shortage is the fact that we have relied heavily for many years on cheap and abundant oil and natural gas. It is now becoming increasingly difficult to produce these fuels in the quantities demanded. In addition to this limitation, the oil exporting countries of the world have decided to take advantage of the situation by raising the price of oil to about four times the price that existed in the early 1970s. Faced with this double problem of decreasing supplies and increasing prices, we must turn to other energy sources. The alternative fuels—coal and nuclear fission and fusion materials—as well as geothermal and solar energy, are available in great abundance. But long development times and various technological advances are necessary before we can alter in any significant way our energy usage patterns. We will continue to burn our precious petroleum reserves until we can begin to liquefy coal or to produce large quantities of electricity from solar energy or to harness geothermal energy on a large scale. We will continue to produce potentially hazardous radioactivity in nuclear fission reactors until we can solve the problems of nuclear fusion or solar energy.

It is virtually certain that we will eventually meet the challenge of our energy problems. At some time in the future, we will have available to do our work abundant and inexpensive energy from fusion power stations, from geothermal heat sources, and from solar radiations. But when will this happy day occur? Probably not for at least 75 or 100 years.

Questions and Exercises

1. How would you classify the energy associated with the water in a high storage tank? How could this energy be used to do work?
2. A pendulum consists of a heavy sphere suspended by a long string from a fixed support. Discuss the energy in a pendulum as it swings back and forth.
3. Classify the energy in the following cases according to the *basic* energy forms: (a) sonic boom, (b) sugar, (c) boiling water, and (d) moving automobile.
4. Some devices are said to "waste" energy. Is such a statement strictly true? What happens to "wasted" energy?
5. The pumping action of the heart gives to the blood some kinetic energy. Where does this energy originate, and what happens to the blood's kinetic energy?

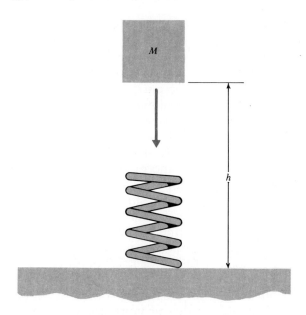

6. A block is dropped from a height h onto a spring that rests on the ground, as shown in the diagram. The moving block compresses the spring and then the spring uncoils and propels the block upward. Trace the conversions of energy in this process. To what height will the block rise?
7. Two automobiles, one of large mass and one of small mass, coast down the same hill, starting from rest (see the diagram). Neglecting friction, which automobile will rise higher on the next hill? Explain.
8. When a stake is driven into the ground, what happens to the energy that was expended?
9. When a box is pushed across a rough floor, work is done

against the force of friction. Is the energy expended stored as potential energy? Explain. What is the difference between doing work against a frictional force and doing work against the gravitational force?

10. A certain coal-burning power plant operates at an efficiency of 40 percent and produces 1000 MW of electrical power. At what rate is energy produced by the burning of the coal? At what rate is energy exhausted by the plant as thermal power that is not converted into electrical power? [Ans. 2500 MW; 1500 MW.]

11. At present, coal costs about $45 per metric ton and oil costs about $11 per barrel. Use the information in Table 4-1 and calculate the cost per kWh of energy in coal and in oil. Compare these figures with the cost per kWh for electrical energy. Why is there such a great difference? [Ans. coal, 0.5 cents/kWh; oil, 0.65 cents/kWh; electricity, 3 cents/kWh.]

12. How many metric tons of coal would have to be burned in order to produce the amount of energy used in the United States in one year? (Use the information in Table 4-1.) [Ans. 2.5 billion.]

13. Is food an expensive source of energy? What would be your annual food bill if it could be purchased at the same rate as electrical energy (3 cents/kWh)? [Ans. About $38.]

14. Comment on the environmental price we must pay for increasing the usage of coal during the years before we have developed cleaner sources of energy.

Additional Details for Further Study

Energy and Power Units

In the metric system, the basic unit of energy or work is the *joule* (J). If a force of 1 newton acts through a distance of 1 meter, the work done (or the energy expended) is 1 joule:

1 N-m = 1 J

Thus, when a 5-kg object is raised through a height of 8 m, the amount of work required is

mgh = (5 kg) × (10 m/s²) × (8 m) = 400 J

and we say that the object has a potential energy of 400 J.

The unit that is used to specify the energy content of foods is the *Calorie* (Cal):

1 Cal = 4186 J

Thus, an 800-Cal meal has an energy content of (800) × (4186) J or about 3.3 million joules.

The definition of *power* is the amount of work done (or energy expended) per unit time. When 1 joule of work is done each second, the power is 1 watt:

1 W = 1 J/s

Thus, if a 5-kg object is raised through a height of 8 m in 10 s, the power required is

$$P = \frac{\text{work}}{\text{time}} = \frac{mgh}{t} = \frac{400 \text{ J}}{10 \text{ s}} = 40 \text{ J/s} = 40 \text{ W}$$

When 1 kilowatt of power is used for 1 hour, the energy expended is 1 kWh:

$$1 \text{ kWh} = \left(1000 \frac{\text{J}}{\text{s}}\right) \times \left(\frac{3600 \text{ s}}{1 \text{ h}}\right) \times (1 \text{ h}) = 3.6 \times 10^6 \text{ J}$$

We often see motors of various kinds rated in terms of *horsepower* (hp). In metric units,

1 hp = 746 W

That is, 1 hp is approximately equal to $\frac{3}{4}$ kW.

Kinetic Energy

Suppose that a constant force F acts on a block with a mass m that is initially at rest. If the force pushes the block across a friction-free horizontal surface, the block will accelerate according to Newton's law, $F = ma$. An amount of work $W = F \times d$ is required to move the block a distance d. What will be the velocity of the block at this position? In Chapter 1 we found two results for an object undergoing constant acceleration, starting from rest. First, the velocity after a time t is $v = at$. Second, the total distance moved after a time t is $d = \frac{1}{2}at^2$. We can use these results and Newton's law in the work equation:

$W = F \times d$
$ = (ma) \times (\frac{1}{2}at^2)$

Regrouping, we can write

$W = \frac{1}{2}m(at)^2$

Now, substituting v^2 for $(at)^2$, we have

$W = \frac{1}{2}mv^2$

This amount of work has been done on the block and the block possesses an equal amount of *kinetic energy:*

kinetic energy $= \frac{1}{2}mv^2$

Notice that kinetic energy is proportional to the *square* of the velocity. That is, if the velocity of an object is *doubled*, its kinetic energy increases *four* times. An automobile traveling at 80 km/h has four times the destructive capacity in a collision as does the same automobile traveling at 40 km/h.

We can use the idea of energy conservation together with the expression for kinetic energy to solve a variety of different problems. For example, suppose that we wish to calculate the velocity with which an object will strike the ground if it falls from a height of 20 m. In its initial position the object has a gravitational potential energy equal to mgh. When it strikes the ground, the object has a velocity v and a kinetic energy equal to $\frac{1}{2}mv^2$. Because all of the initial potential energy has been converted into kinetic energy at the moment of impact, we can equate the two expressions:

$\frac{1}{2}mv^2 = mgh$

Solving for the velocity v,

$v = \sqrt{2gh}$

Notice that the mass of the object cancels and does not appear in the expression for the velocity. This agrees with our knowledge that objects with different masses fall at the same rate.

Substituting $h = 20$ m, we find

$v = \sqrt{2 \times (10 \text{ m/s}^2) \times (20 \text{ m})} = \sqrt{400} \text{ m/s} = 20 \text{ m/s}$

Additional Exercises

1. How much work must be done by an astronaut on the Moon (where $g = 1.6$ m/s^2) to lift a 10-kg object through a height of 2 m? [Ans. 32 J.]

2. If your food intake is 3000 Cal per day and if you convert all of this to bodily energy, at what average power does your body work? [Ans. 145 W or 0.2 hp.]
3. How long will be required for a 2-hp motor to lift 25-kg to a height of 30 m? [Ans. 5 s.]
4. What is the kinetic energy of a 2000-kg automobile moving with a speed of 72 km/h? [Ans. 4×10^5 J.]
5. How much work is required to raise a 4-kg object to a height of 20 m and simultaneously give it a velocity of 10 m/s? [Ans. 1000 J.]
6. An object is thrown upward with a velocity of 30 m/s. To what maximum height will the object rise? [Ans. 45 m.]

How do electric motors work?

5
ELECTRICITY AND MAGNETISM

If you look around your home or almost anywhere in your city, you will find an enormous number of items that contain electric motors. Indeed, most of the heavy work and many of the smaller tasks that are performed in the advanced countries of the world today are carried out by means of electric motors or by internal combustion (gasoline or diesel) engines. In the home, electric motors run refrigerators, vacuum cleaners, clothes and dish washers, hair dryers, food blenders, as well as various kinds of motorized toys. Larger electric motors operate industrial equipment, elevators, water pumps, subway trains, and an increasing number of small automobiles.

One of the most significant features of electric motors is their high degree of reliability. (The same statement can be made regarding gasoline and diesel engines.) How often have you had to repair a device because the electric motor failed? If a mechanical device *does* require repair, it is usually because of a defect in some other part, not the motor itself. Electric motors have been in widespread use for little more than 50 years. During this period, the design and construction of these motors have been perfected to the point that almost any present-day motor, if properly used, will operate for many, many years.

The reason for the long life and low failure rate of electric motors is that they are actually very simple devices. In this chapter we will learn how electric motors work. We will examine the physical principles that are involved in the design of motors, and we will see how a few simple ideas are sufficient to understand the operation of these and other electrical devices. While pursuing the answer to our chapter title question, we will have an opportunity to look into several other aspects of the subject of electricity and magnetism.

Electricity

You have probably heard several terms used in connection with discussions of electricity and of electric motors in particular: voltage and volts, current and amperes, power and watts (or kilowatts), energy and kilowatt-hours. If you examine the metal plate that is attached to most motors, you will see that every motor is designed for a particular voltage (for example,

120 volts), draws a certain current (for example, 6 amperes or *amps*), and has a definite power rating (for example, 1 horsepower). (Remember, 1 hp is approximately equal to 750 W or $\frac{3}{4}$ kW.)

What do these various terms mean? We have already discussed power and energy, and we have defined the units for these quantities. Therefore, let us concentrate on the new electrical terms, *voltage* and *current*.

Voltage has to do with the amount of work that is required to move electrical charges. You have probably learned that all matter is composed of atoms and that atoms consist of electrically charged particles. We will discuss in more detail the structure of atoms in Chapter 10. For now, it is sufficient to know that the outer portion of every atom consists of negatively charged *electrons* and that the central core (or *nucleus*) carries a positive charge. Every atom in the normal condition carries equal amounts of positive and negative charge and so is electrically neutral.

In a bulk sample of a metal, not all of the electrons are attached to the atoms. Some of the electrons are free to move around within the material. (The atoms, however, remain stationary in a solid substance.) The movement of these free electrons is responsible for the conduction of electricity in metals. We say that metals are electrical *conductors*. Other materials, such as glass, plastics, and air, do not contain any substantial number of free electrons. Electricity is therefore not readily conducted through these substances. We call such materials *nonconductors* or *insulators*.

In the normal condition, essentially all everyday matter is electrically neutral. We can cause an object to become electrically charged by arranging to add charge to it or to remove charge from it. This is almost always done by moving electrons, *not* atoms. Therefore, if we wish to place a positive charge on an object, we remove electrons from the object. Similarly, if we wish to place a negative charge on an object, we add electrons (which have been removed from some other object). When we charge an object in this way, we are not *creating* electrical charges. Instead, we are simply separating charges that were initially very close together, so close that their electrical effects cancelled and gave the appearance of no charge at all. Net electrical charge can neither be created nor destroyed. This is one of the fundamental (and never violated) laws of Nature.

One way to separate electrical charge is by chemical reac-

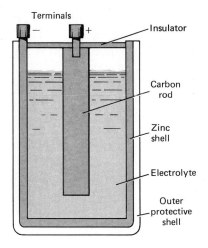

Figure 5-1 *Schematic diagram of a carbon-zinc battery. Chemical reactions cause electrons to be moved from the carbon rod (which becomes positively charged) to the zinc shell (which becomes negatively charged).*

tions, as in a battery. The structure of a typical carbon-zinc dry cell is shown schematically in Fig. 5-1. The outer part of the cell is a zinc shell and in the center there is a carbon rod. The space between the zinc shell and the carbon rod is filled with a spongy, porous material that is saturated with a fluid that is a good electrical conductor. (Such a fluid is called an *electrolyte*.) The chemical reactions that take place in the dry cell are complicated, but the net result is that electrons are transferred from the carbon rod to the zinc shell. Thus, the central terminal (attached to the carbon rod) becomes the positive terminal, and the outer terminal (attached to the zinc shell) becomes the negative terminal. Most batteries of this general type are made to be discarded after the electrical energy has been drained. Other batteries, constructed from different materials, can be recharged. The standard lead-acid automobile battery is a rechargeable battery. A variety of small items, such as electric shavers and electronic calculators, use rechargeable nickel-cadmium batteries.

Various types of electrical and electronic devices can be used to separate charge. Some of these devices are capable of producing huge accumulations of electrons and positive charge on opposite terminals. As we will see in the next section, this means that high *voltages* result. In the atmosphere we have a natural mechanism in thunderclouds that produces charge separation and gives rise to lightning.

Electrical charges have the familiar property that *opposite charges attract* and *like charges repel*. That is, if we charge

Electricity 115

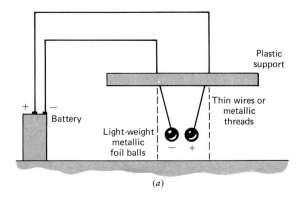

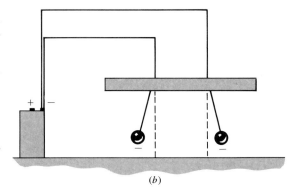

Figure 5-2 *The terminals of a battery or dry cell are designated positive (+) and negative (−). A conductor that is connected by a wire to one of the terminals will acquire a charge of the same sign as that of the terminal. (a) If two lightweight balls are connected to opposite terminals of a battery, they acquire opposite charges and attract. (b) If the balls are connected to the same terminal, they acquire like charges and* repel.

two objects in the same way so that both carry a positive charge (or a negative charge), there will be a mutual repulsive force between the objects. On the other hand, if one object carries a positive charge and the other carries a negative charge, the force between the objects will be attractive. Figure 5-2 illustrates this principle.

By moving the charged balls in Fig. 5-2 closer together or farther apart, it is easy to discover that the magnitude of the electrical force depends on the distance separating the charges. In fact, detailed measurements show that the electrical force depends on distance in the same way that the gravitational force does: namely, inversely as the square of the distance (that is, $1/r^2$).

The electrical force between two objects depends on the

116   *Electricity and Magnetism*

amount of charge that is carried by the objects. Compared to the gravitational force, the electrical force is enormously stronger. By transferring only a relatively small amount of charge from one object to another, the electrical force between the objects can be made much stronger than the gravitational force. For example, by transferring only about 300 kg of electrons from the Earth to the Moon, the electrical force between these bodies could be made as strong as the actual gravitational force! The electrical force between two objects would completely overwhelm the gravitational force if it were not for the fact that most ordinary objects are electrically neutral, or very nearly so.

Voltage

Let us see how we can apply these ideas concerning electrical charges and electrical forces to the concept of voltage. Suppose that we connect two metal plates to the opposite terminals of a battery, as shown in Fig. 5-3. Each plate then acquires an electrical charge of the same sign as that of the terminal to which it is attached. (Why is this so? See Ques-

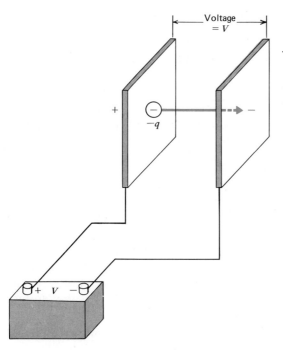

Figure 5-3 *The amount of work required to move the charge* $-q$ *from the positive plate to the negative plate is directly proportional to the voltage V across the plates (or, equivalently, the voltage across the terminals of the battery).*

tion 1 at the end of this chapter.) Suppose next that we place a charge $-q$ very close to the positive plate. We know that the charge will be attracted toward the positive plate and will be repelled by the negative plate. Therefore, to move the charge from the positive plate to the negative plate requires the expenditure of *work*. Clearly, if the magnitude of the charge is increased or if the battery is replaced by a more powerful battery (that is, a battery with a higher voltage), then the electrical forces will be greater and more work will be required to move the charge. The work done is directly proportional to both the charge and to the battery voltage:

work = (charge) × (voltage)

From this expression, we can write

$$\text{voltage} = \frac{\text{work}}{\text{charge}}$$

Thus, we can define the *voltage* between two points as the *work per unit charge* required to move a charge from one of the points to the other (see Fig. 5-3).

The terms "voltage" and "potential difference" are used interchangeably. We measure voltage or potential difference in units of *volts,* abbreviated V. Thus, we say that household electrical lines have a voltage of 120 V or that the potential difference between the terminals of a flashlight battery is 1.5 V.

Look again at Fig. 5-3. If the charge $-q$ is moved from the positive plate to the negative plate, work must be done by some outside agency against the electrical forces acting on the charge. Now, suppose that we start the charged particle at a point near the *negative* plate. In this situation, no outside force is required to move the charge to the other plate because the electrical forces acting on the charge push it toward the positive plate. Therefore, if we were to release the particle at the negative plate, it would be acted upon by the electrical forces and would accelerate toward the positive plate. Upon arriving at the positive plate, the particle would have some velocity and some motional or kinetic energy. In this case, work is done *on* the charged particle and this work appears in the form of the particle's kinetic energy. This energy is supplied by the battery, and the energy stored in the battery is correspondingly reduced. The amount of kinetic energy acquired by the charge in moving from the negative plate to the positive plate is exactly equal to the work required to move the charge from the positive plate to the negative plate. (This is another example of energy conservation.)

The Electric Field

In Fig. 5-3 we have a pair of parallel plates connected to a battery. The charges on the two plates are distributed uniformly over the surfaces. (Can you see why? On the negative plate the free electrons mutually repel one another and therefore move apart. What happens on the positive plate?) In order to calculate the force on a charge in the region between the plates, we could vectorially add the effect of every charge on each plate. Clearly, this would be a difficult procedure. An easier way to treat such cases is in terms of the *electric field*.

Let us begin by looking at an isolated charge. We know that such a charge will exert a force (attractive or repulsive) on any other charge in its vicinity. We can view this action in the following way. Every charge sets up in space around itself a condition to which any other charge will respond by experiencing a force. We call this "condition in space" the *electric field*. We find it useful to say that the charge sets up an electric field and that the field exerts a force on any other charge.

We can map the field around a charge by drawing lines (called *lines of force* or *field lines*) that represent the direction of the force that would be exerted by the field on a *positive* charge. At every point in space the direction of the field line is the same as the direction of the electrical force on a positive charge at that point. Figure 5-4 shows the field lines for an isolated positive charge and an isolated negative charge. Notice that the field lines are directed *away from* the positive charge

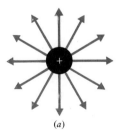
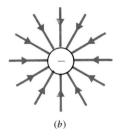

Figure 5-4 (a) *The electric field lines around a positive charge are directed* away from *the charge.* (b) *The field lines around a negative charge are directed* toward *the charge.*

(a)   (b)

because another positive charge in the vicinity would be repelled. On the other hand, the field lines are directed *toward* the negative charge because a positive charge in the vicinity would experience an attractive force.

We can gain further insight from this way of describing the effects of electrical charges. If two charges are close together, the mutual electrical force between them (attractive or repulsive) will be much stronger than if there is a considerable distance between the charges. (Remember, the electrical force depends on $1/r^2$.) We can see this effect in the diagrams of the field lines (Fig. 5-4). The magnitude of the electrical force on another charge will be large where the field lines bunch together, and the force will be small where the lines are sparse. The *strength* of the electric field at any point is proportional to the force that would be exerted by the field on a charge at that point. Thus, the strength of the electric field is large where the field lines converge toward a charge and is small where the field lines have diverged away from the charge.

In Fig. 5-5 we see the electric field in the vicinity of a pair of charges with the same magnitudes but with opposite signs. At

Figure 5-5 *Electric field lines in the vicinity of a pair of charges with the same magnitudes but with opposite signs.*

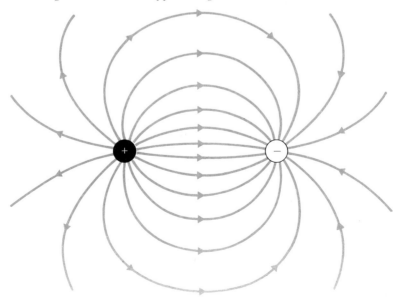

any point, the direction of the field line is the same as that of the vector that represents the sum of the attractive force due to the negative charge and the repulsive force due to the positive charge. Notice that the field lines originate on the positive charge and go smoothly over to the negative charge where they terminate. This is entirely consistent with the idea that a field line represents the direction of the force on a positive charge.

Now we can return to the case of the pair of uniformly charged plates (Fig. 5-3). In Fig. 5-6 we show the field lines for this case. Notice that the lines are all straight and uniformly spaced. We say that such a field is *uniform*. If a charge is placed in a uniform field, it will experience the same force (both in magnitude and direction) regardless of its position within the field.

The *strength* of an electric field is related to the force that is experienced by a charge placed in that field. The force is proportional both to the field strength and to the magnitude of the charge:

electrical force = (field strength) × (charge)

What has all this to do with electric motors? First of all, electric fields within wires cause electrons to move, that is, cause currents to flow, as we will see in the next section.

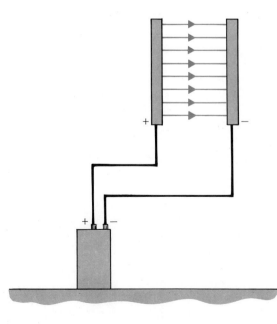

Figure 5-6 *The electric field between a pair of uniformly charged parallel plates is* uniform; *that is, the field lines are all straight and uniformly spaced.*

Electric currents are necessary in the operation of all electric motors. Second, the idea of a *field* is easy to understand for the electric case. We are now prepared to make the extension to *magnetic fields* of the types that are involved in electric motors. Because magnetic fields are produced by electric currents, we must next study some of the features of moving electric charges.

Current

Any sample of conducting material, such as a copper wire, contains many free electrons. These electrons are continually in motion, moving in random directions with various speeds in much the same way that molecules move in a gas. Because the electrons are moving randomly, if we focus on a particular cross section of the wire, we will find that just as many electrons move to the right through this cross section as move to the left. That is, there is no net flow of electrons along the wire.

Suppose that we now connect the ends of the wire to the opposite terminals of a battery. This establishes an electric field within the wire, and the free electrons respond by moving away from the negative terminal and toward the positive terminal (Fig. 5-7). Now there *is* a net movement of electrons through any cross section of the wire. The net flow of charge in a conducting medium constitutes a *current*.

Electric charge occurs in discrete units. The electron carries one such unit of negative charge. (Each nuclear proton carries one unit of positive charge, equal in magnitude to the electron charge.) All charges that are found in Nature are multiples of this fundamental charge. Because the electron charge is so small, we find it convenient to use a large unit in discussions of electrical quantities. The combined charge on 6.25×10^{18} electrons is called 1 *coulomb* (1 C).

Electric current is measured in terms of the rate of flow of electric charge. If there is a net flow of charge amounting to 1 coulomb per second through an imaginary cross sectional surface in a conductor, such as that shown in Fig. 5-7, we say that the current is 1 *ampere* (1 A). That is, 1 ampere = 1 coulomb per second, or 1 A = 1 C/s. In other words, if 6.25×10^{18} electrons flow through a surface each second, the current is 1 A.

In Fig. 5-8 we again show a portion of a conductor in which

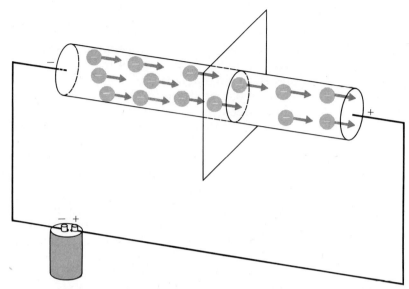

Figure 5-7 *When a conductor is connected to a battery, the free electrons move away from the negative terminal and toward the positive terminal. If a net charge of 1 coulomb passes through the cross-sectional surface shown in .1 second, the current is 1 ampere.*

electrons are moving. Here, the electrons flow from left to right. However, it proves convenient, as we will see, to define the direction of electric *current* to be *opposite* to that of the electron flow, namely, from right to left in this case. The reason for this arbitrary choice is that we usually define elec-

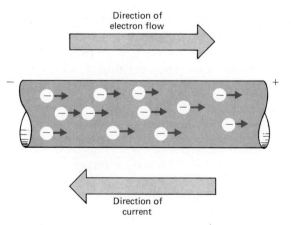

Figure 5-8 *The direction of electric* current *is defined to be the direction in which* positive *charge would flow. The direction of the current in a conductor is therefore opposite to the direction of electron flow.*

trical quantities in terms of the actions of *positive* charges. Thus, in Figs. 5-4, 5-5, and 5-6, we drew the direction of each electric field line to represent the direction of the electrical force on a *positive* charge. Similarly, we define the direction of electric current to be the direction in which a *positive* charge would move. In a solid conductor, only the electrons move; the positively charged atomic nuclei remain stationary. Even though positive charge does not move in a solid conductor, the direction in which it *would* move is opposite to that of the electron flow and this we define as the direction of the current. (Positive charges *do* move in conducting liquids and gases.) Whenever we wish to refer explicitly to the direction of electron motion, we will use the term "electron flow." The term "current" will always refer to the conventional current defined in terms of the motion of positive charge.

Electrical Resistance

How much current will flow through a conductor when it is connected to a battery? As the electrons move along a conductor, they collide frequently with the stationary atoms of the material. Therefore, instead of moving smoothly through the conductor, the electrons are deflected and deviated from their intended paths at frequent intervals. The materials that impede electron flow the least are the best conductors. In this class we have silver, copper, gold, and other metals. We say that these materials have high electrical *conductivity* or low electrical *resistance*. Carbon, sulfur, and other nonmetals have low conductivity or high resistance.

It has been found experimentally that the amount of current that flows in most materials is directly proportional to the voltage that is applied. We define the electrical *resistance* to be the factor connecting voltage and current, and we write

$$\text{current} = \frac{\text{voltage}}{\text{resistance}}$$

or, using symbols,

$$I = \frac{V}{R}$$

This relationship is known as Ohm's law, after its discoverer, Georg Simon Ohm (1787–1854). When I is measured in amperes and V in volts, the electrical resistance R is given in *ohms*. We use the symbol Ω (capital Greek omega) to denote resistance values in ohms.

Electrical wiring is almost always made from copper. A copper wire with a cross-sectional area of 1 mm^2 and a length of 100 m will have a resistance of 1.8 Ω. Therefore, if such a wire is connected to the terminals of a 6-V battery, the current will be

$$I = \frac{V}{R} = \frac{6\text{ V}}{1.8\text{ }\Omega} = 3.33\text{ A}$$

In many types of electrical circuits, the current is controlled by inserting into the circuits specific resistance values. These *resistors* are usually in the form of small cyclindrical pieces of molded carbon. By varying the size and the way in which the molding is carried out, carbon resistors can be made in a variety of resistance values from a fraction of an ohm to many millions of ohms.

Fluid Flow Analogy

In many ways the flow of electrical current through wires is similar to the flow of fluids through pipes. The electrical concepts of voltage, current, and resistance all have their analogies in fluid flow. Look at Fig. 5-9. Here we have a fluid being pumped through a pipe from the bottom of a reservoir to the top. The rate at which the fluid is transferred clearly depends on how powerful the pump is. The "strength" of the pump is analogous to the voltage of the battery in an electrical circuit. And the flow rate of water through the pipe is analogous to the electrical current. Increasing the "strength" of the pump causes more water to flow just as increasing the battery voltage produces an increase in electrical current.

What will happen to the flow rate of the fluid in Fig. 5-9 if we substitute a pipe with a smaller diameter? It is more difficult to force a fluid through a small pipe than through a large pipe. Therefore, decreasing the pipe size will decrease the flow rate. We say that the small pipe has a greater *resistance* than the large pipe. Similarly, if we increase the length of the pipe, the resistance will increase and the flow rate will decrease. With regard to size and length, the electrical resistance of a wire is exactly the same as the resistance to fluid

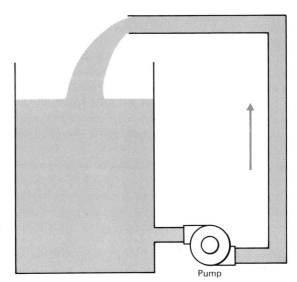

Figure 5-9 *The pumping of a fluid through a "circuit" is analogous to the flow of electrical current in a circuit.*

flow through a pipe. A long, thin wire has a much higher electrical resistance than a short, thick wire made from the same material.

Electric Power

If you have ever touched an electric motor that has been running for a while, you know that an operating motor is *hot*. Every device that has moving parts experiences friction and therefore produces *heat*. But friction is not the only source of

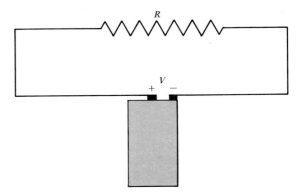

Figure 5-10 *A simple electrical circuit. The sawtooth symbol represents a resistor whose resistance value is* R. *The voltage of the battery is* V. *The current that flows in the resistor causes it to become heated.*

heat in an electric motor. The temperature rise of an operating motor is due also to *electrical* heating. We can account for electrical heating in the following way.

Consider the simple circuit shown in Fig. 5-10, where a resistor R is connected to a battery that has a voltage V. The battery does work in causing a current to flow in the resistor. That is, energy is dissipated in the resistor by the collisions of the electrons with the atoms. Consequently, the resistor becomes hot.

Let us look at this situation in more detail. The battery causes charge to move through the resistor. If a charge q has moved from one battery terminal, through R, and to the other terminal, it has moved through a voltage V. The work required to do this is supplied by the battery:

work = (charge) × (voltage)

If we divide both sides of this expression by *time,* we have

$$\frac{\text{work}}{\text{time}} = \left(\frac{\text{charge}}{\text{time}}\right) \times (\text{voltage})$$

Now, we know that charge per unit time is equal to the *current* I in the circuit. Also, work per unit time, as we found in the preceding chapter, is equal to the *power* P. Therefore, we can express the result as

power = (current) × (voltage)

or

$$P = I \times V$$

That is, the *power* (energy per unit time) expended in the resistor is equal to the current in the resistor multiplied by the voltage across the resistor. If I is measured in amperes and V in volts, then the power P is expressed in watts.

Our analysis here is not limited to the case of electrical resistors. The power expended in any electrical device or circuit is equal to the product IV. If a motor draws 6 A from a 120 V line, then the power required to run this particular motor is (6 A) × (120 V) = 720 W or about 1 hp. Some of this power is used to drive the mechanical equipment attached to the motor and the remainder is consumed in electrical and frictional heating in the motor itself.

The cost of operating an electric motor or other electrical device can be computed by using the power rating stamped on the device. (If the power rating is not given, it can be obtained by multiplying the current rating by the voltage at which the device operates. Most common electrical appliances are designed to operate at normal household voltage, 120 V.) As we saw in the preceding chapter, if we multiply the power rating in kilowatts by the time of operation in hours, we obtain the energy used in kilowatt-hours (kWh). Each kWh costs about 3 cents. The exact charge for electrical energy made by the power company in your area can be found on your monthly electric bill.

Magnetism and Magnetic Fields

Almost everyone has had the experience of observing the interaction of two magnets. One common shape for a magnet is that of a small bar. The two ends of a bar magnet are called *poles* and they have different effects on the poles of other magnets. Generally, we identify one pole as the *N pole* and the other as the *S pole*. (We really should not call these the *north* pole and the *south* pole, for reasons that we will see shortly.) If we bring the N pole of one bar magnet near the N pole of another magnet, we find that the two poles *repel* one another (Fig. 5-11a). Similarly, two S poles also repel one another. However, if we bring an N pole near an S pole, we find a mutual *attraction* (Fig. 5-11b). These results are similar to

Figure 5-11 (a) *Like magnetic poles repel.* (b) *Opposite magnetic poles attract.*

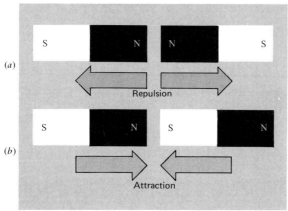

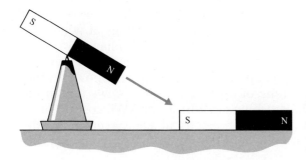

Figure 5-12 *A pivoted magnet acts as a* compass *and aligns itself to indicate the direction toward an opposite magnetic pole.*

those we found for electrical charges. We can summarize our observations by stating: *Like magnetic poles repel* and *opposite magnetic poles attract.*

Suppose that we mount a small bar magnet in such a way that it can pivot freely about its center. If the S pole of another magnet is brought near the pivoted magnet, the latter will turn until its N pole is pointing in the direction of the S pole of the second magnet (Fig. 5-12). That is, the pivoted magnet acts as a *compass* and aligns its N pole to indicate the presence of a nearby S pole.

We all know that a compass magnet that is sufficiently far from other magnets will point toward the Earth's north pole. (Actually, a compass points toward the *magnetic* pole, which is about 1300 km from the *geographic* pole.) The end of the compass magnet that points northward is the north-seeking or N pole. Because the N pole of a compass magnet is attracted toward an S pole, this means that the Earth's north magnetic pole is actually an S pole. It is for this reason that we should always refer to magnetic N and S poles and not to magnetic "north" and "south" poles. Otherwise, we would have the confusing situation that the Earth's north pole is a south pole!

If we move a small compass around a bar magnet, we observe that the compass magnet points in different directions at different points. This is illustrated in Fig. 5-13, which shows the directions of compass magnets at various positions around a bar magnet. In this diagram we see that the compass direction changes smoothly as we proceed from the N pole to the S pole. We can find other positions for the compasses around the bar magnet that also give smooth changes in direction along a path from the N pole to the S pole. We describe this situation by saying that there is a *magnetic field* around the bar magnet and that a compass magnet aligns itself with the direction of the field at every point. That is, the line of

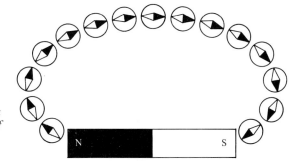

Figure 5-13 *Mapping the magnetic field of a bar magnet with a small compass. (The black end of the compass magnet is the N pole.)*

compass magnets in Fig. 15-13 indicates one of the *magnetic field lines* for the bar magnet. When we sketch several of these field lines, as in Fig. 5-14. we have a portion of a map of the bar magnet's field. The magnetic field of the Earth is very similar to that of a bar magnet (see Fig. 5-15).

In Fig. 5-14 we can see that the field lines of a bar magnet are concentrated near the poles and spread out in space far from the magnet. In analogy with the electrical case, we can say that the strength of the magnetic field is greatest near the poles where the lines bunch together and diminishes as we move away from the magnet.

The magnetic field of a bar magnet is certainly not uniform:

Figure 5-14 *The magnetic field of a bar magnet.*

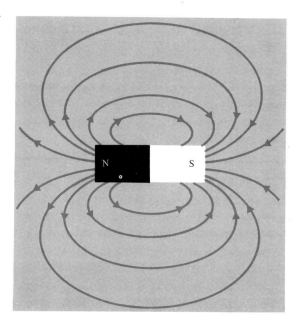

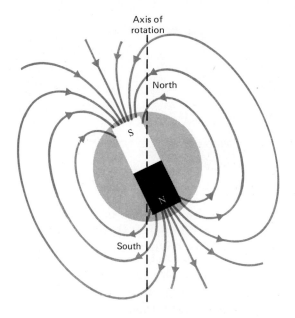

Figure 5-15 *The Earth's magnetic field is like that of a bar magnet with the S pole near the north geographic pole.*

The field lines are not straight and evenly spaced (as they are in the electrical case shown in Fig. 5-6). We can, however, produce a uniform magnetic field by bending a bar magnet into the form of a C, as shown in Fig. 5-16. Here, the surfaces of the N and S poles are parallel and in the gap between the poles the field is uniform. Fields of this general type are found in the simple electric motors that we will discuss shortly.

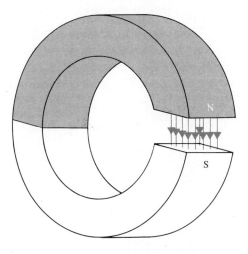

Figure 5-16 *By bending a bar magnet into the shape of a C, a uniform magnetic field in the gap between the poles can be produced.*

A bar magnet consists of a very large number of tiny crystalline pieces of iron. Each of these units (called *domains*) is actually a miniature magnet with N and S poles. If the domains are arranged in random fashion, the individual magnetic effects cancel and the bar has no net magnetism. On the other hand, if the domains are mostly aligned with their N poles in the same direction, the individual magnetic effects add together and the bar has a net magnetism.

The magnetism of a bar magnet is a permanent feature of the material. Left alone, the magnetism will remain indefinitely. However, this magnetism can be destroyed. If the bar is heated to a sufficiently high temperature (for iron the temperature is 1300°C), the domains are jiggled about to the extent that their original alignment is upset and they take up random directions. The net magnetism of the bar therefore ceases to exist.

Magnetic Fields Produced by Electric Currents

The magnetism of a bar magnet is not easy to turn on or off. But there is a simple way in which we can produce temporary magnetism. In the early 1800s it was discovered that an electric current in a wire will influence the direction of a nearby compass magnet. Because a compass magnet changes direc-

Figure 5-17 *Circular magnetic field lines surround a current-carrying wire.*

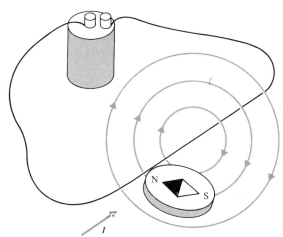

tion only in response to a magnetic field, this observation means that an electric current flowing in a wire sets up a magnetic field in the vicinity of the wire.

This experiment is easy to reproduce. We need only a wire connected to a battery and a compass. If we place the compass directly beneath the current-carrying wire, as in Fig. 5-17, we find that the compass magnet takes up a direction *perpendicular* to the wire. We can trace the field lines by moving the compass around the wire in the same way that we mapped the field of a bar magnet. By doing this, we discover that the field lines are all *circles* that surround the wire. Moreover, the field lines have directions that are given by the fol-

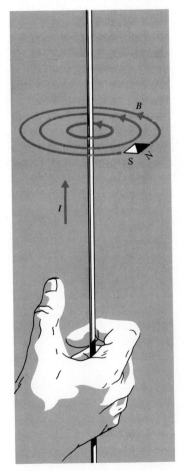

Figure 5-18 *Illustration of the right-hand rule for determining the direction of the magnetic field lines surrounding a current-carrying wire.*

Magnetic Fields Produced by Electric Currents   133

lowing rule: *Grasp the wire in your right hand with your thumb pointing in the direction of current flow; then, your fingers encircle the wire in the same direction as the field lines.* This *right-hand rule* is illustrated in Fig. 5-18.

We can use this right-hand rule to deduce the main features of the magnetic field produced by a current flowing in a loop of wire. Imagine that the straight wire in Fig. 5-17 is bent into a loop. Next, use the right-hand rule at various positions around the loop. You will find, as shown in Fig. 5-19, that the field lines inside the loop all have the same direction. Outside the loop, the lines spread out and resemble the field of a bar magnet. If we had many loops (that is, a wire *coil*), each loop would contribute an equal amount to the overall field. (Can

Figure 5-19 *The field lines for a current-carrying loop of wire. Notice that the field resembles that of a bar magnet. Another type of right-hand rule is in operation here. If the fingers of the right hand are curled in the same sense that current flows in the loop, the thumb points in the same direction as the magnetic field lines that pass through the loop. (Check that this is consistent with the application of the previous right-hand rule.)*

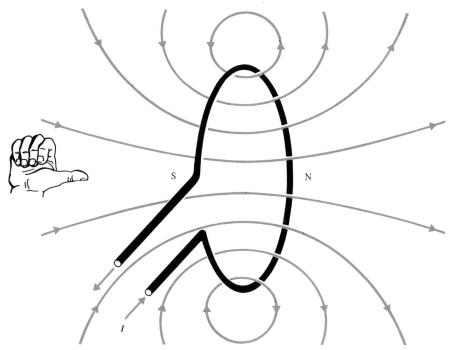

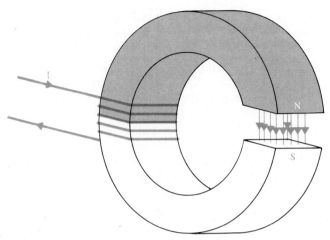

Figure 5-20 *An electromagnet is made by passing a current through a coil of wire that surrounds a bar of iron. Check that the polarity is correct by using the right-hand rule given in Fig. 5-19.*

you see why?) Therefore, passing a current through a coil of wire is one way to produce a strong magnetic field.

We can now introduce a piece of iron into a current-carrying coil of wire to construct a very useful device called an *electromagnet*. If the iron is shaped into a C, we have the arrangement shown in Fig. 5-20. Here, the strong field produced inside the coil is concentrated and guided by the iron to the gap between the N and S poles. The C-shaped iron need not be a permanent magnet; the field produced by the coil will align the magnetic domains in the iron and a strong field will exist for as long as the current flows in the coil. Electromagnets of one type of another are used in a variety of situations because they are simple to construct, because the field can be turned on and off, and because the field strength can be changed by increasing or decreasing the current in the coil. Many types of electric motors use electromagnets instead of permanent magnets.

Forces Due to Fields

The operation of an electric motor requires *forces* to be exerted on the rotor. Therefore, we must now enquire how these forces arise. An electric field will exert a force on any charged particle in that field. Thus, the electric field set up within a wire when the ends are connected to a battery exerts a force on each of the free electrons in the wire and causes a current to flow.

What about magnetic fields? Do they also exert forces on charged particles? There are two important differences in the behavior of a charged particle in a magnetic field compared to an electric field. First, *any* charged particle, whether it is stationary or moving, will experience a force due to an electric field. That force, as we have already pointed out, is proportional to the strength of the field and to the magnitude of the charge on the particle:

electrical force = (electric field strength) × (charge)

In a magnetic field, however, a charged particle will experience a force *only if it is moving*. A charged particle at rest in a magnetic field will experience no magnetic force whatsoever. In analogy with the expression above for the electrical force, we can write for the magnetic force on a moving charged particle,

magnetic force = (magnetic field strength)
$$\times \text{(charge)} \times \text{(velocity)}$$

Thus, in addition to being proportional to the strength of the magnetic field and to the charge, the magnetic force on a moving charged particle is also proportional to the velocity of the particle.

The second important difference between electric and magnetic forces involves the *direction* of the force. The two cases are illustrated in Fig. 5-21. In each case the field lines are directed from left to right and the (positively) charged particle enters each field along the same direction. When the particle enters the electric field (Fig. 5-21a), the field exerts a force just as we expect, namely, the positively charged particle is pushed toward the negative plate. That is, the direction of the electrical force on a positively charged particle is the same as the direction of the electric field lines.

Figure 5-21b shows that the magnetic force does *not* act in the same direction as the magnetic field lines. In fact, the magnetic force is *perpendicular* to both the field direction and the direction of motion of the particle. Another (and final) right-hand rule specifies the direction of the magnetic force. Refer to the lower portion of Fig. 5-21b. If you curl the fingers of your right hand in the direction that carries the velocity vector

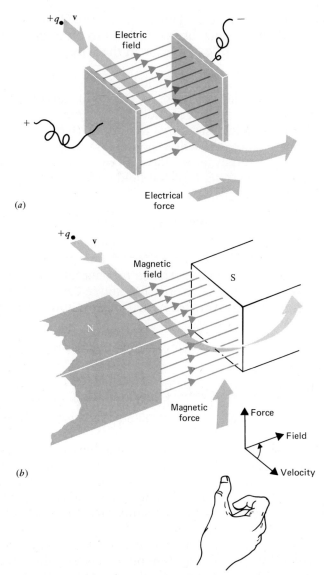

Figure 5-21 (a) *The direction of the electrical force on a positively charged particle is the same as the direction of the field lines.* (b) *The direction of the magnetic force on a moving charged particle is perpendicular to the direction of the field lines and to the direction of the velocity. The force direction is given by a right-hand rule, as illustrated.*

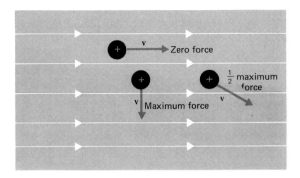

Figure 5-22 *The magnetic force on a moving charged particle depends on the angle between the direction of motion and the field direction.*

toward the field direction, then your thumb points in the direction of the magnetic force. Remember, this rule applies (as do all or our right-hand rules) for the effect on a particle that carries a *positive* charge. (In short, this right-hand rule says: Curl from **v** toward the field to find the direction of **F**.)

The magnitude of the force that a magnetic field exerts on a moving charged particle depends on the direction of the particle's motion with respect to the field direction. A particle that moves *perpendicular* to the field lines, as in Fig. 5-21b, experiences the maximum force. On the other hand, a charged particle that moves *directly along* a field line will experience no force at all. These two cases are illustrated in Fig. 5-22. If we change the angle between the velocity vector and the field direction from 90° toward 0°, we find that the magnetic force gradually diminishes, becoming zero when the angle reaches 0°. The intermediate case in which the magnetic force is half the maximum value is also shown in Fig. 5-22.

Because our concern is with the operation of electric motors, we are really interested, not in the behavior of individual charged particles in magnetic fields, but in the behavior of current-carrying wires. The extension from particles to wires is easy to make because a stream of moving charged particles is exactly equivalent to a current. In Fig. 5-23a we show a wire that passes through a magnetic field. What will happen when the switch is closed and a current flows through the wire? Tracing the current flow from the positive terminal of the battery to the negative terminal, we see that the direction of the current in the region of the field is the same as that

138   *Electricity and Magnetism*

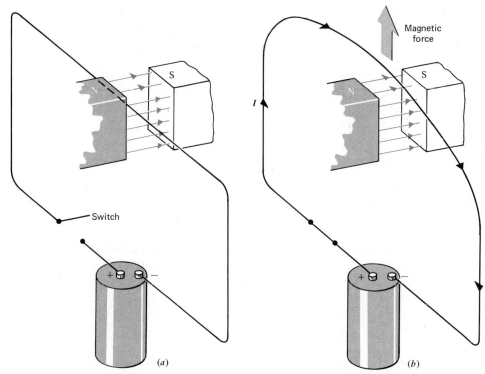

Figure 5-23 *The magnetic force on a current-carrying wire is the same as that on a stream of moving charged particles.*

of the moving charged particle in Fig. 5-21b. Remembering that the direction of an electric current is the same as the direction in which a positive charge would move, we immediately conclude that the magnetic force in this case will be upward, just as in Fig. 5-21b. That is, the magnetic force attempts to push the current-carrying wire upward and out of the field, as shown in Fig. 5-23b. Here we have the essential feature of the effect of a magnetic field on a current-carrying wire that permits us to understand how electric motors operate.

Electric Motors

Now that we understand how current-carrying wires behave in magnetic fields, we are in a position to discuss the opera-

Electric Motors 139

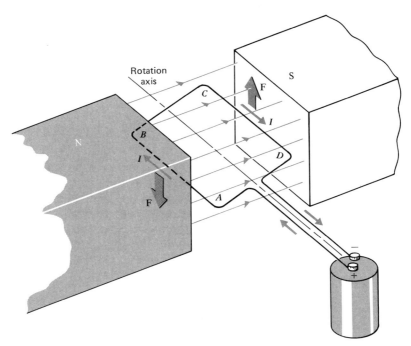

Figure 5-24 *The magnetic forces on a current-carrying loop of wire in a magnetic field produce a rotation of the loop. The modification shown in Fig. 5-25 is necessary to convert this system into a true electric motor.*

tion of electric motors. Actually, this next step is very easy. Let us begin by examining the forces on a current-carrying loop of wire in a magnetic field. A diagram of this situation is shown in Fig. 5-24. The current I flows through the rectangular wire loop in the direction $ABCD$. What are the forces on the various segments of the loop? First, look at the segment AB and apply the right-hand rule to determine the direction of the magnetic force. We find that this force is *downward*. Next, look at the segment CD. The same procedure shows that the force on this segment is *upward*. These two forces, acting together, produce a rotation of the loop around the axis shown in the diagram. (The forces on the segments BC and AD are both *outward* along the rotation axis and therefore cancel.) Thus, a current-carrying loop of wire in a magnetic field tends to *rotate*. This is the essential feature of an electric motor.

We have not yet completed our story. Look again at Fig. 5-24. What happens when the loop rotates so that it is parallel to the faces of the magnetic poles? In this position, the force on the highest part of the loop (segment *CD*) is upward, and the force on the lowest part (segment *AB*) is downward. Therefore, when the rotation has progressed to this position, there is no longer a tendency for the magnetic forces to produce rotation.

If we are to have a useful electric motor, we must arrange for the magnetic forces to act in such a way that rotation continues. In order to do this, we must provide a scheme that reverses the direction of the current each time the loop moves to a vertical position. In this way, the force on the segment *CD*, for example, will be *upward* when the segment moves upward, and the force will reverse and become *downward* during the next half cycle when the segment moves downward. Thus, the rotation will be maintained.

The current from a battery flows always in the same direction. A change in the direction of the current through the loop therefore requires some mechanical modification of the system. Figure 5-25 shows how this can be done. Here, we introduce between the battery and the loop a metal cylinder that is split into halves that are insulated from one another. Each of the two wires leading to the loop is connected to one of the cylinder halves. The circuit through the battery is completed by sliding contacts that touch the cylinder halves. Every time the loop rotates to the vertical position, each contact slips from one half of the cylinder to the other, thereby reversing the direction of the current through the loop for the next half cycle. Figure 5-25 represents, in a schematic way, the simplest type of battery-powered electric motor.

We have remarked that the current from a battery flows always in the same direction. Such current is called *direct current* or DC. Motors of the type illustrated in Fig. 5-25 are called *DC motors*. You probably know that ordinary household current is *alternating current* or AC. This type of current is reversed in direction at regular intervals. Household current in the United States completes 60 complete cycles of reversal each second; we call this *60-cycle AC*. (The European standard is 50 cycles per second.) In the next section we will see how alternating current is produced.

The requirement for a continuously rotating electric motor is that the direction of the current in the loop be reversed every half cycle. The use of alternating current meets this requirement in an automatic way without the necessity of intro-

Electric Motors 141

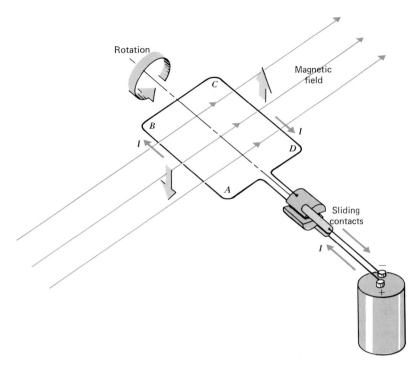

Figure 5-25 *By introducing a split metal cylinder between the battery and the loop, the current through the loop is reversed every half cycle. This permits the loop to rotate always in the same direction. All real electric motors have coils of wire, instead of a single loop, in order to enhance the magnetic forces.*

ducing a split cylinder and sliding contacts. If the two wires leading to the loop in Fig. 5-24 are connected to an AC power line, the loop will rotate in synchronism with the current reversals in the line. Such a device is an *AC motor*.

If an AC motor is constructed as in Fig. 5-24, the loop (or *rotor*) will undergo one complete rotation for each reversal cycle of the current. That is, when connected to an ordinary household power line, the rotation rate will be 60 revolutions per second or 3600 revolutions per minute (rpm). Different rotation rates can be arranged by constructing multisegment rotors and by using several pairs of magnetic poles.

If you take apart an electric motor and examine its construction, you may find that it does not look like Fig. 5-24 or 5-25. However, if you trace the wires and connections, you will find that the essential features are the same as those in the diagrams.

Induction and Electric Generators

How do we generate the electric power that is transmitted to our homes to run our appliances, lights, and heaters? What kind of equipment is needed to produce alternating current? Actually, an electric generator is simply the reverse of an electric motor. When an electric motor is operated, electric energy is consumed and mechanical work is done (and heat is produced). On the other hand, in an electric generator, mechanical work is done (by some outside agency) and electric energy is produced. The "outside agency" that provides the input to the generator can be the water flowing from behind a dam or steam from a boiler heated by burning coal or oil or by nuclear fission.

The basic idea of an electric generator can be seen in Fig. 5-26. Here, a segment of a wire circuit is being moved downward through a magnetic field. The vector **v** indicates this mo-

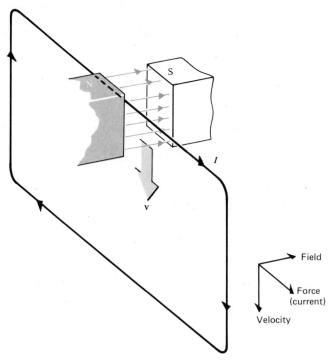

Figure 5-26 *The motion of a wire through a magnetic field induces a current to flow in the wire. The direction of current flow is given by the right-hand rule.*

Induction and Electric Generators 143

tion. The wire contains free electrons, and so the movement of the wire means that charged particles are in motion in the magnetic field. When charged particles move in a magnetic field, they experience a force. If we apply the right-hand rule to the situation in Fig. 5-26, we see that the force (on a positive charge) is along the wire in the direction shown. Thus, a current is induced to flow in the wire. (As always, the moving charged particles are actually electrons and they move through the wire in the direction opposite to that of the current I.) This process is called *electromagnetic induction* and was discovered by the great English scientist, Michael Faraday (1791–1867) in 1831.

Figure 5-26 illustrates the principle of induction and the conversion of mechanical work into electrical energy, but such a system is not a practical way to generate electricity. Figure 5-27 shows, in a schematic way, the basic features of an electric generator. Note that this diagram is almost the same as Fig. 5-24. But in Fig. 5-27 an external agency is causing the wire loop to rotate as indicated by the velocity vectors **v**. The segment that is moving downward has induced

Figure 5-27 *By rotating a wire loop in a magnetic field, an alternating electric current can be produced. Use the right-hand rule and verify that the direction of induced current is correctly shown for both segments of the wire.*

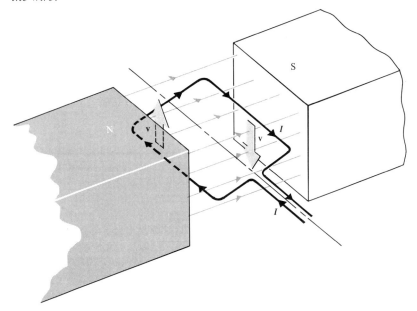

144   *Electricity and Magnetism*

in it a current that has the same direction as in the downward-moving wire in Fig. 5-26. As the loop rotates, the direction of current flow reverses every half cycle. (Compare the discussion of electric motors.) Therefore, Fig. 5-27 represents an *AC generator*. By using a split cylinder and sliding contacts, we could modify this system to produce DC. Can you see how?

Questions and Exercises

1. Figure 5-3 shows that a plate attached to the positive terminal of a battery acquires a positive charge and a plate attached to the negative terminal acquires a negative charge. Why is this so? To develop your explanation, remember the following facts. (a) Like charges repel; (b) the connecting wires and plates are made of conducting materials; (c) only the free electrons—*not* the atoms—can move around within conductors.
2. Two identical copper spheres carry charges with equal magnitudes but opposite signs. Explain in detail what will happen if the spheres come into contact.
3. If you shuffle across a carpet and then touch a doorknob, you will experience a mild shock. This is due to the fact that your body carried an electric charge. How do you suppose your body became charged?
4. Make a sketch similar to Fig. 5-5 for the case of two equal positive charges.
5. Is there an electric field in the vicinity of a battery? Explain.
6. Can two field lines ever cross? Explain.
7. Two wires have the same lengths but one wire has twice the diameter of the other. Which wire do you suppose has the greater electrical resistance? Explain your reasoning.
8. Figure 5-11 shows the attraction and repulsion of magnetic poles. In Fig. 5-11a, the S pole of the left-hand magnet should be attracted toward the N pole of the right-hand magnet. Similarly, the S pole of the right-hand magnet should be attracted toward the N pole of the left-hand magnet. In spite of these two attractive forces, there is a *net* repulsion between the two magnets. Why do you suppose this is true?
9. If you suspend a compass magnet on a horizontal axis, you will find that it will point downward at a certain angle. Explain why this is so. (Make a sketch to demonstrate your answer.)

10. It is believed that huge electrical currents flowing in the molten iron core of the Earth are responsible for the Earth's magnetism. Make a sketch showing the direction in which the *electrons* must be moving to account for the Earth's polarity.
11. A current flows in a wire that is oriented in a north-south direction. A compass magnet located above the wire points *west*. In what direction is the current flowing in the wire?
12. An electron moves directly along the axis of a wire loop that carries a steady current. Describe how the electron will react to the magnetic field. (Refer to Fig. 5-19.)
13. The movement of free electrons along a wire constitutes a current. Surrounding such a wire there is a magnetic field but no electric field. Why?
14. A friend tells you that within a certain black box he has established a field. What simple experiments could you do to determine whether the field within the box is an electric field or a magnetic field?
15. Explain how you could build a DC motor without using a permanent magnet.
16. Explain why electric motors are always constructed with coils of wire consisting of many loops instead of single loops as shown in Figs. 5-24 and 5-25.

Additional Details for Further Study

Coulomb's Law

The expression for the force exerted by one electrical charge on another is very similar to that for the gravitational force exerted by one mass on another. Newton's law of universal gravitation takes the form

$$F_G = G \frac{m \times M}{r^2}$$

The expression for the electrical force between a charge q and a charge Q is

$$F_E = K \frac{q \times Q}{r^2}$$

This is called *Coulomb's law*. Whereas the gravitational force is always attractive, the electrical force is attractive only when q and Q have opposite signs; if q and Q have the same signs, the force is repulsive.

We choose to measure electrical charge in units called *coulombs* (C). In these units, the electron carries the charge

$$e = -1.60 \times 10^{-19} \text{ C}$$

With charge measured in coulombs, the constant K must be determined experimentally so that F_E is the force in newtons when the distance r is given in meters. The value is

$$K = 9.0 \times 10^9 \text{ N-m}^2/\text{C}^2$$

For example, the force between a charge $q = +4 \times 10^{-5}$ C and a charge $Q = -8 \times 10^{-6}$ C separated by a distance of 0.8 m is

$$F_E = (9.0 \times 10^9 \text{ N-m}^2/\text{C}^2)$$
$$\times \frac{(+4 \times 10^{-5} \text{ C}) \times (-8 \times 10^{-6} \text{ C})}{(0.8 \text{ m})^2}$$
$$= -4.5 \text{ N}$$

The negative sign means that the charges carry opposite signs and therefore that the force is attractive.

Voltage, Current, and Electrical Heating

How does the electrical power expended in a resistor R change as we change the current through the resistor? The power expended is

$$P = I \times V$$

Now, we can write Ohm's law in the form

$$V = I \times R$$

Substituting this expression for V into the equation for the power, we find

$$P = I \times (I \times R) = I^2 \times R$$

That is, the power delivered to a resistor R varies as the *square* of the current through the resistor. If the current through R is *doubled,* then the power loss, which appears in the form of heat, is *quadrupled*. This effect is sometimes referred to as the "*I*-squared-*R*" heating of a resistance element.

The loss of power in a wire due to the resistance of the wire is an important consideration in the design of power transmission systems. Suppose that it is desired to transmit a certain power *P* through a particular set of wires. Should a high voltage or a low voltage be applied to the wires in order to minimize the power loss? Because the loss in the wire is equal to I^2R, clearly we want the current *I* to be as small as practicable. (We cannot do very much about decreasing the resistance *R* of the wires.) At the delivery end of the system, the user receives power equal to the voltage across the wires multiplied by the current through the wires, $P = IV$. If *I* is to be small, then *V* must be made very large to keep the power *P* at the desired level. Therefore, power transmission lines are always operated at high voltages. Almost all long-distance transmission lines operate at 200 000 V or more, and some operate at 600 000 V.

The very high voltages of transmission lines permit the efficient delivery of electric power, but the voltages are far too high to be used directly. Therefore, the high voltages are "stepped down" in stages to the normal household value of 120 V. This "stepping down" is done in several steps, at power substations, by devices called *transformers*. A transformer is a device that alters the voltage and the current in a circuit while maintaining the product $I \times V$ constant (except for the relatively small power loss in the windings of the transformer). Transformers are used at the generator site to step up the voltage for transmission and at the delivery end to step down the voltage before the power is used by the consumer. This step-up, step-down system is essential for the effective transmission of electrical power. Because transformers operate only with AC, not DC, all power transmission lines carry alternating current.

Additional Exercises

1. Two electrical charges that are separated by a certain distance repel one another with a certain force *F*. If the separation distance is reduced to one-fifth of the original value, what will be the new force between the charges? [Ans. 25*F*.]

2. The gravitational force on a 1-kg object near the surface of the Earth is 10 N. What must be the separation of two 1-C charges if the electrical force between them is equal to 10 N? [Ans. 30 km.]

3. What power (in megawatts) is transmitted by a system that operates at 400 000 V and carries a current of 3 A? [Ans. 1.2 MW.]

4. A 3-Ω resistor is connected to a 6-V battery. At what rate is electrical energy drained from the battery? [Ans. 12 W.]

How do refrigerators work?

6

HEAT AND THERMODYNAMICS

One of the most common appliances in American households is the *refrigerator*. A refrigerator is both a useful and a remarkable device. It is "useful" because it provides lowered temperatures that assist in preserving foods that would quickly spoil if left at room temperature. A refrigerator is "remarkable" because it appears to operate counter to the natural order of things. We know, for example, that when we burn a fuel, as in a gas heater, energy is released and there is an increase in the temperature of the air. That is, the stored chemical energy in the fuel is converted into thermal energy in the air and *heating* takes place. This seems only natural. But when energy—usually electrical energy—is released in a refrigerator, *cooling* takes place. How can the release of energy in a refrigerator produce an effect that is so different from that in the usual situation?

In this chapter, we will study some of the basic principles of *thermodynamics,* and we will see how these ideas are involved in the operation of refrigerating devices.

The Laws of Thermodynamics

Let us begin by examining a natural process that we have all observed. Suppose that we bring together two blocks that initially have different temperatures. Let us further suppose that we insulate these blocks so that they do not exchange heat with the surroundings. What will happen to the temperatures of the blocks? We know from our experience that the hotter block will become cooler and that the cooler block will become hotter. After a period of time, both blocks will have the same temperature. In this process, thermal energy is transferred from the hotter block to the cooler block. This transfer continues until a uniform temperature is reached, at which point no further energy transfer is possible. Nature always works this way. The spontaneous flow of heat always proceeds from a hotter object to a cooler object so that the temperature tends to become equalized; the reverse never happens. This is a fundamental rule of Nature and one of the important laws of thermodynamics. In a refrigerator, however, heat flows from *cool* objects (within the refrigerator) to *warm* objects (outside the refrigerator), and the cool objects become even cooler. How does this come about?

Another basic principle of thermodynamics is the conservation of energy, a principle that we have previously studied. In discussing thermodynamic systems, however, one must be careful to include the internal or thermal energy of each part of the system and to take account of any changes in this internal energy. When the principle of energy conservation is framed in this way, it is called the *first law of thermodynamics*. There is no new physics in the first law: It is simply a general way of stating the familiar energy principle. The rule that the spontaneous flow of heat is always from hotter to cooler objects *is* a new physical idea. There is nothing in the energy conservation principle—or in any other law of Nature—that specifies for us the direction of heat flow. If energy were to flow spontaneously from a block of ice to a surrounding volume of water, this could occur in complete accord with energy conservation. But such a process never happens. This idea is the substance of the *second law of thermodynamics*. Clearly, a refrigerator, which is a physical system, must obey not only the first law (energy conservation) but the second law as well.

To see why the second law is not violated by a refrigerator, we must be careful in our statement of the law. The second law of thermodynamics says that heat never flows *spontaneously* from a cooler to a hotter object. Or, alternatively, heat can flow from a cooler to a hotter object only as the result of work done by an outside agency. We now see the distinction between an everyday spontaneous process, such as the flow of heat between water and ice, and the flow of heat from the inside to the outside of a refrigerator. In the water-ice system, the exchange of energy takes place spontaneously and the flow of heat always proceeds from the water to the ice. The water gives up energy and becomes cooler while the ice receives energy and melts. In a refrigerator, on the other hand, the exchange of energy is not spontaneous. Work provided by an outside agency is necessary to reverse the natural flow of heat and to cool the interior at the expense of further heating the warmer surroundings.

The operation of a refrigerator depends on the compression and expansion of a gas and on the way in which this gas is converted from the gaseous state to the liquid state and back again. Therefore, to understand how a refrigerator works, we need to discuss the behavior of gases under changing conditions of pressure, volume, and temperature. *Volume* is a familiar concept and needs no further elaboration. We measure volume in cubic meters (m³), in cubic centimeters (cm³), or, sometimes, in liters (1 l = 1000 cm³). The ideas of pressure and temperature, however, require commentary.

Pressure

If you push on an object, you are exerting a force. This force is localized; for example, it might be concentrated at the end of a stick. When you blow up a balloon, you are also exerting a force. In this case, however, the force is not localized. In fact, the force exerted through the pump nozzle (Fig. 6-1) is transmitted to the interior surface of the balloon so that any small area of the surface receives the same force as any other equal area. We can readily see that this is the case. If you blow up a spherical balloon (whose thickness is uniform), the shape of the balloon remains spherical. That is, every square centimeter of the surface receives the same outward force and the balloon increases in size while retaining its shape.

All gases and liquids behave in the same way as the gas in a balloon. If a force is exerted on one part of a gas or a liquid, the force is transmitted by the material and exerts a uniform push on the walls of the container and on any object within the substance. Instead of describing this situation in terms of *force*, it is much more convenient to emphasize the *force per unit area* exerted on the container walls or other objects. This

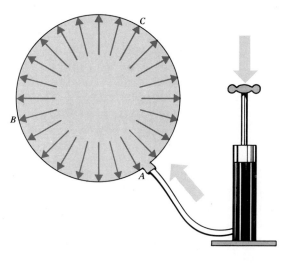

Figure 6-1 *The pump pressure is transmitted undiminished to every part of the interior surface of the balloon. The force per square centimeter at A, B, and C is exactly the same.*

we call the *pressure:*

$$\text{pressure} = \frac{\text{force}}{\text{area}}$$

or, using symbols,

$$P = \frac{F}{A}$$

Looking again at Fig. 6-1, we can see that if the force on 1 cm² of surface at B were greater than the force on 1 cm² of surface at C, then a bulge would develop at B and this part of the balloon would expand differently. But this does not happen, and we conclude that the pressure on the balloon surface is everywhere the same. Indeed, the pressure on the surface at B or C, or any other point, is exactly the same as the pressure at the nozzle A. Thus, we can make the following statement: *The pressure exerted on a gas or a liquid at one position is transmitted undiminished to every part of the substance.* This is known as *Pascal's law* after its discoverer, Blaise Pascal (1623–1662), a French scientist.

Notice that Pascal's law is stated in terms of *pressure,* not *force.* It is *not* true that a force exerted on a gas is transmitted unchanged. In fact, the *total* force exerted by the gas on the surface of the balloon in Fig. 6-1 is much greater than the force at the nozzle.

It is important to realize the significance of the definition of pressure in terms of the force *per unit area.* Suppose that we apply a force F to a piston that closes a container of gas (or liquid), as in Fig. 6-2. If the piston has an area A, the pressure at every point within the gas is $P = F/A$ (Fig. 6-2a). Now, suppose that we reduce the area of the piston to $\frac{1}{4}A$ but maintain the applied force at the value F (Fig. 6-2b). Because the area is *smaller* and the force is the *same,* the pressure is *greater* than in the first case. In fact, we now have a pressure of $4P$. Thus, we see that huge pressures can be generated by the application of even very small forces; all that is necessary is that the force be applied to a small area. This fact has numerous applications in all types of hydraulic systems.

The air in the atmosphere is pulled downward toward the Earth by gravity. This results in a pressure—called *atmospheric pressure*—that is exerted on the surface of the Earth and on all of the objects around us. The magnitude of this atmospheric pressure is actually very large. Suppose that you

154 Heat and Thermodynamics

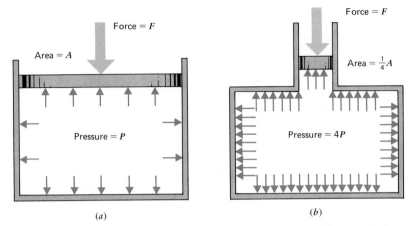

Figure 6-2 *A given force F produces a greater pressure when applied to a smaller area.*

extend your hand with your palm upward. The atmospheric pressure that acts downward on your hand is approximately the same as would be exerted by a 120-kg block! (See Fig. 6-3.) The total downward force due to atmospheric pressure on an area of 1 m² is the same as the weight (that is, the gravitational force) due to a mass of 10 000 kg acting on the same area. The normal atmospheric pressure at the surface of the Earth is called *1 atmosphere* and is abbreviated 1 atm. If, in Fig. 6-2a, the force F produces a pressure of 2 atm within the container, then, in Fig. 6-2b, this same force produces a pressure of 8 atm.

How do we survive under the crushing load of atmospheric pressure? Why does a tabletop not disintegrate due to the downward force exerted by the air above it? The point is that any object immersed in a fluid experiences pressure over its entire surface, not simply the upper surface. The table in Fig. 6-4 does not collapse even though there is a tremendous downward force on the upper surface because there is an equal upward force on the lower surface. In the same way, the inward pressure on our bodies is balanced by the outward pressure from within.

If atmospheric pressure is removed from one side of a surface (for example, by means of a vacuum pump), then the atmospheric pressure on the surface exposed to the air will cause the surface to bend or even rupture. This effect is easy

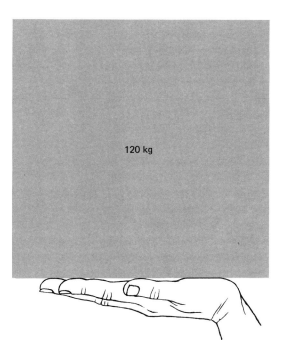

Figure 6-3 *The total downward force due to atmospheric pressure on your hand is approximately the same as the gravitational force due to a 120-kg block.*

to demonstrate with a metal can and a vacuum pump. Before connection to the pump (Fig. 6-5a), the atmospheric forces on all surfaces of the can are balanced. However, when the pump exhausts the air from within the can, there is no longer any outward pressure and the external atmospheric pressure collapses the can (Fig. 6-5b). Notice, in both parts of this

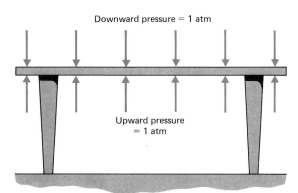

Figure 6-4 *A table top does not collapse due to atmospheric pressure because there is an equal atmospheric pressure on the lower surface.*

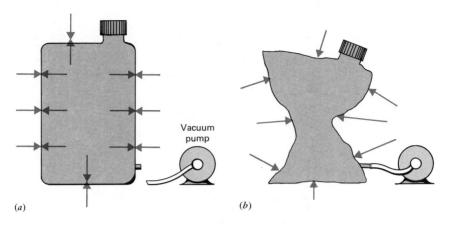

Figure 6-5 *Exhausting the air from within a sealed can removes the outward force and atmospheric pressure causes the can to collapse.*

example, that the forces due to atmospheric pressure (or to any gas or liquid pressure) are always directed *perpendicular* to the surface. (If a gas or a liquid *could* exert a nonperpendicular force on the surface of an object, then there would be a tendency for the object to move. Clearly, a static gas or liquid cannot spontaneously cause an object to move.)

Temperature

Temperature is a familiar concept; it specifies the degree of "hotness" or "coldness" of an object. From our discussion at the beginning of this chapter, we know that temperature determines the direction of heat flow: The spontaneous flow of heat is always from an object at a high temperature to one at a lower temperature. Although this is an important idea, it does not provide a practical means for *measuring* temperature. In order to do this, we need to take advantage of some other physical effect that depends on temperature.

A common type of temperature-measuring device, or *thermometer,* relies on the fact that most substances expand when the temperature is raised and contract when the temperature is lowered. Most household thermometers utilize this property of liquid mercury or alcohol. The liquid is sealed within a tube that has a small diameter and a bulb reservoir at one end (Fig. 6-6). When the bulb is heated, the liquid expands and the

Temperature 157

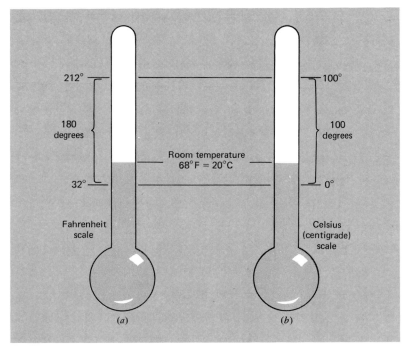

Figure 6-6 *Comparison of the Fahrenheit and Celsius temperature scales. The freezing point of water is designated 32° F or 0° C, and the boiling point of water is designated 212° F or 100° C.*

level within the tube rises. A scale is marked beside the tube. On the *Fahrenheit* temperature scale, the freezing point of water is designated 32°F and the boiling point of water is designated 212°F. The corresponding points on the *Celsius* (or *centigrade*) scale are 0°C and 100°C. These two temperature scales are compared in Fig. 6-6. The United States is now almost alone in its retention of the Fahrenheit scale for everyday usage.

There is a difference of 100 degrees between the freezing and boiling points of water on the Celsius scale, whereas the difference is 180 degrees on the Fahrenheit scale. Thus, the Celsius degree is a larger unit than the Fahrenheit degree; the ratio is 180/100 or 9/5. We can convert ordinary room temperature (68°F) to degrees Celsius in the following way. The temperature 68°F is 36 F degrees above freezing. This corresponds to 36 ÷ 9/5 or 20 C degrees above freezing or 20°C. The correspondence between 68°F and 20°C is shown in Fig.

6-6. (Notice the difference between the terms *36°F* and *36 F degrees*. "36°F" means a *particular temperature* referred to the Fahrenheit scale, whereas "36 F degrees" means a *temperature interval* or *difference* of 36 degrees on the Fahrenheit scale.)

Another temperature scale, one which is important in thermodynamics, is the *absolute* or *Kelvin* scale. On this scale, we measure temperatures, not from an arbitrary point such as the freezing point of water, but from the *absolute zero* of temperature. *Absolute zero* is the lowest conceivable physical temperature; no physical object or system can ever exist at a lower temperature. Absolute zero has been closely approached in the laboratory, but this temperature has never been reached (and *can* never be reached). The value of absolute zero on the Celsius scale is $-273°C$. On the absolute temperature scale we call this point $0°K$ and we measure temperature in *degrees Kelvin*. The size of the Kelvin degree is exactly the same as the Celsius degree. Therefore, a temperature on the absolute scale has a value that is 273 degrees higher than the corresponding Celsius temperature. That is, $0°C = 273°K$, and room temperature is $20°C = 20° + 273° = 293°K$. We will return to the discussion of absolute zero when we discuss the behavior of gases at low temperatures.

The idea of *temperature* is connected with the internal motions of a substance. All matter is composed of atoms and molecules, and these particles are continually in motion. This is easy to accept for gases because it is not difficult to imagine gas molecules moving around at random within a container. Molecular motion is also important in liquids and solids. In a liquid the moving molecules slide past one another and the substance *flows*. In solids, on the other hand, the atoms and molecules do not move *away* from their positions, but they do vibrate back and forth *around* these positions.

The motion of the molecules in a substance depends on the temperature of that substance. The higher the temperature, the more rapid is the motion. In a gas, for example, we can specify the temperature in terms of the average speed of the gas molecules. This association of temperature and molecular motion is important because it provides us with a connection between one aspect of the bulk or *macroscopic* properties of matter (namely, the temperature) and the small-scale or *microscopic* behavior (namely, molecular motion).

The Behavior of Gases Under Changing Conditions

In a refrigerator, the working gas is sealed in a system of tubes and is driven through the cooling cycle by a pump. During this cycle, the gas is subjected to compression and expansion. How does a gas respond to changes in pressure? The first important discovery relating to the behavior of gases under changing conditions of pressure was made in 1662 by Robert Boyle (1627–1691), an English chemist. Boyle's experiments consisted of measuring the pressure and the volume of a quantity of gas confined in a cylinder such as that shown in Fig. 6-7. Boyle began by measuring the pressure P and the volume V of the confined gas (Fig. 6-7a). When he increased the force on the piston so that the gas pressure increased to $2P$, he found that the volume was decreased to $\frac{1}{2} V$ (Fig.

Figure 6-7 *Robert Boyle discovered that when the pressure of a confined gas is increased by a certain factor, the volume is decreased by the same factor (as long as the temperature is maintained constant).*

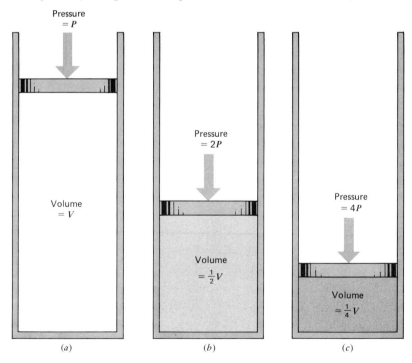

6-7b). A further increase in the pressure to $4P$ resulted in a volume decrease to $\frac{1}{4} V$ (Fig. 6-7c). All of these measurements were carried out at constant temperature. If we multiply the pressure by the volume, we find for the three cases:

(a) $P \times V$
(b) $2P \times \frac{1}{2}V = P \times V$
(c) $4P \times \frac{1}{4}V = P \times V$

That is, the product, (pressure) × (volume), is always the same. We can therefore state

(pressure) × (volume) = constant
$$\text{(for constant temperature)}$$

or

$P \times V$ = constant   (for T constant)

This rule of gas behavior is known as *Boyle's law*.

As we have already mentioned, the average speed with which gas molecules move depends on the temperature of the gas. If the temperature of the gas remains constant, as specified in the statement of Boyle's law, then the average molecular speed also remains constant. Realizing this, we can easily argue from the standpoint of molecular motion why Boyle's law is true.

Look at Fig. 6-8a. Here we focus on the motion of one molecule of the gas that is confined in the box. The molecule moves with a certain speed, and during a certain interval of time, it experiences a number of collisions with the container walls. For example, in Fig. 6-8a, the molecule strikes the right-hand wall *three* times. Now imagine that one wall of the container is moved inward so that the volume of the container is reduced to one-half its former value, as in Fig. 6-8b. If the temperature in the two cases is the same, the molecule will

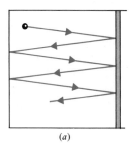
(a)

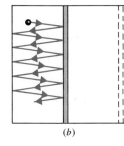
(b)

Figure 6-8 *A molecule in a gas at a certain temperature bounces back and forth between the container walls. In* (a) *the container has a volume V. In* (b) *the volume is reduced to $\frac{1}{2} V$ and the molecule strikes the wall twice as frequently as in* (a). *Thus, the* pressure *of the gas in* (b) *is twice that in* (a). *Boyle's law is therefore satisfied.*

move through the same distance during equal time intervals. In Fig. 6-8b we see that this results in *six* collisions with the wall. That is, decreasing the volume by *half* causes the frequency of collisions to *double*. It is easy to see that if the volume had been reduced to one-third, the number of collisions would have increased by a factor of three, and so forth.

Each time that a molecule strikes and rebounds from a container wall, it exerts a push on that wall. The sum of all the molecular pushes per unit area of the wall is just the *pressure* of the gas. Thus, we conclude that the product, (pressure) × (volume), remains constant for constant temperature. This conclusion, based on the motion of the molecules in a gas, is the same as Boyle's law.

In all of the preceding discussion, we considered only cases in which the temperature remained constant. That is, each system was assumed to be in contact with some large reservoir that was able to supply or absorb heat in order to maintain a constant temperature for the substance being studied. For example, we could investigate the pressure-volume relationship for a container of gas at the constant temperature of 0°C by immersing the container in a large water-ice bath. Now, we need to expand our discussion to include the possibility of temperature changes.

Suppose that we maintain a certain quantity of gas at a constant pressure. How does the volume of the gas depend on the temperature? This question was answered in 1802 by the French scientists Jacques Charles (1746–1823) and Joseph Louis Gay-Lussac (1778–1850). Independently of one another, they discovered that as the temperature of a gas is reduced (while maintaining constant pressure), the volume is also reduced. In fact, the graph of volume versus temperature is a straight line, as shown in Fig. 6-9. By extending this straight line, we find that at a temperature of $-273°C$ the volume of the gas should be reduced to *zero*. Actually, no real gas behaves in exactly this way. At some temperature, the gas condenses to a liquid. But for all temperatures at which the substance is a *gas*, the straight-line relationship holds and its extension (the dashed portion of the graph in Fig. 6-9) indicates that zero volume would occur at $-273°C$. Because there is no physical significance for a temperature below this value, we call this point the *absolute zero* of temperature. Using absolute zero as a starting point, we can measure temperature in degrees Kelvin on the absolute temperature scale, as we have

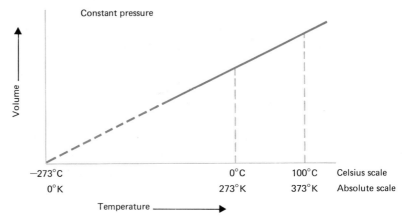

Figure 6-9 *At constant pressure, the volume of a gas decreases uniformly with temperature. Extending the volume-temperature graph beyond the range of measurements, we identify* $-273°$ *C as the* absolute zero *of temperature.*

previously indicated. Absolute zero is 0°K and the freezing point of water is 273°K.

Using the absolute temperature scale, we can state the discovery of Charles and Gay-Lussac in the following way: *At constant pressure, the fractional change in the volume of a gas is the same as the fractional change in its temperature.* That is, if the temperature is doubled, the volume is also doubled; if the temperature is reduced by one-fourth, the volume is also reduced by one-fourth. We can express the Charles–Gay-Lussac law as

$$\frac{\text{volume}}{\text{temperature}} = \text{constant} \quad \text{(for constant pressure)}$$

or

$$\frac{V}{T} = \text{constant} \quad \text{(for } P \text{ constant)}$$

This relationship is called the *Charles–Gay-Lussac law*.

We must mention an important point about the law of Charles and Gay-Lussac. The statement of the law in terms of fractional changes of volume and temperature, or in the form $V/T = $ constant, is true only if temperature is measured on the absolute scale. For example, if the temperature of a gas is increased from 20°C to 40°C, the volume does *not* double.

Moreover, on the Celsius scale (or on the Fahrenheit scale), *negative* temperatures are possible. But to insert a negative temperature into the Charles–Gay-Lussac law is meaningless. (This would require a *negative* volume!) In order for the law to have physical meaning, the temperature must never become negative; that is, we must use the absolute scale in which all temperatures are measured from absolute zero at 0°K.

The two laws of Boyle and of Charles and Gay-Lussac can be combined into a single statement concerning the behavior of gases under changing conditions of pressure, volume, and temperature. The combined statement is

$$\frac{(\text{pressure}) \times (\text{volume})}{\text{temperature}} = \text{constant}$$

or

$$\frac{P \times V}{T} = \text{constant}$$

Look carefully at this equation and see how it does represent both laws. For example, if the temperature remains constant, we can incorporate T into the "constant" on the right-hand side; then, we have $P \times V =$ constant, which is Boyle's law. Similarly, if the pressure remains constant, the equation reduces to the statement of the Charles–Gay-Lussac law.

The above relation connecting pressure, volume, and temperature is closely followed by all real gases. There are, however, small deviations, and we say that only an *ideal* gas would follow the relation exactly. Consequently, the statement $PV/T =$ constant is called the *ideal gas law equation*. The gas helium is almost an ideal gas. Helium obeys the ideal gas law down to a temperature very close to absolute zero (4°K), at which point it becomes a liquid. By measuring the pressure in a constant volume of helium, we can determine temperatures down to 4°K. We call such an instrument a *helium gas thermometer*.

Heat and the Transfer of Energy

Next, we come to the role of energy exchanges in the behavior of gases. Suppose that we have a certain fixed volume of gas and we wish to increase the temperature and the pressure. How can we do this? We know that we can increase the temperature by placing the gas container over a flame or on a hot stove. The ideal gas law tells us that increasing the temperature of a constant volume of gas will result in an increase in the pressure. (This is why a sealed can, such as an aerosol or spray container, should never be placed in a fire; the pressure could increase to the point at which the can will explode.) What has happened to cause the increase in temperature and pressure?

We have already learned that the temperature of a gas is determined by the speed of the gas molecules. If the molecular speed rises, so does the temperature. Then, the more rapidly moving molecules strike the container walls more frequently and this means an increase in pressure, as indicated in Fig. 6-10. (Compare the discussion of pressure and Boyle's law from the molecular viewpoint.) When the speed of a molecule is increased, its motional or kinetic energy is increased. Therefore, to increase the temperature of a gas means that energy must be supplied to the gas from some outside source such as a flame or a hot stove element. In this way, the internal or thermal energy of the gas is increased and we say that *heat* has been supplied to the gas. Similarly, if we *remove* heat from a gas that is maintained at constant volume, the pressure and the temperature will decrease.

Next, we consider a different kind of situation. Instead of supplying energy to a gas by heating the gas over a flame, sup-

Figure 6-10 *The input of heat to a gas that is maintained at constant volume results in an increased temperature and an increased pressure.*

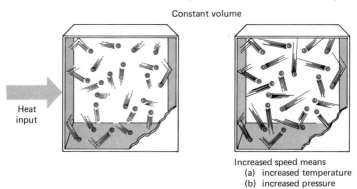

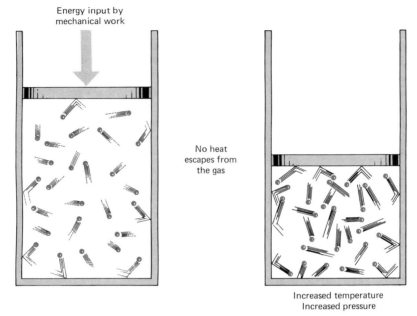

Figure 6-11 *Mechanical work done on a gas results in an increase in temperature and an increase in pressure.*

pose that we do mechanical work on the gas by compressing it. Figure 6-11 shows a cylinder with a movable piston. If the piston is pushed downward by some outside force, mechanical work is done on the gas. Let us imagine that the gas container is well insulated and that the mechanical compression takes place rapidly. Under these conditions, no heat will escape from the gas. That is, the total amount of work done on the gas remains as internal energy of the gas. We call this an *adiabatic* process. An increase in the internal energy must take the form of an increased speed of the gas molecules. We know that this means an increase in temperature and pressure. Thus, the sudden compression of a quantity of gas causes both the temperature and pressure to rise. This is exactly what happens, for example, in a bicycle tire pump. When you push down on the handle, the gas in the pump cylinder is suddenly compressed. The increased pressure acts to inflate the bicycle tire, and the increased temperature is readily apparent if you touch the cylinder wall.

What is the result of a sudden *expansion* of a gas? This process is just the reverse of that just considered. Look at the right-hand diagram in Fig. 6-11. If this high-pressure gas is allowed to expand suddenly, work is done *by* the gas in moving the piston. This work is done at the expense of the internal energy of the gas. Therefore, both the pressure and the temperature suddenly drop. This is what happens when you open the valve of a spray can. The gas suddenly expands and you can feel the cooling of the can. Or, perhaps you have seen a carbon dioxide (CO_2) fire extinguisher in action; when the valve is opened, the compressed gas expands so rapidly that the cooling actually freezes some of the gas into tiny solid particles—a kind of "snow."

We can also arrange for a compression to take place in which heat is removed from the gas so that the temperature remains constant. Then, the only effect of the compression is an increase in pressure. We call this an *isothermal* process. Similarly, an expansion can be made to occur in which the only effect is a decrease in pressure. Usually, an isothermal process takes place slowly so that there is sufficient time for heat transfer to be effective in maintaining a stable temperature. Can you see why a very rapid compression or expansion of a large amount of gas cannot be an isothermal process?

Changes of State

The cooling cycle of a refrigerator involves not only changes in pressure, volume, and temperature, but also changes of the working substance from gas to liquid and from liquid to gas. We are all familiar with the fact that materials exist in three possible conditions: solid, liquid, and gas. These are referred to as the three *states of matter*. The particular state in which a substance exists depends on the temperature and the pressure. At normal atmospheric pressure, water exists as a liquid only between the temperatures of 0°C and 100°C. At temperatures below 0°C, water is a solid (ice), and at temperatures above 100°C, water is a gas (steam). The same is true of all substances. Even iron will liquefy at 1535°C and will vaporize at 3000°C.

Water can exist both as a liquid and as a solid at the temperature of 0°C, and it can exist both as a gas and a liquid at the temperature of 100°C. In fact, water can be changed from the liquid state to the solid state or from the gaseous state to the liquid state without any change in the temperature. How does this happen? The answer is to be found by examining the internal molecular structure of substances in the various states in which they exist.

In solids, the molecules are closely packed and exert strong attractive forces on one another. These forces are so strong in the solid state that the molecules are bound in permanent positions and cannot move about. In liquids, the intermolecular forces are weaker and the molecules are free to slide past one another. Finally, in gases, the forces between the molecules are extremely weak—almost zero—and, consequently, gas molecules move around at random, completely free of the other molecules.

When a substance is changed from a solid to a liquid, the strong molecular binding forces must be broken. To tear the molecules apart so that they can move relative to one another requires that work be done on the substance. That is, energy in the form of heat must be supplied to a solid to convert it into a liquid. If just the right amount of energy is supplied, the only effect will be the breaking of the bonds and the solid will be changed into a liquid without any increase in temperature. Of course, if more energy is supplied, the liquid will experience a rise in temperature.

The change from solid to liquid requires the *input* of energy. Conversely, the change from liquid to solid *releases* energy. That is, in the process of freezing, energy is given up by water to some cooler object and ice is formed. For example, if water at 0°C is exposed to air at -10°C, energy will be spontaneously transferred from the water to the air. As a result, the water will be frozen and the air will be warmed.

The amount of energy that is required to convert a substance from the solid to the liquid state, or the amount of energy that is released when the substance changes from the liquid to the solid state, is called the *heat of fusion* of the substance. For example, the amount of energy required to convert 1 kg of ice at 0°C to water at the same temperature is equal to the energy required to raise 1 kg of water from 0°C to 80°C. We usually measure heat in units called *Calories* (Cal); 1 Cal is the heat required to raise 1 kg of water by 1 C degree. Thus, in these units the heat of fusion of water is 80 Cal/kg.

The same kinds of processes take place in liquid-to-gas and gas-to-liquid conversions. If energy is supplied to a liquid, it can be vaporized. Or, if energy is extracted from a gas, it can be liquefied. The amount of energy involved in one of these conversions at the boiling point of the substance is called the *heat of vaporization* of the substance. For water, the amount of energy required to convert 1 kg of liquid water at 100°C to gaseous steam at the same temperature is 540 Cal. Notice that the heat of vaporization for water is much greater than the heat of fusion. That is, it is much easier to break the solid state bonds to change ice into water than it is to separate the molecules from one another completely so that water becomes steam. In fact, for most substances, the heat of vaporization is several times greater than the heat of fusion.

Direct changes between the solid and the gaseous states take place for a few substances in certain pressure ranges. If you have ever observed a block of *dry ice* (solid carbon dioxide), you have probably noticed that the material gradually disappears without any sign of melting to a liquid. Dry ice has the property that it passes directly from the solid to the gas at normal atmospheric pressure; this process is called *sublimation*.

If you look back over our discussion of changes of state, you will see that we can summarize much of the information as follows. An input of energy is required when a substance undergoes a change from a state in which the molecules have low mobility to a state in which they have higher mobility. That is, energy is required to produce the following changes of state: solid to liquid, liquid to gas, and solid to gas. Also, energy is released in the opposite processes, which involve a change from a state of high molecular mobility to a state of lower molecular mobility. This summary is illustrated in Fig. 6-12.

We remarked at the beginning of this section that the state of a substance depends on the temperature and the pressure. At normal atmospheric pressure, water will change from a liquid to a gas at 100°C if sufficient heat is added. At other pressures, however, the change of state will occur at different temperatures. You are probably familiar with the fact that on a mountaintop, water boils at a temperature lower than 100°C. This is a result of the decreased atmospheric pressure at high altitudes. (If you live in a sea-level city and like eggs that are cooked three minutes in boiling water, you will need to lengthen the cooking time if you move to Denver, Colorado.) Also, in a pressure cooker, in which the pressure is greater than normal atmospheric pressure, water will remain liquid at temperatures above 100°C.

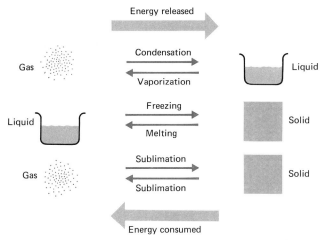

Figure 6-12 *Energy changes involved in changes of state. (Notice that the gas-to-solid and solid-to-gas transformations are both called* sublimation.*)*

The pressure also affects the amount of energy required to vaporize a liquid and the amount of energy released in the condensation of a gas. That is, the heat of vaporization depends on pressure (as does the heat of fusion). Can you see why this is so? Suppose that we have a cup of water at 20°C (room temperature) in a closed container at normal atmospheric pressure. Now, by means of a vacuum pump, we reduce the pressure in the container to $\frac{1}{40}$ of normal atmospheric pressure. Next, we add heat to the water and observe the temperature. When the temperature rises by only 1.5 C degrees to 21.5°C, we would see the water actually *boil*. At reduced pressure a liquid boils more readily than at high pressures, and the energy required to vaporize a liquid decreases as the pressure decreases. Similarly, at high pressures, the energy required to vaporize a liquid is greater than at normal atmospheric pressure.

As we will see, the dependence of the heat of vaporization on pressure is an important aspect of the operation of refrigerators.

The Refrigeration Cycle

The only changes of state that are important in the operation of refrigerators are liquid-to-gas conversions and gas-to-liquid conversions. These changes of state take place at the boiling (or vaporization) point of the substance. (Sometimes we refer to this temperature as the *condensation point* when the change involves the conversion of a gas to a liquid.) In a refrigerator we want the liquid-to-gas conversion (that is, the vaporization) to take place at a temperature below 0°C so that the heat absorbed during this process by the working substance will lower the temperature of the surroundings below the freezing point of water. In this way we ensure that the temperature in the freezing compartment is sufficiently low to produce ice. Also, we want the gas-to-liquid conversion (that is, the condensation) to take place at a temperature above room temperature so that the heat rejected by the working substance in this process can be absorbed spontaneously by the surrounding cooler air.

If we have the working substance of a refrigerator sealed in a system of tubes, how can we arrange for the vaporization to take place at a *low* temperature and the condensation to take place at a *high* temperature? Remember, the boiling (or vaporization) point is the same as the condensation point only if we have the same pressure for both processes. The vaporization can be made to take place at a lower-than-normal temperature by using a reduced pressure. Similarly, the condensation can be made to take place at a higher-than-normal temperature by using an increased pressure. Therefore, as the working substance moves through the system of tubes in a refrigerator, we must arrange for the pressure to be low in the refrigeration compartment and high in the part that is exposed to air. This pressure difference is maintained by a *compressor* that is driven by an electric motor. The energy supplied to the electric motor produces the work necessary to drive the refrigerator's working substance (the *refrigerant*) through its cycle. The most common refrigerant in household refrigerators is *Freon,* although sulfur dioxide and ammonia have also been used.

With these facts about changes of state in mind, we are now in a position to describe the operation of a refrigerator. A schematic of a refrigeration system is shown in Fig. 6-13. The compressor is at the top, the refrigeration coils (located in the freezer compartment of the refrigerator) are at the left, and the cooling coils (which are exposed to air) are at the right. On the down stroke of the piston in the compressor, the valve C to the refrigeration coils is forced closed and the valve A to

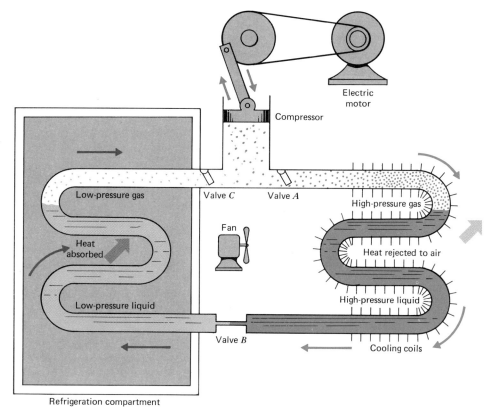

Figure 6-13 *Schematic diagram of the refrigeration cycle in a typical household refrigerator.*

the cooling coils remains fully open. The refrigerant enters the cooling coil as a gas at high pressure and at a temperature above room temperature. As it flows through the cooling coils, the gas gives up heat to the cooler air which is forced across the coils by an electrically operated fan. Notice that these coils have fins attached to conduct the heat away from the gas and to facilitate cooling. When the gas cools to about room temperature, it condenses and becomes a liquid (still at high pressure). During every cycle of the system, a certain amount of heat is rejected to the surrounding air.

In the next step, the high-pressure liquid is forced through the small-diameter valve B. To the left of this valve, the pressure in the system is lower, and so the refrigerant expands as it passes through valve B. (This valve is called a *throttling valve;* the diameter of the opening is so small that it severely constricts the flow and permits a pressure difference to exist across the valve.) The expansion of the refrigerant in the low-pressure region causes the temperature to drop and some of the liquid vaporizes. At the low pressure existing in the refrigerant coils, the liquid is easily vaporized. As the liquid passes through the coils, it absorbs heat from the refrigeration compartment until it is completely vaporized. Thus, there is a certain amount of heat absorbed by the working substance from the refrigeration compartment during each cycle.

The upward stroke of the compressor piston pulls closed valve A and opens valve C. The low-pressure, low-temperature gas is drawn into the compressor cylinder, ready to begin a new cycle.

We can show in a diagram (Fig. 6-14) how the pressure and the volume of the refrigerant change as we proceed through a cycle of the system. Let us begin at point A in Fig. 6-14, which corresponds to the high-temperature, high-pressure gas that has just emerged from the compressor and has entered the cooling coils through valve A. During the phase $A \to B$, several things happen. As we see in the diagram, the volume decreases, but the pressure remains constant. This results from the condensation of the gas to liquid form. The condensation process releases energy and the cooling coils reject the heat to the surrounding air so that the temperature remains approximately constant. Thus, $A \to B$ is essentially an isothermal condensation at constant pressure.

Next, $B \to C$ represents the effect of the refrigerant passing through the throttling valve (valve B in Fig. 6-13). During this

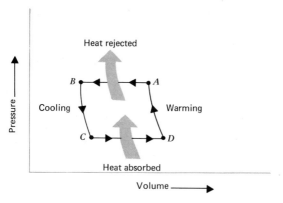

Figure 6-14 *Pressure-volume diagram of a refrigeration cycle.*

process, the pressure decreases significantly but the volume change is relatively small. This part of the cycle takes place rapidly so that there is almost no opportunity for the transfer of heat. Therefore, there is a large temperature decrease. Thus, $B \rightarrow C$ is essentially an adiabatic expansion with a large temperature decrease.

The part of the cycle $C \rightarrow D$ represents the vaporization of the refrigerant in the refrigeration coils. This process takes place at constant pressure (as does the condensation $A \rightarrow B$), and heat is rejected to the surrounding air. This part of the cycle is approximately isothermal.

Finally, $D \rightarrow A$ represents the adiabatic compression of the gas in the compressor cylinder. During this process both the pressure and the temperature increase. The refrigerant has now been returned to the starting point and a new cycle can begin.

During each cycle of the system, a certain amount of work is done by the driving motor on the refrigerant. At the same time, a certain amount of heat is absorbed from the refrigeration compartment and a certain amount of heat is rejected into the surrounding air. Conservation of energy tells us that

(heat rejected) = (heat absorbed) + (work done)

In this equation we see that the amount of heat rejected to the air must be greater than the amount of heat absorbed from the refrigeration compartment. The heat rejection process involves the condensation of the refrigerant in the cooling coils. For each kilogram of the refrigerant that is condensed, the amount of heat rejected is equal to the heat of vaporization of the substance. Also, the heat absorption process involves the vaporization of the refrigerant and the absorption of an amount of heat equal to the heat of vaporization of the refrigerant. Because the same physical quantity—the heat of vaporization of the refrigerant—is involved in both processes, how can this agree with the energy conservation equation? We must remember that the heat of vaporization of a substance depends on the *pressure*. If the pressure is low, the heat of vaporization is low; if the pressure is high, the heat of vaporization is high. Thus, the heat rejected by the cooling coils (in the high-pressure part of the system) is greater than the heat absorbed from the refrigeration compartment (in the low-pressure part of the system).

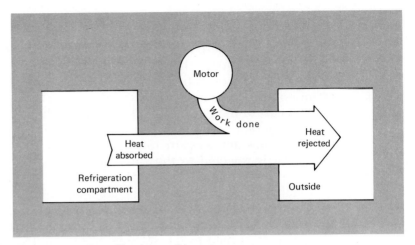

Figure 6-15 *Heat flow in a refrigerator.*

Heat Pumps

We can represent the flow of heat in a refrigerator as in Fig. 6-15. Here, the work done by the motor adds to the heat absorbed from the refrigeration compartment to equal the heat rejected to the air outside the refrigerator. We can construct a similar diagram to represent the cooling of a house by an air-conditioner (Fig. 6-16a). The refrigeration compartment becomes the interior of the house, and the air outside the refrigerator becomes the entire outdoors. An interesting feature of this system becomes apparent if we reverse the direction of heat flow and use the backwards running air-conditioner to *heat* the house (Fig. 6-16b). This system, which can cause heat to flow in either direction, is called a *heat pump*.

In Fig. 6-16b, notice that the amount of heat added to the house is *greater* than the amount of work done by the motor. If we were to use the energy that operates the heat-pump motor to power an electric heater in the house, we would have a *smaller* heating effect. That is, a heat pump is a more efficient heater than a direct conversion of the energy into heat. Heat pumps are beginning to be used widely in the construction of new homes. However, heat pumps do not operate efficiently if the outside temperature is too low. Consequently, these devices are useful only in moderate climates.

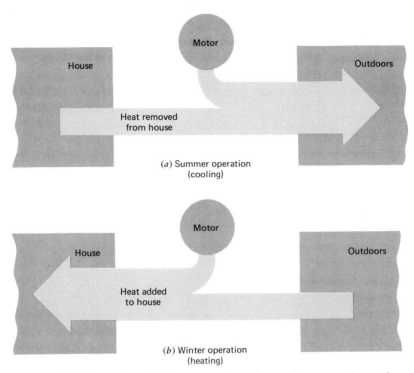

Figure 6-16 *Operation of a heat pump. (a) The cooling operation is the same as that of a refrigerator (Fig. 6-15). (b) Reversing the heat flow causes heat to be added to the house. This process is more efficient than using the same amount of energy to heat the house directly.*

Questions and Exercises

1. The expansion of glass for a particular temperature increase is much less than that of mercury or alcohol. Why is this fact important in the construction of thermometers?
2. By using a hydraulic jack, you can lift the entire weight of an automobile with a relatively small force on the jack handle. Explain why this is so.
3. What would happen to an astronaut making a "space walk" if his space suit were suddenly punctured?
4. In one type of nuclear power reactor the water that passes through the reactor core (the coolant water) is actually at a temperature considerably above 100°C. Explain why this is

possible. Does your explanation suggest any problems that might arise in the design of nuclear reactors?

5. Could you cool a closed room by leaving open the door of a refrigerator in the room? Explain.

6. If you want a cup of boiling water, you could pump away the air above the water until the water boils. Would this cup of boiling water be suitable for making a cup of hot coffee? Explain.

7. *Evaporation* is a process in which the work necessary to remove molecules from the surface of a liquid is supplied by the internal energy of the liquid. If a liquid container were well insulated so that no heat could flow into or out of the liquid, would evaporation tend to warm or cool the liquid?

8. If you have ever used "moth balls," you may have noticed that they tend to become smaller with time. Why?

9. Could a refrigerator be made to work with a (hypothetical) refrigerant whose heat of vaporization does not depend on pressure? Would such a refrigerator obey the second law of thermodynamics?

10. In a refrigerator the amount of heat removed from the refrigeration compartment can actually be greater than the amount of work done on the system by the electric drive motor. Does this seem reasonable in view of the fact that energy must be conserved for the entire system (refrigerator + electrical energy input + surrounding air)?

Additional Details for Further Study

Using the Ideal Gas Law

The ideal gas law states that the pressure, volume, and temperature of a quantity of gas are related by

$$\frac{PV}{T} = \text{constant}$$

Suppose that we have a gas initially specified by the quantities P_1, V_1, and T_1. The gas undergoes some change and the new values become P_2, V_2, and T_2. In each of the two conditions of the gas we have $PV/T = $ constant, and the "constant" has exactly the same value in each case. Therefore, we can write

$$\frac{P_1 V_1}{T_1} = \frac{P_2 V_2}{T_2}$$

Let us see how we use this relation in a particular case. If you have ever seen a photograph of the launching of a high-altitude balloon, you have probably noticed that the gas bag is floppy and only partially filled at release. Why is this so? We can answer this question by calculating the volume occupied by the balloon gas at its highest altitude.

The initial conditions of the balloon gas are as follows:

(a) The initial pressure P_1 is normal atmospheric sea-level pressure. Let us call this pressure P_0 (= 1 atm); $P_1 = P_0$.

(b) A typical high-altitude research balloon designed to carry a 1000-lb load of equipment to an altitude of 100 000 ft will be filled (at sea level) with about 400 m³ of helium; $V_1 = 400$ m³.

(c) The initial temperature will be close to room temperature, 20°C; $T_1 = 293°K$.

The conditions at maximum altitude are as follows:

(a) At an altitude of 100 000 ft, the atmospheric pressure is only about 1 percent of sea-level pressure; $P_2 = 0.01 P_0$.

(b) The volume at altitude is the quantity we are seeking; $V_2 = ?$

(c) At 100 000 ft, the temperature is usually about $-50°C$; $T_2 = 223°K$.

We now have specified five of the six quantities appearing in the ideal gas law equation. Solving for V_2, we have

$$V_2 = \frac{P_1}{P_2} \times \frac{T_2}{T_1} \times V_1$$

Substituting the known values,

$$V_2 = \frac{P_0}{0.01 P_0} \times \frac{223°K}{293°K} \times 400 \text{ m}^3$$
$$= 30\ 440 \text{ m}^3$$

or, approximately 76 times the initial volume.

Thus, we see that the huge expansion of the gas under the conditions at high altitude requires that the gas bag be only partially filled at launch.

Specific Heat

Suppose that you place a 1-kg container of water and a 1-kg block of iron in a refrigerator. Suppose also that the refrigeration system removes heat from each substance at the same rate. (That is, both the water and the iron lose the same number of Calories per minute.) If you examine the samples some time later (but before they have both been reduced to the temperature of the refrigeration compartment), you will find that the temperature of the iron block is *lower* than that of the water. Similarly, if you have two 1-kg samples of water and iron at the same temperature and if you add the same amount of heat to each, you will find that the iron has a *higher* temperature than the water.

This behavior of water and iron is typical of all materials. Every substance has its own particular way of responding to a heat input (or heat loss) in terms of a temperature change. Remember, *temperature* is a measure of molecular motion and the different molecular compositions and structures of different materials means that they will exhibit different temperature changes for the same heat change. We measure these differences in terms of a quantity called the *specific heat* of the substance.

The specific heat of a particular substance is the amount of heat in Calories required to raise the temperature of that substance by 1 C degree. Recall that the unit of heat, the *Calorie*, is defined to be the amount of heat that will raise the temperature of 1 kg of water by 1 C degree. Therefore, the specific heat of water is 1 Cal/kg-C deg. (Read this as "1 Calorie per kilogram per Celsius degree.") Some other specific heat values are given in Table 6-1.

How do we use the specific heat value (which we represent by the symbol c) to predict the temperature rise of a substance

Table 6-1 Specific Heats of Some Materials

Substance	Specific Heat (Cal/kg-C deg)
Air	0.17
Aluminum	0.22
Copper	0.09
Iron	0.12
Water	1.00

when a certain amount of heat is supplied? We can write down the formula with just a little thought. The amount of heat Q that must be supplied to a sample to raise its temperature from T_1 to T_2 must be proportional to the temperature change, $T_2 - T_1$. Furthermore, if we increase the mass M of the sample, then we must also increase the amount of heat to produce the same temperature change; that is, Q is proportional to M. Finally, a material with a large value of the specific heat requires more heat input to produce a particular temperature change; that is, Q is proportional to c. Combining these statements, we can write

heat transferred = (specific heat) × (mass)
$$\times \text{(temperature change)}$$

or

$$Q = cM(T_2 - T_1)$$

where

Q = heat transferred in Cal
c = specific heat in Cal/kg-C deg
M = mass in kg
$T_2 - T_1$ = temperature change in C deg

How much heat is required to raise the temperature of 2 kg of iron from 20°C to 80°C? From Table 6-1, we find for iron, $c = 0.12$ Cal/kg-C deg. Therefore,

$$Q = (0.12 \text{ Cal/kg-C deg}) \times (2 \text{ kg}) \times (80°C - 20°C)$$
$$= 14.4 \text{ Cal}$$

If we supplied this same amount of heat to a 2-kg sample of water, the temperature change would be

$$T_2 - T_1 = \frac{Q}{cM} = \frac{14.4 \text{ Cal}}{(1 \text{ Cal/kg-C deg}) \times (2 \text{ kg})}$$
$$= 7.2 \text{ C deg}$$

Water (which has an exceptionally large value of specific heat) requires a much larger heat input compared to other substances to produce a particular temperature change. That is, water is a good medium for the storage of heat. For this reason, water is the most commonly used fluid in heating and cooling systems.

Additional Exercises

1. A sample of gas has a volume of 15 m³ when at a temperature of 27°C and a pressure of 2 atm. If the temperature is increased by 200 C deg and the pressure is increased to 6 atm, what will be the new volume? [Ans. 8.3 m³.]

2. A sealed tank of nitrogen gas has a volume of 0.2 m³. When the temperature is 300°K, the tank pressure is 20 atm. What will be the pressure if the building that houses the tank catches fire and the tank temperature increases to 1200°K? Is there any danger involved (apart from the fire hazard)? [Ans. 80 atm.]

3. How much heat is required to change 4 kg of water at 20°C to steam at 100°C? [Ans. 2480 Cal.]

4. A bar of copper and a bar of aluminum, each with a mass of 2 kg and both at the same temperature, are each supplied with 20 Cal of heat. Which bar will experience the greater temperature change and by how much? [Ans. The copper bar; 65.6 C deg.]

5. A certain room has dimensions of 5 m × 4 m × 3 m. How much heat is required to raise the air in this room from 20°C to 23°C? (The density of air is approximately 1 kg/m³.) [Ans. 30.6 Cal.]

What is a sound wave?

WAVES, SOUND, AND NOISE

The most ancient and most universal method of personal communication involves the use of *sound*. We all take for granted the use of speech and other sounds to convey information to others. But what *is* sound? How are sounds produced and how are they transmitted through the air? Actually, the phenomenon of sound has many characteristics in common with several other familiar effects—the vibration of a piano or violin string, the propagation of waves across an expanse of water, and even the propagation of radio waves and light.

In this chapter we will be concerned with wave motions of various types. We will begin by studying the simplest type of wave motion—waves on strings and springs. The basic ideas that we develop in this discussion will lead us to the case of sound waves and will permit us, in the next chapter, to look briefly at the wave character of light.

Wave Pulses on a String

Suppose that you grasp one end of a long, extended string (or rope) and then move the end quickly up a few inches and then down to the initial position. What will be the effect of this motion on the string? You have probably performed this simple experiment with a rope or a garden hose. The sudden motion of the end of the string produces a kink that moves along the string at a certain speed. This is an elementary type of wave motion. The kink is a *wave pulse,* and this pulse propagates along the string.

Let us examine in more detail how the string reacts to the sudden motion of one end. For simplicity, we will assume that the string is very long and consists of a number of discrete particles connected together, as shown in Fig. 7-1. In Fig. 7-1a we see the end particle being moved upward. As soon as this motion begins, a force is exerted through the string on the second particle and it also begins to move upward, lagging slightly behind the end particle. This is shown in Fig. 7-1b. Now, the second particle, in turn, exerts a force on the third particle and it begins to move (Fig. 7-1c). By this time, the end particle has reached the limit of its upward motion; in Fig. 7-1c the end particle is momentarily at rest. Next, the end particle begins to move downward and the force exerted by the

Wave Pulses on a String

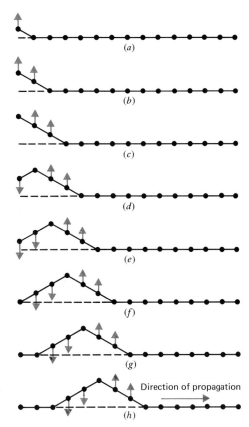

Figure 7-1 *The sudden up-and-down motion of one end of a string produces a wave pulse that propagates along the string.*

end particle on the second particle causes the latter to stop (Fig. 7-1d) and then to move downward (Fig. 7-1e). In Fig. 7-1f the particles are in several different states of motion. The end particle has returned to its original position and is stationary. The fourth particle has reached the maximum upward excursion and is momentarily at rest. The second and third particles, however, are moving downward, while the fifth and sixth particles are moving upward.

This type of activity of the particles continues along the string. Each particle is pulled upward and then downward by the action of the particle on its left. Similarly, each particle exerts first an upward and then a downward force on the particle to its right. In this way, the wave pulse propagates along the string. Wave motion is the *collective* motion of the particles that make up the medium (in this case, the string) through which or along which the wave propagates.

Look again at the motion of the end particle in Fig. 7-1. As this particle is pulled upward and then downward (by your hand), it exerts a force on the second particle. The elastic forces in the string resist this pull, and so work must be done to move the end particle. The energy imparted to the string by the work that is done remains in the string in the form of the kinetic energy of the moving particles and the potential energy of the stretched parts of the string. Notice that this energy propagates along the string. No part of the string moves very far; each particle moves only a short distance upward and then returns to its original position. Even so, energy is transported along the string at a steady rate. This is a feature of all types of waves: Waves transport *energy,* not *matter.*

Although no part of the string experiences any net displacement, we can easily follow the motion of the top (or the *crest*) of the wave pulse. By measuring the time required for the crest to move a certain distance, we can determine the *wave speed* for pulses on the string. The value we find will be valid only for the particular string used; strings of different sizes and compositions will exhibit different wave speeds. Indeed, we can change the wave speed on a particular string by altering the tautness of the string. However, the wave speed does not depend on the size of the pulse (the height of the crest), nor does it depend on how rapidly the pulse is imparted to the string.

Traveling Waves on a String

In Fig. 7-1, only a single pulse moves along the string. Now, instead of stopping the end particle after the up-and-down motion, suppose that you continue to drive the end of the string by a regular series of motions. Figure 7-2a shows what happens. Each time the end particle is moved upward from its original position (which is on the dashed line) to the maximum upward displacement and then downward to the original position, a pulse is formed just as in Fig. 7-1. Furthermore, each time the end particle is moved downward from its original position to the maximum downward displacement and then upward to the original position, a similar pulse is formed, but in the opposite sense. Because the up-and-down motion of the end particle is uniform, a series of pulses of identical shape is formed and these pulses fit smoothly together to produce a regular pattern (Fig. 7-2a).

Wavelength, Period, and Frequency 185

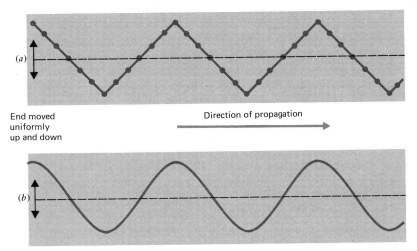

Figure 7-2 (a) *If the end of the string is moved uniformly up and down, a series of pulses (identical to that in Fig. 7-1) is formed.* (b) *In a real case, the string will exhibit smoothly curved (sinusoidal) pulses instead of triangular pulses.*

Each individual pulse (whether above or below the dashed line) behaves in exactly the same way as the pulse in Fig. 7-1. In particular, each pulse propagates along the string with the same speed. Thus, the entire pattern moves uniformly to the right in Fig. 7-2a without changing its shape. This train of identical pulses constitutes a *wave*. Because the pulses all move together, retaining the overall shape, we call this a *traveling wave*.

If you actually attempt to make a traveling wave in the manner we have discussed, you will find that the individual pulses are not triangular, as we have shown schematically in Figs. 7-1 and 7-2a. Instead, the string will be smoothly curved, as illustrated in Fig. 7-2b. Waves of this type usually have a shape that is the same as the sine function of trigonometry. Accordingly, we call these *sinusoidal* waves.

Wavelength, Period, and Frequency

In Fig. 7-3 we show a typical sinusoidal wave form together with some of the terms we will use to describe these waves. The upward pulse is called the wave *crest*, and the downward pulse is called the wave *trough*. The size of the wave is specified in terms of the *amplitude*, which is the maximum excur-

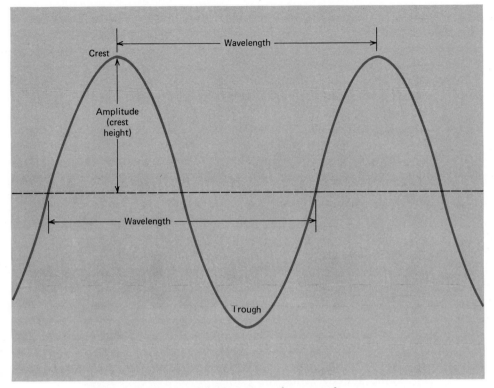

Figure 7-3 *Some of the important features of a wave.*

sion of the wave from the nonmoving condition (the dashed line). Notice that the crest amplitude is the same as the trough amplitude. The distance along the string that is required for the wave pattern to repeat itself is called the *wavelength* of the wave. The wavelength can be determined by measuring the distance between successive crests or between any two successive corresponding features of the wave (such as the distance between points of zero displacement, as shown in the figure).

If you compare Fig. 7-3 with Fig. 7-2, you will see that as the driven end of the string is moved through one complete cycle of the up-and-down motion, the wave moves forward by a distance equal to one wavelength. The time required for this motion is called the *period* of the wave. (The period of the wave motion is the same as the period of the driving force.)

The speed with which a wave propagates is related to the

Wavelength, Period, and Frequency

period and the wavelength of the wave. We know from the discussion in Chapter 1 that *speed* is defined to be *distance per unit time:*

$$\text{speed} = \frac{\text{distance}}{\text{time}}$$

For a wave, the distance moved forward during a time equal to one period is one wavelength. Therefore, we can write

$$\text{wave speed} = \frac{\text{wavelength}}{\text{period}}$$

The period of a wave is the *period* of time required for the passage of two successive wave crests past a particular point. If the period of a wave is $\frac{1}{10}$ s, then 10 crests will pass a particular point in 1 s. This latter number we call the *frequency* of the wave. The frequency specifies how *frequently* wave crests pass a given point. The period and the frequency are reciprocally related:

$$\text{frequency} = \frac{1}{\text{period}}$$

The period, being a time, is measured in *seconds*. Frequency, therefore, is measured in a unit we might call "per second." Indeed, we sometimes speak of "the number of wave vibrations per second." Current practice, however, is to measure frequency in terms of an equivalent unit called the *hertz* (Hz), named in honor of Heinrich Hertz (1857–1894), a German scientist who made great contributions to the study of electrical waves. A wave that has a period of $\frac{1}{10}$ s has a frequency of 10 Hz; if the period is 0.001 s, the frequency is 1000 Hz or 1 kHz; and so forth.

We can use the relation connecting the frequency and the period to rewrite the expression for the wave speed. Substituting *frequency* for 1/*period*, we have

wave speed = (wavelength) × (frequency)

This is a fundamental relation that is valid for all types of wave motion. If we designate the wavelength by λ (Greek *lambda*) and frequency by f, we can express the wave speed as

$$v = \lambda f$$

None of the terms we have introduced—wavelength, period, frequency, or amplitude—is limited to the case of waves on strings. Indeed, we use these terms to describe wave motions of all types.

Transverse and Longitudinal Waves

In the waves we have been discussing, the individual particles that make up the string vibrate back and forth while the wave pulses move along the string. Waves that propagate in this manner are called *transverse* waves. That is, the motions of the individual particles are transverse (or perpendicular) to the direction of propagation of the wave.

Now suppose that instead of a string we have a coiled wire spring, such as a "slinky" toy. We could vibrate one end up and down to produce a transverse wave just as we did using the string. For a coiled spring, however, a different kind of wave motion is possible. Instead of vibrating the end of the coil up and down, let us now vibrate the end *back and forth* along the direction of the coil. By this action, we first push the coils near the end closer together and then pull them farther apart than in the normal condition. Thus, we form a *compressional pulse*, and this pulse propagates along the spring in the same way that a transverse pulse propagates along a string. Figure 7-4 shows how this compressional pulse is formed.

If we move the end of the spring back and forth in a uniform way, we cause a series of pulses to propagate along the spring, as shown in Fig. 7-5a. Because the particles that make up the spring vibrate back and forth *along* the direction of propagation of the wave, we call waves of this type *longitudinal waves*.

The same general type of wave motion can be set up in a column of air. Figure 7-5b shows an air-filled tube that has a movable piston in one end. Moving this piston back and forth produces a series of regions in which the air is alternately compressed to a higher-than-normal pressure and expanded to a lower-than-normal pressure. As the piston is moved back and forth, these high- and low-pressure regions propagate along the tube. Figure 7-5c shows that the pressure in the tube follows the same sinusoidal pattern typical of many types of wave motion. Notice that it is the *pressure* of the air, not the

Transverse and Longitudinal Waves 189

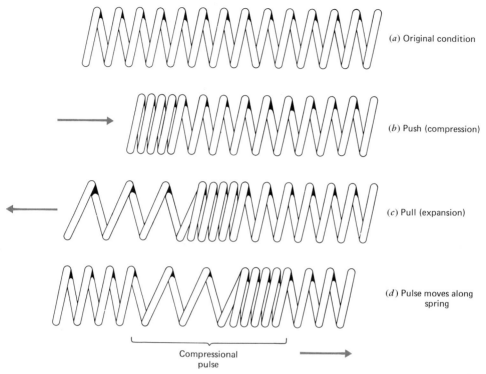

Figure 7-4 *Formation of a compressional pulse in a spring by first pushing and then pulling on one end. This is a* longitudinal *pulse.*

distance through which the air molecules move, that is important in this case.

It is important to remember that, in a medium through which a wave propagates, the individual particles do not move very far. This is true for waves on strings and springs, and it is also true for compressional waves in air. Air is an elastic medium, and when it is compressed by the action of the piston, the air expands into the low-pressure regions. During this process, the air molecules collide with one another, thereby causing the high-pressure region to shift along the tube. A series of compressional pulses moves along the air column, but the individual air molecules vibrate back and forth around their original positions.

How do we know whether the waves that propagate in a particular medium or system will be transverse or longitudinal

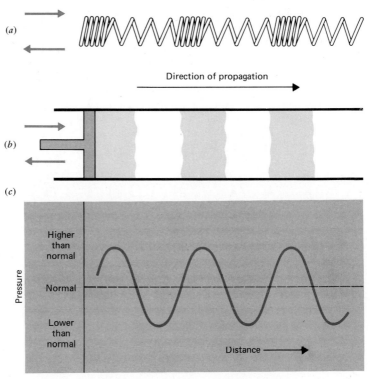

Figure 7-5 *Longitudinal (compression-expansion) waves* (a) *on a spring and* (b) *in an air-filled tube.* (c)*The pressure in the tube follows a sinusoidal pattern.*

waves? The character of the waves in a system is determined by the nature of the elastic forces that exist among the components of the system. As soon as wave motion begins in a system, the particles or molecules are displaced from their normal positions. In the displaced condition, the particles are subjected to forces due to the other particles in the system. These forces tend to restore the displaced particles to their original positions. This is exactly what happens in the various simple cases we have examined so far. In Fig. 7-1, for example, the elastic forces in the string tend to pull any displaced particle back to its original position. After the pulse has passed, the particles occupy the same positions that they did before being set into motion. In Figs. 7-4 and 7-5a, the elastic forces act longitudinally but they still tend to restore the spring coils to their original positions. A spring can sup-

Transverse and Longitudinal Waves 191

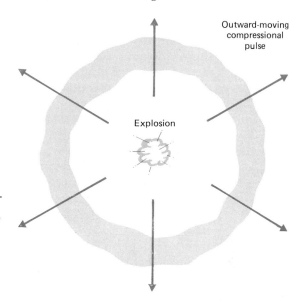

Figure 7-6 *An explosion in an unconfined expanse of air results in an outward-moving compressional pulse. The direction of motion of every part of this pulse is directly away from the site of the explosion. That is, the pulse is a longitudinal pulse.*

port either transverse or longitudinal waves because the structure of a spring is such that restoring forces can act both along and perpendicular to the direction of the spring. A string, on the other hand, is a floppy structure and cannot be compressed. Consequently, only transverse waves will propagate along a string.

What about compressional waves in air? In Fig. 7-5b, we see the air molecules bunched up at various positions along the tube. When the molecules are crowded together, they tend to push one another back to the regions of lower pressure. Because the air is confined in the tube, no transverse expansion of the compressed air is possible, and the restoring forces act along the direction of the tube. The compressional waves in a confined column of air are longitudinal waves. Suppose, however, that the air is not confined in a tube. Will transverse waves than be possible? Figure 7-6 shows a compressional pulse that is set up in an unconfined expanse of air by an explosion. The energy released in the explosion does work on the surrounding air and compresses it to a high pressure. In attempting to expand, the air molecules move outward to the low-pressure region. This is the only direction of motion available to the molecules. If they attempted to move transversely, they would encounter a region of equally high pressure. Therefore, even though the compressional pulse is spherical in shape, the pulse is still longitudinal. That is, every part of the compressional pulse moves directly away from the site of the explosion. Compressional pulses and waves in air are always longitudinal. There are no restoring forces in a gaseous (or liquid) medium that can support transverse wave motion.

In a solid material, such as a block of metal or a piece of rock, restoring forces can act both along and perpendicular to the direction of propagation of a wave. Solids can therefore support both transverse and longitudinal waves. In fact, both types of waves can be propagated simultaneously through solids. When an earthquake takes place, the forces exerted on the Earth's rocky material are complex combinations of lateral and longitudinal pushes and pull. Therefore, the waves that radiate outward from the site of an earthquake are both transverse and longitudinal waves. These two types of waves actually have different wave speeds through the Earth. By detecting and analyzing the complicated wave patterns caused by earthquakes, much has been learned about the Earth's interior.

Sound

The delicate membranes within the ear are sensitive to changes in air pressure. If a compressional wave in air is incident on the ear, the pressure differences between the compressed (high-pressure) and the expanded (low-pressure) parts of the wave arriving at regular intervals cause the ear membranes to vibrate. These vibrations provoke a nervous response and we *hear* a sound. If the frequency of the wave is too high or too low, however, the ear membranes cannot respond properly, and there is no sensation of hearing. The range of frequencies to which the human ear is sensitive is from about 16 Hz to about 20 000 Hz. We call this the *audible* range of frequencies.

There are large differences among persons in their ability to hear sounds, especially high-frequency sounds. Generally, a person's high-frequency response deteriorates with age. A young person may be able to hear 20 000-Hz sounds, but when this same person reaches middle age, he may be unable to hear sounds with frequencies above 12 000 or 14 000 Hz. Similarly, persons who are exposed to very loud sounds for long periods of time (for example, persons who work near high-powered aircraft or jack hammers, and rock performers or fans) will often experience impaired hearing, with the most pronounced effect in the higher frequencies.

Generally, we use the term *sound* to mean those compressional waves in air that have frequencies between 16 Hz and 20 000 Hz. If the frequency is above 20 000 Hz, we call these waves *ultrasonic waves*. (Even though these waves are above the auditory range of most humans, some animals and some insects respond to waves with much higher frequencies.) If the frequency is below 16 Hz, we call these waves *infrasonic waves*. In humans, the response to a 10-Hz wave, for example, is one of *feeling* the wave with the entire body instead of *hearing* with the ear.

There are many ways to produce sound waves. We have already shown a simple way in Fig. 7-5b. The vibrating piston causes a sound wave to propagate along the tube. The frequency of the wave is the same as the frequency of the piston. Each time a compressional pulse reaches the end of the tube, the air outside is compressed and the wave is propagated into the surrounding space (Fig. 7-7). In the same way, vibrations

Figure 7-7 *A sound wave is produced by the vibrations of a piston in a column of air. At the end of the tube, the sound wave is radiated into the surrounding space.*

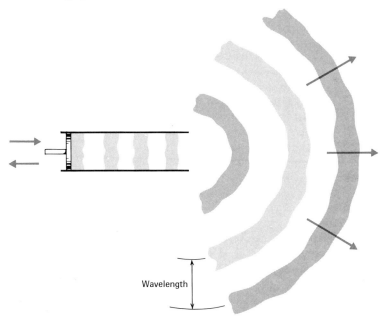

of the human voice box cause compressions of the air, and sound waves are radiated from the mouth.

Sounds are produced by musical instruments in several ways. For example, a vibrating drum head, or a vibrating violin string, or a vibrating clarinet reed causes the compressions in the nearby air that result in a propagating sound wave. In fact, any mechanism capable of forcing the air to vibrate with a frequency in the audible range can produce sound.

The Speed of Sound

When you see a lightning stroke during a thunderstorm, you do not hear the thunder until several seconds after the flash. When you hear a jet aircraft overhead and attempt to locate it by looking in the direction from which the sound is coming, you find that your line of sight falls a considerable distance behind the aircraft (Fig. 7-8). Both of these effects are due to the fact that sound travels through air with a speed that is extremely slow compared to the speed of light. Thus, the light from a lightning flash reaches your eyes almost instantaneously, but if the stroke is a mile away, there will be a delay of about 5 s before the sound of the thunder reaches your ears.

At sea level and at a temperature of 0°C, the speed of sound in air is approximately 330 m/s or 1100 ft/s. The speed is essentially the same for all frequencies of sound, but the speed does depend on the pressure and the density of the air. At an altitude of 40 000 ft or 12 km (where many jetliners fly), the pressure and density conditions are such that the speed of sound is about 13 percent smaller than at sea level.

Aircraft speeds are frequently stated in units of the speed of sound using *Mach numbers*. A speed of Mach 1 corresponds to the speed of sound (750 mi/h at sea level); Mach 2 corresponds to twice the speed of sound; and so forth. Because of the variation of the speed of sound with altitude, an aircraft flying at Mach 0.9 near sea level actually moves faster than an aircraft flying at Mach 0.9 at a high altitude.

The speed of sound in air and other gases is limited by the fact that the moving molecules must collide with one another in order to propagate the compressional wave. In liquids and solids, in which the molecules are closer together and interact

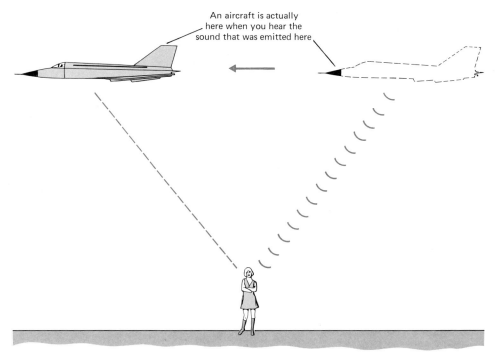

Figure 7-8 *Because the speed of sound is so much slower than the speed of light, the sound from a rapidly moving aircraft appears to come from a position far behind the actual position.*

more strongly with one another, the speed of sound is substantially greater than it is in a gas. Table 7-1 gives the speed of sound in several different materials.

We can calculate the wavelengths of sound waves with various frequencies by using the relation connecting wave speed, wavelength, and frequency, and substituting the speed

Table 7-1 *Speed of Sound in Various Materials*

Material	Speed (m/s) at 0°C
Air	330
Lead	1210
Sea water	1450
Iron	4480
Granite	up to 6000

of sound in air for the wave speed. We have

$$\text{wavelength} = \frac{\text{wave speed}}{\text{frequency}} = \frac{330 \text{ m/s}}{\text{frequency in Hz}}$$

For the audible sound with the lowest frequency,

$$\text{wavelength} = \frac{330 \text{ m/s}}{16 \text{ Hz}} = \text{approximately 20 m}$$

which is a little more than 60 ft. For higher frequencies, the wavelength is smaller. For an 11 000-Hz sound,

$$\text{wavelength} = \frac{330 \text{ m/s}}{11\,000 \text{ Hz}} = 0.03 \text{ m} = 3 \text{ cm}$$

which is a little more than 1 in.

Standing Waves

We have been discussing waves that travel on very long strings or springs or through large expanses of air. Ideally, these waves continue to propagate away from the source forever. Consider, now, a short string, one end of which is attached to a rigid support, such as the wall of a room. If you grasp the free end of the string and move it up and down, a wave will propagate along the string. When this wave reaches the opposite end, it will be *reflected* from the fixed support. (Remember, the wave carries energy and this energy cannot suddenly disappear at the fixed end—it must go *somewhere*. The only possibility is for the wave to be reflected in much the same way that a ball would bounce off the wall.)

We now have *two* waves traveling along the string—the direct wave moving away from the driven end and the reflected wave moving in the opposite direction. How does the string respond to two different wave motions at the same time? The answer to this question is actually very simple. Each wave moves along the string completely independent of the other!

To see what this means in a simple case, look at Fig. 7-9. Here, we examine the way in which individual pulses combine on a string. In Fig. 7-9a, we have two identical upward (or positive) pulses approaching one another. The sequence (1)–(5) shows the time development. Each pulse moves along the string as if the other pulse were not there. When the pulses

Standing Waves 197

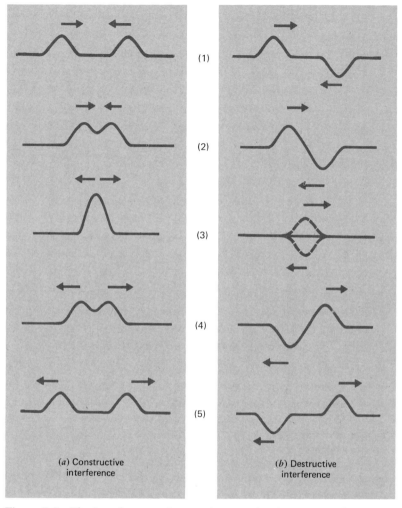

Figure 7-9 *The interference of two pulses moving in opposite directions along a string. In each case, the amplitudes combine. In* (a) *the pulses reinforce one another and the interference is* constructive. *In* (b) *the pulses cancel one another and the interference is* destructive.

meet, their amplitudes add together. At the instant the peaks coincide (Fig. 7-9a3), the displacement of the string is the sum of the displacements due to the two pulses. When the pulses overlap, we say that they *interfere*. If the amplitudes add together, as in this case, we call this *constructive interference* (or we say that the waves are *in phase*). After the interference has taken place, the pulses continue on as before. Notice that the pulses are *not* reflected from one another when they meet. If the pulses have identical shapes, as in this case, it might appear that a reflection has taken place. We could check that this does not happen by using pulses with different shapes or amplitudes.

In Fig. 7-9b, we have a positive pulse and a negative pulse. Each pulse has the same shape, so that when they overlap completely, cancellation results. At this instant, the string has zero displacement (Fig. 7-9b3). We call this *destructive interference* (or we say that the waves are *out of phase*).

Let us return to the case of the two waves moving along the string. We now see that the direct and reflected waves will interfere with one another, sometimes constructively and sometimes destructively. If the string is being driven at some random frequency, the wave motion on the string will be a jumbled pattern of constructive and destructive interference. By changing the frequency, we discover that there are certain frequencies at which the direct and reflected waves interfere to produce regular patterns of vibration. At these frequencies, certain positions along the string remain stationary (these points are called *nodes*), while the rest of the string vibrates in a smoothly changing way. In these situations, we no longer need to drive one end of the string. Once the wave pattern has been established, we can fix the driven end to a wall and the vibrations will continue. We call these patterns *standing waves*. (Of course, in real cases, frictional effects will eventually damp the motion.)

The lowest frequency at which a standing wave can be set up produces a pattern with nodes only at the fixed ends of the string. This situation is shown in Fig. 7-10a. This frequency is called the *fundamental frequency* for the particular string. The next lowest frequency for standing waves is twice the fundamental frequency (Fig. 7-10b); in this case there are three nodes and the wavelength of the wave is exactly equal to the distance between the supports. Additional standing waves all have frequencies that are integer multiples of the fundamental frequency (Figs. 7-10c, d). The waves with frequencies $2f_0$, $3f_0$, $4f_0$, and so forth, are called *harmonics* of the fundamental. In music, we refer to the waves with these frequencies as *overtones*. The frequency $2f_0$ corresponds to the first overtone or second harmonic; the frequency $3f_0$ corresponds to the second overtone or third harmonic; and so forth.

Superposition and Musical Sounds

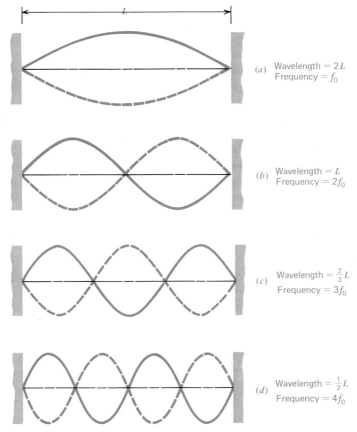

Figure 7-10 *Standing-wave patterns on a string with fixed ends. The fundamental* (a) *and the first three overtones are shown.*

Any type of wave motion can exist in standing-wave patterns if proper reflective surfaces are provided. For example, standing sound waves can exist in a room because of the reflection of the sound from the room walls. If your living room has a length of 22 ft, the lowest frequency standing sound wave that can exist in the room is a 25-Hz wave. This frequency is near the lower limit of the audible range.

Superposition and Musical Sounds

An electronic tone generator (or audio oscillator) is an instrument that can produce electrical oscillations with various pure frequencies. If you attach the output of such a device to a loudspeaker and adjust the controls for a signal with a frequency of 440 Hz, you will hear a tone corresponding to A on the musical scale. However, if you ask a violinist or a pianist to play the note A, you will hear quite different and distinctive sounds. What is the difference between the A note produced by a tone generator and the A note produced by a musical instrument?

When a violin string is stroked with a bow, the string is set into vibration. If the note A is played, the string vibrates with a frequency of 440 Hz. But in addition to this type of vibration, the string also vibrates at other frequencies, corresponding to various overtones of the fundamental frequency. This complicated motion of the string produces a sound that is characteristic of the violin. If the A key of a piano is struck, the string vibrates at a frequency of 440 Hz plus other frequencies. The relative amounts of energy contained in the various overtones are different for the violin compared to the piano. We say that the quality or *timbre* of the sound is different in the two cases. Thus, an A note from a violin is never confused with an A note from a piano.

We can readily observe the standing-wave patterns on a long string or rope, but the vibrations of a violin string or a piano string are too rapid and too complex to be analyzed visually. An easy way to "see" the vibrations of a musical in-

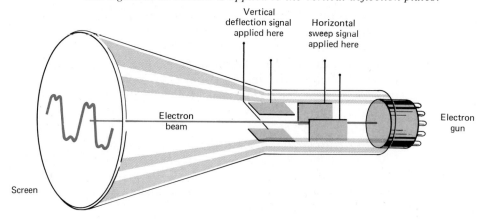

Figure 7-11 *Schematic diagram of a cathode-ray tube of the type used in oscilloscopes. The electron beam is swept horizontally across the screen by a varying voltage applied to the horizontal deflection plates. The signal to be studied is applied to the vertical deflection plates.*

strument or other sound source is to examine the wave patterns as a function of *time* instead of *distance*. We can do this by using an *oscilloscope,* an instrument that converts electrical signals into a visual display. The sound to be studied is picked up by a microphone which produces an electrical signal that corresponds in frequency and intensity to the sound. The microphone output is the input for the oscilloscope. The heart of an oscilloscope is a *cathode-ray tube* (Fig. 7-11), which is similar to a television picture tube. A beam of electrons is produced within the tube and when these electrons strike the fluorescent screen on the tube face, they produce a visible spot of light. As the electron beam is scanned

Figure 7-12 *The combination or superposition of two waves, the fundamental* (a) *and the third harmonic* (b).

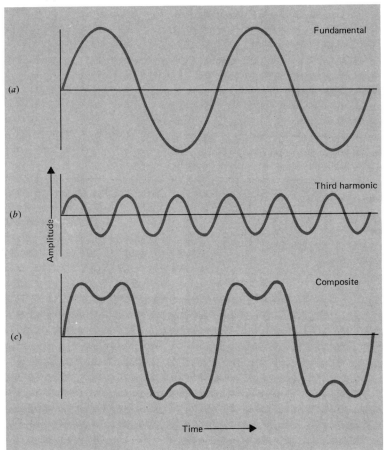

across the screen, the input signal causes up-and-down deflections of the beam and a curve is traced on the screen. If the sound picked up by the microphone consists of a pure frequency, the signal displayed on the oscilloscope face will be a sinusoidal pattern corresponding to the particular frequency of the sound (Fig. 7-12a).

Suppose, now, that instead of a sound with pure frequency, we have a sound that consists of two different frequencies—for example, a certain fundamental frequency and its third harmonic. In this case, the input signal consists of the frequencies f_0 and $3f_0$. The third harmonic is shown in Fig. 7-12b. The display that will appear on the oscilloscope is the sum or *superpositon* of these two signals. This composite signal is shown in Fig. 7-12c (see also the signal in Fig. 7-11).

The sound of any musical instrument consists of a fundamental and several different harmonics. Such sounds are more complex than the signal shown in Fig. 7-12c. The wave pattern of violin playing the note *A* is shown in Fig. 7-13.

The sound that we hear and identify as a violin *A* note can be synthesized by combining the pure notes generated by individual vibrating strings or electrical tone generators. If the mixture of the fundamental and harmonics is the same as that produced by the violin, the resulting tone will sound exactly the same as that from a violin. Indeed, electric organs synthesize tones in just this way. Such organs can generate tones that duplicate a variety of different instruments by combining pure notes in different ways. It is difficult, however, to produce exactly the correct mixture of pure notes, and an artificial violin tone from an electric organ sounds slightly different

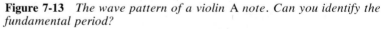

Figure 7-13 *The wave pattern of a violin* A *note. Can you identify the fundamental period?*

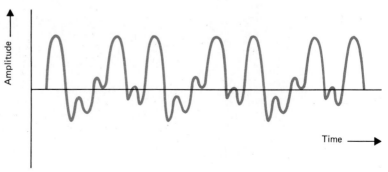

from a real violin tone. The ear is quite sensitive to the presence of extra harmonics or to those that should be present but are missing.

Loudness

There are two qualities of every sound that we hear: the *pitch* (or frequency) and the *loudness* (or intensity). The *intensity* of a sound wave is measured in terms of the amount of energy per second (that is, the *power*) the wave delivers to a unit area of a surface.(Accordingly, sound intensity is measured in watts per square meter, W/m^2.) The frequency of a sound wave depends on the rate of vibration of the wave; the intensity of a sound wave depends on the amplitude of the wave. These two aspects of a wave are independent. That is, we can have a high-frequency or a low-frequency sound wave with the same intensity. Or, we can have a loud or a soft sound with the same frequency. (See Fig. 7-14.)

The human ear is sensitive over an incredibly large range of sound intensity. The loudest sound that the ear can tolerate has an intensity about a *trillion* (10^{12}) times that of the softest sound the ear can perceive! Because this range is so large, a special scale has been devised which compares sound intensities by factors of 10. The unit of sound intensity is called the *bel* (B), in honor of Alexander Graham Bell (1847–1922), the inventor of the telephone. If a sound has an intensity of one bel (1 B) greater than another sound, the ratio of the two in-

Figure 7-14 *The frequency and the intensity of a sound wave do not depend on one another.*

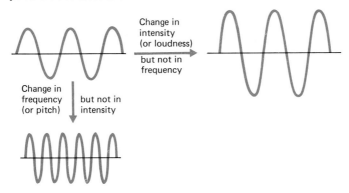

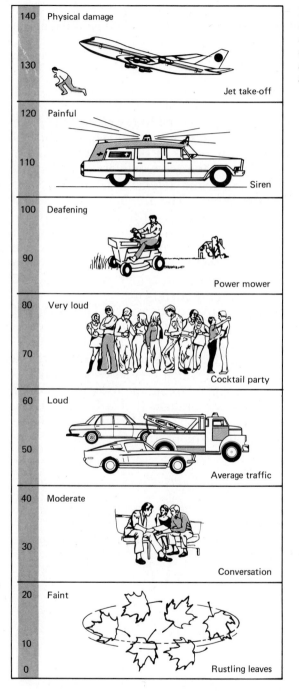

Figure 7-15 *The decibel scale of sound intensity and some common sounds. Each difference of 10 dB corresponds to a factor of 10 in sound intensity.* (Newsweek, *February 7, 1972, p. 45.*)

tensities is 10. A difference of 2 B means an intensity ratio of 100; a difference of 3 B means an intensity ratio of 1000; and so forth.

A bel is a rather large unit, and so we usually refer to sound intensities in terms of the *decibel* (dB): 10 dB = 1 B. A difference of 20 dB (= 2 B) means a sound intensity ratio of 100; and so forth. A factor of 2 in intensity corresponds to a difference of approximately 3 dB. The threshold of hearing corresponds to a sound intensity of about one-trillionth (10^{-12}) of a watt per square meter; this intensity is designated 0 dB. The loudest tolerable (but painful) sound has an intensity of about 1 W/m², or 120 dB. Figure 7-15 illustrates some of the sound intensities to which we are exposed, from the gentle rustle of leaves to the ear-splitting roar of a powerful jet aircraft.

Noise

We are all familiar with the sound patterns of *speech* and *music*. But what are the characteristics of *noise?* Generally, we classify as noise any sound that is disagreeable to the ear or that interferes with our hearing of desirable sounds. The din of a power lawn mower is certainly noise, but so is the music from a radio when it prevents normal conversation.

In today's power-driven world we are assaulted on all sides by increasing noise levels. The list of noise makers is almost endless: trucks, aircraft, jack hammers, motor bikes, lawn mowers, and on and on. Even in the home we must contend with blaring radios and television sets, washing machines, air conditioners, and other appliances. In rural areas, which we ordinarily consider to be quiet, tractors and farm machines disturb the residents. It has been estimated that city noise (on average) has doubled in the last 15 years, and that some downtown locations have experienced noise intensity increases of up to 18 dB.

What effect does noise have on our lives? It has been demonstrated that prolonged, high-level noise has pronounced physiological and psychological effects. A person who is exposed to loud noises will exhibit constriction of blood vessels in the skin, dilation of the pupils of the eye, increased heart rate, and tightening of the intestines. Psychological distress can also result from high noise levels: A person can become annoyed, irritable, and subject to emotional disorders. Exposure for long periods of time to noises that produce these effects can cause increased susceptibility to viral infection and ulcers; some persons will become increasingly prone to heart attack and stroke. These physiological effects are all in addition to partial (or, in some cases, almost complete) loss of hearing. Probably half of the population of the United States has suffered some impairment of hearing as the direct result of exposure to excessive noise levels.

Although we are beginning to make some headway in controlling other forms of pollution, very little progress has been made in the area of noise pollution. We can always elect not to swim in polluted water, and a steady breeze and a rain shower will help to cleanse the air, but noise pollution seems always to be with us. Many states and local communities have antinoise ordinances, but these vary widely in standards and in degree of their enforcement. Federal regulations are slow in being instituted. Noise pollution will probably be with us for many years, and its effects will be increasingly evident.

The Doppler Effect and Sonic Booms

Have you ever noticed the change in pitch of the siren on an ambulance or police car when it races past you? As the vehicle moves toward you, the pitch is high, and as it moves away from you, the pitch is low. Just as the vehicle passes, the pitch suddenly changes, and you hear the familiar *whee-oo* sound. This phenomenon is called the *Doppler effect*, after the Austrian scientist Christian Johann Doppler (1803–1853), who studied the way in which the frequency of a sound wave depends on the motion of the source.

Suppose that we have a stationary sound source that is emitting waves with a definite frequency. These waves radiate outward as expanding spherical shells. Each compressional pulse is separated from the previous one and the following one by exactly one wavelength (Fig. 7-16a). This is a very simple wave pattern, and a listener would hear a tone of the same frequency at any position around the source.

Now, let us consider the case in which the source is moving *toward* a listener, as in Fig. 7-16b. Each time a new compressional pulse is formed, the source is farther to the right than for the previous pulse. Therefore, the wave pattern is a series of expanding spherical shells, but the center of each shell is

The Doppler Effect and Sonic Booms   207

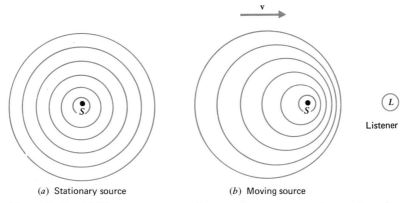

(a) Stationary source   (b) Moving source

Figure 7-16 (a) *The wave pattern for a stationary source is a series of concentric spherical waves.* (b) *If the source is moving, the waves bunch together in the direction of motion and the listener will hear a sound with a pitch higher than that for the stationary source.*

displaced a certain distance to the right of the center of the preceding shell. Thus, the waves become bunched together in the direction of motion of the source.

What does the listener hear in these two cases? When the source is stationary, the waves arrive at the listener's position at a certain rate, and the listener hears a sound with a frequency corresponding to that rate. When the source is moving toward the listener, however, the time interval between the arrival of successive compressional pulses is *shorter* than in the previous case. Consequently, the listener hears a sound with a *shorter period*, that is, a *higher frequency*. If the source were moving *away from* the listener, he would hear a sound with a *lower frequency*. Can you see why?

Figure 7-16b shows what happens when the speed of the source through the medium is *less* than the wave speed in that medium. Next, let us suppose that we increase the speed of the source until it is just *equal* to the wave speed. In this case (Fig. 7-17a), the source keeps pace with the outgoing spherical waves in the direction of motion. The waves can never move ahead of the source and they continue to pile up. As more and more waves are formed, the "pile-up" region extends farther and farther from the source in the direction perpendicular to its motion. The result is a wave front that is a sheet or a plane, with the source located at its center.

Finally, let us see what happens when the speed of the

208   Waves, Sound, and Noise

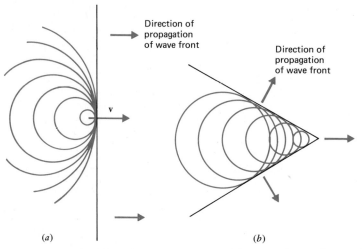

Figure 7-17 (a) *When the source moves with a speed exactly equal to the wave speed, the waves pile up and form a plane that extends perpendicular to the direction of motion of the source.* (b) *When the source speed exceeds the wave speed, a conical wave front is formed which trails behind the source. This front is a* shock wave.

source *exceeds* the wave speed in the medium. In this case (Fig. 7-17b), the source runs ahead of the outgoing waves, and the pile-up of the waves produces a wave front (actually, a cone-shaped wave front) that makes a certain angle with the direction of motion. The faster the source moves, the greater is the degree to which the wave front is bent back. A simple example of this type of wave pattern is the *bow wave* that is produced when a boat moves through water with a speed greater than the speed of waves on the water surface.

When an object, such as a bullet or an aircraft, moves through the air with a speed greater than the speed of sound, the motion is said to be *supersonic*. (At speeds below the speed of sound, the motion is *subsonic*.) Along the cone-shaped wave front that is produced by a supersonic object, the air is highly compressed. This moving sheet of high-pressure air is called a *shock wave*. Actually, there are *two* shock waves produced by a supersonic object—one from the forward part of the object and one from the rear. Figure 7-18 shows the shock waves produced by a supersonic aircraft. The graph at the bottom of the diagram indicates how the air pressure varies between the shock waves. The air pressure

The Doppler Effect and Sonic Booms 209

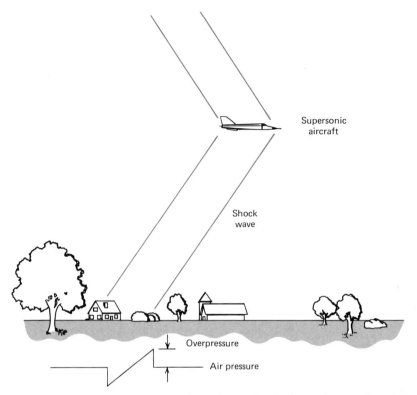

Figure 7-18 *A supersonic aircraft produces shock waves that are heard as a* sonic boom. *The graph at the bottom of the diagram shows that the air pressure rises sharply along the shock wave formed by the forward part of the aircraft. The pressure then falls below normal atmospheric pressure and again rises sharply along the shock wave formed by the rear of the aircraft.*

rises sharply along the leading shock wave, which is due to the forward part of the aircraft. The pressure then falls, becoming less than normal atmospheric pressure. The pressure again rises sharply along the trailing shock wave, which is due to the rear of the aircraft. This second rise brings the pressure back to normal.

If you are unfortunate enough to find yourself along the ground track of a supersonic aircraft, you will hear and feel the effect of the rapid pressure changes due to the shock waves. We refer to the sound produced by a supersonic aircraft as a *sonic boom*. Actually, you will hear only a single *crack* or *boom* when the shock waves pass because the time between the two sudden pressure increases is typically only about $\frac{1}{50}$ of a second.

If the increase in pressure along the shock wave (this is called the *overpressure*—see Fig. 7-18) is sufficiently great, the sudden change can cause damage to structures and to people. After any tests of supersonic aircraft over inhabited areas, the government is usually swamped with complaints about broken windows and cracked plaster. Because of frictional effects in the air and because of the spreading out of the shock wave with distance, the intensity of a sonic boom is much less for high-flying aircraft than for those at low altitudes. Even so, a supersonic transport (SST) in normal operation will produce a 60-km-wide track in which sonic booms can be heard. For this reason, SSTs will operate at supersonic speeds only on flights over water or over desolate land areas.

Notice that shock waves are produced by an aircraft (or any other object) when moving at any speed greater than the speed of sound. Thus, the occurrence of sonic booms is not limited to the case of an aircraft passing through the speed of sound (sometimes referred to as "breaking the sound barrier").

Questions and Exercises

1. Focus on one of the particles near the middle of the string in Fig. 7-1. Describe the motion of this particle as a function of time. Do the same for a particle near the middle of the string in Fig. 7-2a. Make sketches to illustrate these motions.
2. Argue that any elastic medium can support wave motion. Can sound waves be propagated through vacuum?
3. Sound waves do not propagate for great distances through air—eventually, they "die out." What happens to the sound energy when this occurs?
4. If you are interested in how close to you a thunderstorm is, describe a way in which you could easily estimate the distance.
5. Suppose that you are standing in water with your head tilted so that one ear is under water. While standing in this awkward position, someone detonates a small charge of dynamite on the surface of the water a mile away. Describe what you will hear.

6. If you have ever watched a strip-mining operation or a road crew at work when an explosive charge was set off, you may have noticed that you felt a ground tremor before you heard the explosion. Why is this so?
7. The speed of sound in air does not depend on the frequency of the sound waves. If this were not the case, would there be any difficulty in listening to speech or music that originated some distance away? Explain.
8. What is the wavelength of the musical tone *A* in air? [Ans. 0.75 m.]
9. Why does a violinist "finger" the strings of his instrument?
10. The amount of sound energy that reaches your ear each second when listening to a "live" rock band is about a million times greater than that for normal conversation. What is the difference in decibels between these two sound levels? [Ans. 60 dB.]
11. If, by some misfortune, a high-speed rifle bullet were to whiz by close to you, you would hear two sounds—first, a sharp *crack* and, a short time later, a more dull *thud* or *boom*. Explain the origin of these sounds.
12. The speed of a supersonic object can be determined by measuring the angle of the cone of shock waves. (Remember, we know the speed of sound in air.) Make a sketch to show how this can be done by drawing the cones that correspond to motions with speeds of Mach 2, Mach 3, and Mach 4.

Additional Details for Further Study

Water Waves

Have you ever noticed how a small object floating on water bobs up and down when a wave passes? This reminds us of the situation we studied in Fig. 7-2. As a transverse wave travels along a string, the individual particles move up and down. Water waves appear to follow this same pattern. But water waves are not true transverse waves. If we follow closely the motion of a floating object as a water wave passes, we would see the object move through a small circular orbit instead of vibrating exactly up and down. The motion of surface particles due to the passage of a water wave is shown in Fig.

212   Waves, Sound, and Noise

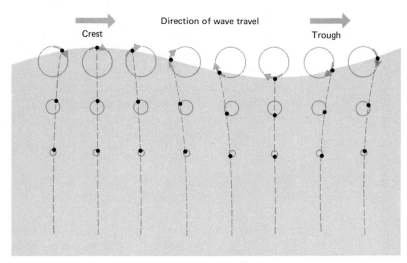

Figure 7-19 *The circular motion of particles due to the passage of a water wave. Notice how the motion diminishes with depth.*

7-19. Notice that the particles at the crest move in the direction of the wave propagation, whereas the particles in the trough move in the opposite direction. Notice also that the circular motion extends below the surface and that the orbit size decreases with depth. At a depth equal to one-half wavelength, the orbit diameter is about $\frac{1}{25}$ of that of the surface particles; at a depth equal to one wavelength, the orbit diameter is about $\frac{1}{500}$ of that of the surface particles. This shows that the violent condition of the ocean due to a storm is really a surface phenomenon—at depths of a few tens of meters or greater, the sea remains calm.

Because the wave motion of a water wave extends below the surface, a wave that approaches the shore begins to "feel" the bottom when the water depth becomes less than about one-half wavelength. The drag exerted by the bottom on the wave causes the wave speed to decrease. The waves then begin to pile up: The wavelength decreases and the wave height increases, as shown in Fig. 7-20. Near the shore, the wave height increases to the point that it becomes unstable. The crest of the wave is moving forward more rapidly than the rest of the wave, and the crest falls forward, creating a *breaker*.

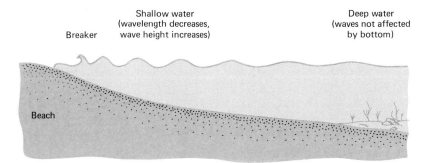

Figure 7-20 *As a wave moves inshore, the decreasing depth gradually influences the wave speed, eventually causing a breaker to form.*

Characteristic Frequencies

If you gently tap the rim of a crystal glass, you will hear a ringing tone. Or, if you suspend a length of pipe from a string and strike the pipe with a hammer, you will again hear a ringing sound, but one that is different from the sound of the glass. Any elastic object can be set to vibrating if struck a sharp blow. Each object has its own particular frequency (or set of frequencies) of vibration. We call these the *characteristic* or *natural* frequencies of vibration for the object, and we say that the object *resonates* at these frequencies. If these characteristic vibrations take place with frequencies in the audible range, a sound will be heard. Because the mixture of characteristic frequencies is different for every object, the sounds emitted are distinctive. For example, whether a pie tin or a half dollar has been dropped on the floor is easily determined by means of the sound.

A simple example of the occurrence of characteristic frequencies is the case of the resonant tones emitted by long, open pipes, such as those found in various wind instruments. (In some cases, the pipes of wind instruments are bent into curved shapes, but, nevertheless, they are still open pipes.) Figure 7-21 shows the way in which the displacement of the air molecules varies in an open pipe that is driven by three different frequencies. These wave patterns are similar to those for standing waves on a string (see Fig. 7-10), but there is one significant difference. There are always nodes at the fixed

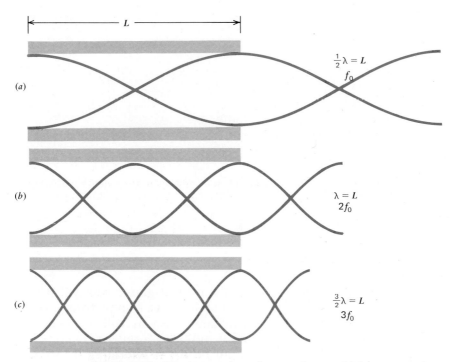

Figure 7-21 *The lowest three frequencies at which an open pipe can vibrate. If the pipe is driven with a mixture of frequencies at one end, it will resonate only at the characteristic frequencies. The curves show the magnitudes of the back and forth displacement of the air molecules. Where the curves cross (the nodal points), the molecules are stationary. Where the curves have maxima (or minima), the molecules experience the greatest back and forth displacement.*

ends of a vibrating string. For an open pipe, however, there are always *anti-nodes* at the ends. That is, the ends of a vibrating string remain at rest, whereas the air molecules at the ends of an open pipe are always moving. The frequency f_0 (Fig. 7-21a) is the fundamental frequency of the pipe. The fundamental and the various harmonics constitute the characteristic frequencies of the pipe.

Additional Exercises

1. The speed with which a water wave propagates in deep water depends only on the wavelength and not on the water depth.

Experiment and theory show that this speed is

$$v_{deep} = 1.25\sqrt{\lambda}$$

The speed is given in meters per second when the wavelength λ is given in meters. (*Deep* water for a particular wave is water with a depth greater than about one-half of the wavelength of the wave.) In shallow water, on the other hand, the wave speed depends only on the depth and not on the wavelength:

$$v_{shallow} = 3.13\sqrt{d}$$

The speed is given in meters per second when the depth d is given in meters. (*Shallow* water for a particular wave is water with a depth less than about $\frac{1}{25}$ of the wavelength of the wave.) Now, suppose that a wave with $\lambda = 100$ m is propagating across deep water. What is the wave speed and what is the wave frequency? If this wave moves inshore, what will be the wave speed at a depth of 1 m? The wave must have the same *frequency* in shallow water as in deep water. (Can you see why?) What will be the wavelength when the depth is 1 m? [Ans. 12.5 m/s, 0.125 Hz; 6.26 m/s; 25 m.]

2. What is the frequency of the second harmonic that can be set up in an open pipe with a length of 20 m? [Ans. 33 Hz.]

3. Make a series of sketches similar to those in Fig. 7-21 for the case of a pipe that is open at one end and closed at the other end. Express the wavelengths in terms of the length of the pipe, and write down the frequencies in terms of the fundamental frequency. (Can you see why there must be a node at the closed end?) This is the case of organ pipes. What must be the length of an organ pipe whose fundamental frequency is 16 Hz, the lower limit of the audible range? [Ans. 5.16 m or about 15 ft.]

4. Show that an open pipe can support both *even* and *odd* harmonics but that a pipe that is closed at one end can support only the *odd* harmonics.

How do cameras work?

8

LIGHT AND OPTICAL INSTRUMENTS

Photography—although it is little more than a hundred years old—is an important aspect of modern culture. Millions of photographs are taken each day, many of which appear in magazines, newspapers, and books. Nearly everyone has some kind of collection of photographs of events or scenes or persons of interest to him. Modern technology has provided us with a wide variety of photographic equipment, from complex and versatile professional instruments to the virtually foolproof "Instamatic" cameras and "pictures-in-a-minute" systems. The professional photographer or the serious amateur has available dozens of different types of black-and-white and color films as well as hundreds of different types of lenses to fit every situation.

Although you have probably taken many photographs yourself, have you ever wondered how your camera really works? How does the lens system produce an image on the film? What is the function of the aperture control? How does the film capture the scene that you see in the viewfinder?

In this chapter we will study the properties of light and how they relate to the operation of cameras and other optical instruments. We will learn how lenses focus light and produce images. We will discuss the wave properties of light and how these give rise to the phenomena of *refraction, diffraction,* and *color*. And we will describe the physical principles that govern the basic operation of cameras and films.

Light Rays—Reflection

Probably the most common statement we hear about light is that light travels in straight lines. We see this verified in a number of situations. For example, a flagpole blocks the passage of sunlight and casts a sharp shadow on the ground. And the light beam from a flashlight travels in the direction in which the flashlight is pointed. We account for both of these observations in terms of the straight-line propagation of light. (Later in this chapter we will find that in certain situations we must modify our idea about the straight-line propagation of light. However, for most of our discussion here, we will not be concerned with these *diffraction* effects.)

Light Rays—Reflection 219

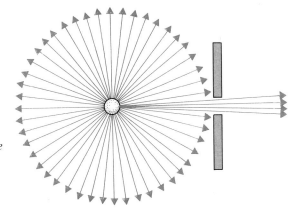

Figure 8-1 *The light from a source is represented as a collection of straight* rays. *The opening in the screen allows the passage of a* bundle *of rays.*

Because light travels in straight lines, we can simplify the description of many situations by representing the light emitted from a source as a collection of *rays*. In diagrams, we draw each ray as a straight line, and we include only enough rays to define the situation clearly (Fig. 8-1). We understand that the light actually fills the entire region exposed to the source.

An important aspect of light that is easy to describe by means of light rays is the phenomenon of *reflection*. It was known even in ancient times that when a light ray is incident on a flat polished surface, the ray will be reflected in such a way that the angle of incidence is exactly equal to the angle of reflection (Fig. 8-2). Notice that the angle of incidence ϕ_i and the angle of reflection ϕ_r are both measured from the dashed line that is perpendicular to the surface of the mirror.

Hero of Alexandria, in the second century B.C., explained the equality of the angles of incidence and reflection as a result of the axiom that a light ray, traveling from one point to another by means of a reflection, takes the *shortest* possible path between the points. Look at the ray that connects points *A* and *B* in Fig. 8-2. Can you see how $\phi_i = \phi_r$ only for the shortest path from *A* to *B*? (See Exercise 1 at the end of this chapter.)

When you look into a mirror, objects that are actually in *front* of the mirror appear to be *behind* the mirror. That is, you see the *image* of an object in a position different from the true position of the object. Figure 8-3 shows how such a mirror image is formed. Trace the various rays that originate at the

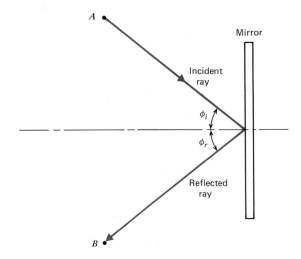

Figure 8-2 *When a light ray is reflected, the angle of incidence ϕ_i is equal to the angle of reflection ϕ_r.*

tip A of the object. Each ray is reflected from the mirror surface with $\phi_i = \phi_r$ (check this). When the observer views the reflected rays, they seem to diverge from a point A' located behind the mirror. The rays from every other point on the object (these rays are not shown in the diagram) also appear to originate behind the mirror. The observer therefore sees a complete image of the object. Notice that this image is located just as far behind the mirror as the object is in front of the mirror.

Light Waves—Refraction

Light is a wave phenomenon. In the preceding chapter, we learned about waves and how they are propagated. Light, as we experience it, seems to have nothing in common with sound waves or waves on strings. Indeed, light waves differ in several important respects from the waves we have studied. Sound waves and waves on strings require some material medium to support the wave motion. However, we know that light from the Sun and the stars reaches us through the vacuum of space. That is, light waves can propagate without the benefit of any material medium. The same is true of radio waves; we communicate regularly by means of radio waves with spacecraft far outside the Earth's atmosphere.

Wave motion involves the vibration of *something*. How,

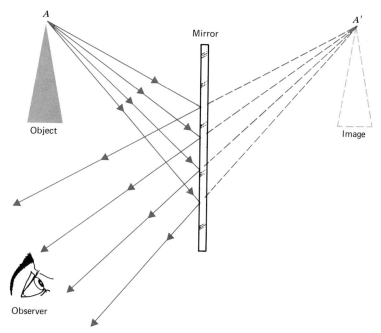

Figure 8-3 *Construction to determine the position of a mirror image. Each ray from* A *has* $\phi_i = \phi_r$.

then, can radio and light waves propagate through vacuum? In Chapter 5 we learned about electric and magnetic fields. These fields can exist in vacuum. When the electric and magnetic field amplitudes vary regularly with time and through space, we have an *electromagnetic wave*. Although no *matter* vibrates, as in sound waves and waves on strings, the regular vibration of the electromagnetic field constitutes a type of wave.

Light and radio waves are both electromagnetic waves. So are television signals, microwaves, infrared and ultraviolet radiation, and X rays. All of these radiations have the same basic character; they differ only in *frequency* and *wavelength*. We give different names to the waves because we detect them and use them in different ways. But apart from frequency and wavelength, there are no essential differences among them. For example, if the human eye and optic nerve system were constructed in a slightly different way, we would be able to see infrared or ultraviolet radiation, and what we call *visible* light would be *invisible!*

Another important difference between sound waves and light waves is the speed of propagation. Sound waves travel through air with a speed of 330 m/s. The speed of light is about a million times greater, 300 000 000 m/s (3×10^8 m/s). All types of electromagnetic waves travel through vacuum with this same speed. (When an electromagnetic wave travels through a material medium, the speed is less than the value for vacuum. As we shall see, light moves more slowly through glass or water than through air or vacuum.)

The wavelengths of sounds that we hear range from about 20 m to about 1.6 cm. The wavelengths of electromagnetic waves that we see (that is, light waves) are extremely small, ranging from about 0.0000004 m (4×10^{-7} m) to about 0.00000075 m (7.5×10^{-7} m). This is the wavelength range of visible light. The wavelengths of infrared radiation, microwaves, and radio and television signals are all *longer* than those of visible light. The wavelengths of ultraviolet radiation and X rays are *shorter* than those of visible light.

What effects do we see that are due to the wave character of light? First, let us look at what happens when a light ray in air

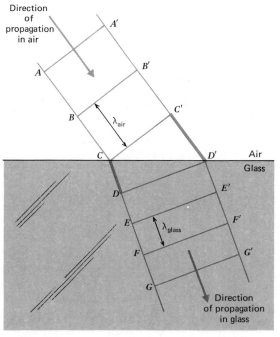

Figure 8-4 *Because light travels more slowly in glass than in air, the advancing wavefronts are retarded as they enter the glass. This produces a change in direction of the wave propagation. The wave is refracted.*

enters a piece of glass. The speed of light in air is essentially the same as in vacuum. But in a medium such as glass, the speed of light is substantially less. (We can view this as due to a kind of "resistance" that the glass offers to the propagation of light.)

Figure 8-4 shows the effect of the change in wave speed when light passes from air into glass. We do not use a ray picture here; instead, the line segments, AA', BB', ..., GG', represent the wave fronts of the light wave. Each segment is spaced one wavelength from its neighbors. Look at the wavefront CC'. The end C has reached the glass surface, but the end C' is still moving through air. As the end C advances to D, the end C' advances to D'. Because the wave is retarded in the glass, the distance CD is less than the distance $C'D'$. This results in a change in direction of wave propagation in the glass compared to the air. We say that the wave is *refracted* at the air-glass boundary. Notice also that the wavelength of the wave is smaller in the glass than in the air. (Can you see why this is so? The *speed* of the wave is smaller in the glass, but the *frequency* is the same. Therefore, what must happen to the *wavelength?*)

In Fig. 8-5 we again show the refraction of light at an air-glass boundary, but here we use a ray picture. Notice that the direction of each ray is the same as the direction of propagation of the corresponding wave in Fig. 8-4. The direction of a

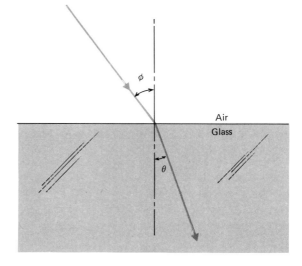

Figure 8-5 *Ray diagram corresponding to the wave picture of Fig. 8-4 for the refraction of light at an air-glass boundary.*

ray is always perpendicular to the wave front of the wave. Now that we understand how the wave character of light produces refraction, we will no longer use the wave diagram to describe these effects but will use the simpler ray diagram, as in Fig. 8-5.

When a light ray travels from air into glass, the ray is bent (refracted) *toward* the perpendicular to the surface. That is, in Fig. 8-5 the angle θ is *less* than the angle ϕ. When the ray travels in the opposite direction, from glass into air, the direction of bending is *away* from the perpendicular. In Fig. 8-5 we can simply reverse the directions of the arrows to obtain the correct description of what happens when the ray travels from glass into air.

The amount of bending that a light ray experiences when it crosses the boundary between two media depends on how much the speed changes. If the speed change is small, the refractive effect is small; if the speed change is large, the refractive effect is large. We measure the ability of a medium to produce refraction in terms of a number called the *index of refraction,* usually denoted by n. This number is the ratio of the speed of light in vacuum to the speed of light in the medium:

$$\text{index of refraction} = \frac{\text{speed of light in vacuum}}{\text{speed of light in medium}}$$

Because the speed of light in any medium is always smaller than the speed in vacuum, the index of refraction is always greater than 1. Different types of glass have different values of n, but most are near $n = 1.5$. For water, $n = 1.33$. Thus, glass is more effective than water in bending light that is incident from air.

Lenses

In order to take a photograph, we must have some way to project onto the film an image of the desired scene. This is accomplished with a *lens* or a system of lenses. To see how a lens forms an image, we need to examine the way light rays pass through pieces of glass of various shapes.

First, let us look at a light ray that passes through a slab of glass with parallel sides. Figure 8-6 shows a light ray entering such a piece of glass at point A. The ray undergoes refraction

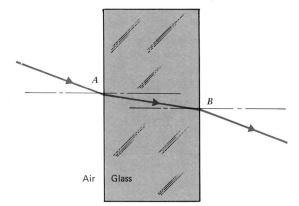

Figure 8-6 *When a light ray passes through a piece of glass with parallel sides, it undergoes equal amounts of refraction at A and B, and therefore emerges in the same direction as the incident ray.*

and bends toward the perpendicular. When the ray emerges from the glass at B, the ray is bent away from the perpendicular. The amount of bending in each case is the same, so that the emergent ray has the same direction as the incident ray. The net result is that the ray has been shifted by a small amount but without any change in direction. If we had many parallel rays incident on the glass, we would not see any effect of the glass on the transmitted beam. Indeed, whenever we look through a glass window, we see a natural and undistorted scene (if the glass has accurately parallel faces).

When a light ray passes through a piece of glass whose faces are not parallel, we find a different effect. Figure 8-7 shows what happens in such a case. The light ray labeled I enters the glass at point A and is refracted. Upon emerging at B, the ray is again refracted. Because the faces are not parallel, the emergent beam has a direction different from that of the incident ray. Any other ray that is parallel to ray I, such as ray II in the diagram, will be deviated by the same angle and will emerge from the glass parallel to ray I.

What use can we make of the deviation of a light ray by a

Figure 8-7 *A light ray that passes through a triangular-shaped piece of glass undergoes a deviation in direction.*

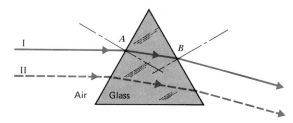

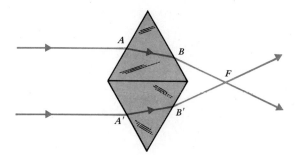

Figure 8-8 *By using two triangular-shaped pieces of glass, we can bring two parallel rays together at* F.

piece of glass with nonparallel faces? Suppose that we have two pieces of triangular-shaped glass and place them together as in Fig. 8-8. The ray that enters the upper piece will be deviated downward just as in Fig. 8-7. The ray that enters the lower piece will be deviated by the same amount, but upward. The two rays come together at F. In this way we have brought two parallel rays to a *focus*. We have not yet produced a useful lens, however, because other pairs of parallel rays that enter the pieces of glass will emerge parallel to the rays shown. Such ray pairs will converge at points different from F. (Try another pair of rays and convince yourself that this is indeed the case.)

If we want to bring an entire bundle of rays together at a single point, we need a piece of glass with *curved* faces. If the faces have the shape of circular arcs, we have a true lens, as shown in Fig. 8-9. Most lenses are circular, so that each face is actually a segment of a spherical surface. Notice how the angle of incidence increases with the distance of the ray from the axis. Therefore, the rays incident far from the axis are deviated by amounts greater than those incident near the axis. In this way, all of the rays in the bundle are brought to a focus at F. If the parallel rays are incident from the right, instead of the

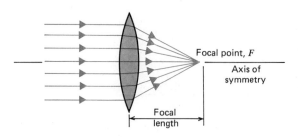

Figure 8-9 *A lens with spherical surfaces brings parallel rays to a focus.*

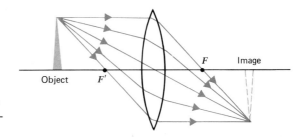

Figure 8-10 *The light rays from an object are brought together by a lens to form an image of the object.*

left, they will converge at the other focal point of the lens, located to the left of the lens.

The light rays that pass through a lens or a lens system are *reversible*. That is, a light ray that follows a certain path through a lens will follow that same path if the ray is projected through the lens in the opposite direction. In Fig. 8-9, if we place a source of light at the focal point F, the light will emerge to the left of the lens as a parallel bundle of rays.

The formation of an *image* by a lens is shown in Fig. 8-10. Several of the rays from the tip of the object are followed through the lens. These rays all converge at a point to produce an image of the tip.

If we know the positions of the focal points of a lens, it is a simple matter to construct the image of an object. Figure 8-11 shows how this is done. Look first at the ray OAI. This ray is incident on the lens parallel to the lens axis; therefore, the ray emerges in a direction that causes it to pass through the focal point F. Next, look at the ray OCI. This ray passes through the focal point F'; therefore, it emerges from the lens in a direction parallel to the lens axis. Finally, look at the ray OBI. This ray passes through the center of the lens where the faces are essentially parallel. Consequently, this ray is not deviated

Figure 8-11 *Three easy-to-draw rays that locate the position of the image.*

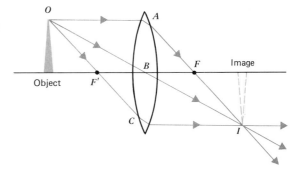

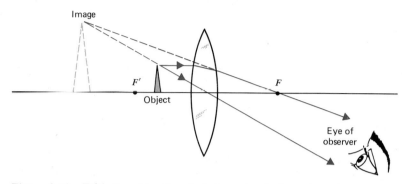

Figure 8-12 With the object located closer to the lens than the focal point, an enlarged virtual image is seen by the observer.

in direction. (We draw the ray *OBI* as a straight line because the lens is thin and the displacement of the ray is very small. Compare Fig. 8-6.) We need only two of these three rays to locate the position of the image; the third ray can be used as a check.

Notice in Figs. 8-10 and 8-11 that the rays from the object actually converge at the position of the image. If we were to place a white card at this position, we would see the projected image of the object. For this reason, we call such images *real images*.

Figure 8-12 shows the formation of a different kind of image. Here, we place the object close to the lens. That is, the object position is *inside* the focal point, whereas in Figs. 8-10 and 8-11 the object was located *outside* the focal point. Two rays are drawn in Fig. 8-12: The parallel ray passes through the focal point F, and the straight-line ray passes through the center of the lens. These rays do not converge to the right of the lens; and no real image is formed in this case. Even so, an observer to the right of the lens can see an image because the lens of the eye focuses the rays, and to him the rays appear to diverge from an image located behind the lens. This image, although it can be seen, cannot be projected onto a screen. We call such images *virtual images*. Notice in Fig. 8-12 that the observer sees an *enlarged* image. Therefore, a simple lens with the object located inside the focal point acts as a *magnifier*.

The lenses we have been considering are thicker at the position of the axis than at the outer rim. Such lenses are called *convex* lenses (or *converging* lenses). Figure 8-13 shows a lens that is thicker at the outer rim than at the center. This type of

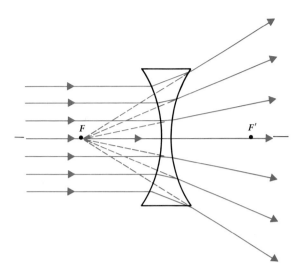

Figure 8-13 *A concave lens causes light rays to diverge.*

lens is called a *concave* lens. Parallel rays that are incident on a concave lens are refracted away from the axis. For this reason we also call these lenses *diverging* lenses. In Fig. 8-13 notice that the diverging rays would appear to an observer on the right of the lens to originate at the point F. We call this the focal point of the lens even though rays are never actually brought to a focus.

Figure 8-14 shows the formation of an image by a concave lens. Again we draw only the parallel ray and the straight-line ray. An observer to the right of the lens would see a virtual image in the position indicated. A concave lens alone always produces a virtual image, regardless of the location of the object. (Verify this by constructing the image for an object located *outside* the focal point.)

Figure 8-14 *A concave lens always produces a virtual image.*

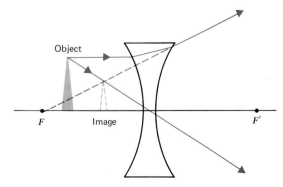

Optical Instruments

Figure 8-12 illustrates the simplest of optical instruments—a convex lens that acts as a magnifier. Using a single lens in this way, we can view small objects with magnifications up to 10 times or so. For larger magnifications, we need a more complex instrument that consists of a *system* of lenses. Figure 8-15 shows a two-lens *microscope*. The object to be viewed is placed just outside the focal point F'_o of the first lens, which is called the *objective*. This lens produces a real image at a position between the objective and the second lens (called the *eyepiece*). This intermediate image is located inside the focal point F'_e of the eyepiece and acts as the object for the eyepiece. Consequently, the eyepiece forms an enlarged virtual image that can be viewed by the observer's eye.

Laboratory microscopes usually employ more complicated lens systems than the simple two-lens arrangement shown in Fig. 8-15. It is common to find a half dozen lenses or so in such instruments. Nevertheless, the operating principle is the same. Each lens produces an image that acts as the object for the next lens, and the final image is an enlarged virtual image that can be seen by the observer. By changing the eyepiece and objective, magnifications in the range from about 10 to about 1000 are possible with many modern microscopes.

The features of a *telescope* are very similar to those of a microscope. In the simplest arrangement there are two lenses,

Figure 8-15 *Ray diagram for a simple two-lens microscope. Notice that the final image is* inverted; *that is, the orientation of the image is opposite that of the object.*

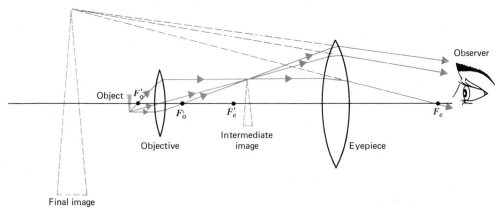

the objective and the eyepiece, as shown in Fig. 8-16. The primary difference between the lenses in a telescope and a microscope is that the telescope objective has a very long focal length, whereas the microscope objective has a very short focal length.

Figure 8-16 shows three parallel rays from a distant object entering the objective lens. These rays are brought to a focus and form an intermediate image. As in a microscope, this image acts as an object for the eyepiece, which forms an enlarged virtual image that can be seen by the observer. Objects at different distances can be brought into sharp focus for the observer by changing the distance between the objective and the eyepiece.

Notice that the image formed by the two-lens telescope in Fig. 8-16 is *inverted*. For astronomical studies, an inverted image is quite satisfactory. For terrestrial observations, however, such an image would be annoying, to say the least. Therefore, in all telescopes used for ordinary, nonastronomical viewing, additional optical elements are included to produce a final image that is erect.

Microscopes and telescopes produce *virtual* images that can be viewed because of the focusing action of the eye. In a camera, however, we require that a *real* image be formed on

Figure 8-16 *Ray diagram for a simple two-lens astronomical telescope. Telescopes for terrestrial viewing incorporate additional optical elements so that the final image is erect.*

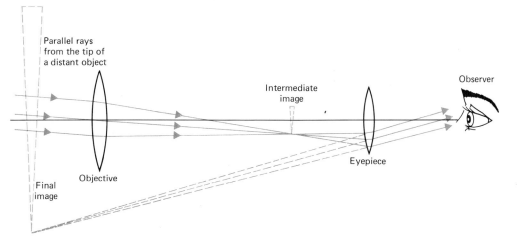

the film so that it can be rendered developable. Therefore, for the remainder of this chapter we shall concentrate on lens systems that produce real images. In addition to cameras, the lens systems in microscopes and telescopes can be arranged to produce real images for photographic purposes.

Cameras

Many cameras in use today have complicated lens systems that consist of several pieces of glass (usually different *kinds* of glass) cut to various curvatures and cemented together or held in specific arrangements by rings. Later in this chapter we will discuss the reasons for employing lenses of this type. However, we can examine the essential features of all cameras by considering first a simple camera with a single lens, as in Fig. 8-17.

The camera lens forms a real image of the object on the plane where the film is located. The bellows arrangement allows different lens-to-film distances so that objects at different distances can be brought into sharp focus. (In cameras with multiple-element lenses, focusing is accomplished either by moving the entire lens system or by moving one or more of the lens elements with respect to the others.) Behind the lens is a *shutter* that can be opened for a fraction of a second (or held open for long exposures) to admit sufficient light from the object to produce a developable image on the film.

The amount of light that reaches the film for a particular shutter opening time depends on the *focal length* of the lens and on its *area* (which is proportional to the square of the lens *diameter*). The ratio of the focal length of a lens to its diameter is called the *speed* or the *f-number* of the lens:

$$f\text{-number} = \frac{\text{focal length}}{\text{lens diameter}}$$

The larger the diameter of a lens with a particular focal length, the smaller will be the *f*-number and the greater will be the amount of light that can reach the film. It is difficult to construct a lens that has a diameter greater than its focal length; most camera lenses therefore have *f*-numbers greater than 1 (written as $f1$). Many 35-mm cameras have an $f2$ lens as standard, with an $f1.4$ lens as an extra-cost option. (A "35-mm camera" is one that accepts film with a width of 35 mm.)

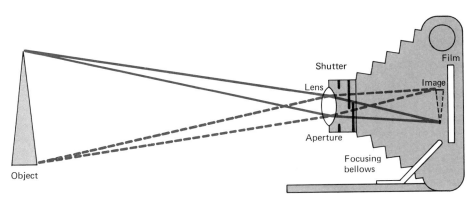

Figure 8-17 *The essential features of a simple camera.*

If two lenses have the same *f*-number, even though they may have different focal lengths and diameters, the degree of darkening of the film will be the same for the same shutter opening times.

Many cameras have an *aperture* of variable diameter located behind the lens (see Figs. 8-17 and 8-18). By changing the size of the aperture, the effective *f*-number of the lens can be changed. The amount of light reaching the film can be controlled by adjusting either the aperture diameter or the shutter speed (or both). For a dimly lit object, a large aperture (small *f*-number) and a long exposure time are required; for a brightly lit object, a small aperture (large *f*-number) and a short exposure time are required.

The aperture settings on a camera are usually indicated by markings that represent changes by a factor of the square root of 2 in *f*-number. Each such change in *f*-number means a change by a factor of 2 in the amount of light admitted. (Why?)

Figure 8-18 *The variable aperture of a camera controls the* f-*number.*

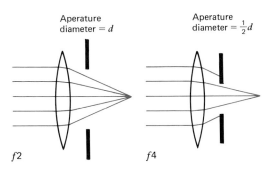

A typical lens for a 35-mm camera has a focal length of 50 mm and a diameter of 25 mm. (Lengths having to do with cameras and films are usually given in *millimeters*.) The f-number for such a lens with the aperture "wide open" is 50 mm/25 mm = $f2$. By "stopping down" the aperture, a range of f-numbers can be obtained: $f2, f2.8, f4, f5.6, f8, f11, f16,$ and $f22$. Changing the aperture by one position (one "stop") corresponds to changing the amount of light admitted by a factor of 2 (if the exposure time remains the same). Shutter speeds are also usually changed in steps of a factor of 2 (or approximately 2). A typical 35-mm camera will have shutter speeds of $\frac{1}{30}, \frac{1}{60}, \frac{1}{125}, \frac{1}{250}, \frac{1}{500}, \frac{1}{1000}$, and $\frac{1}{2000}$ of a second. Sometimes, slower speeds are included: $\frac{1}{15}, \frac{1}{8}, \frac{1}{4}, \frac{1}{2}$, and 1 s.

By changing aperture settings and shutter speeds, it is possible to accommodate a wide range of lighting conditions. But these adjustments are made for other reasons as well. Clearly, if you wish to photograph a rapidly moving object, then a fast shutter speed is necessary to prevent blurring of the image. The aperture size also controls what is called *depth of field*. The rays that pass through a lens near its axis will be in focus for a wide range of object distances. The rays that enter near the outer rim of a lens, however, will be in focus for only a narrow range of object distances. Therefore, if a large aperture is used (for example, $f2$ or $f2.8$), the range in which objects are in sharp focus (the depth of field) is small. If a small aperture is used (for example, $f16$ or $f22$), the depth of field is large. If a subject is located in front of some background that you do not want in the photograph, this background can often be blurred out by using an aperture setting that gives a narrow depth of field.

Most cameras of recent manufacture have some sort of "automatic" feature. These cameras are equipped with a light-sensitive element (a *photoelectric* device) that produces a tiny electric current in a special circuit when exposed to light. The amount of current is proportional to the intensity of the light. This current is used to control either the aperture setting or the speed of the shutter so that a proper exposure of the film is made for the particular lighting of the subject.

Camera Lenses

Many types of modern cameras (with the exception of the simplest models) are designed to accept interchangeable lenses. This feature allows the photographer to select a lens with the focal length appropriate for a particular subject. The "standard" lens for a 35-mm camera has a focal length of 50 mm (or 55 mm). Most manufacturers provide a line of lenses with both larger and smaller focal lengths. Some typical values are 21 mm, 28 mm, 35 mm, 85 mm, 100 mm, 135 mm, 200 mm, and 400 mm. Lenses are also available outside this range, but such lenses are more difficult to manufacture and are therefore usually more expensive.

If you attach a series of lenses with different focal lengths to a camera, you will readily see that the size of the image of a particular object depends on the focal length of the lens. If the focal length is short, the image size will be small; if the focal length is long, the image size will be large. To see why this is so, look at the ray diagrams in Fig. 8-19. Here we have two cameras, one with a long-focal-length (or *telephoto*) lens and one with a normal lens. The distance from the object to the lens is the same in each case, but the image formed by the telephoto lens is much larger than that formed by the normal

Figure 8-19 *Schematic comparison of the image sizes formed by a normal camera lens and a long focal-length (or* telephoto) *lens.*

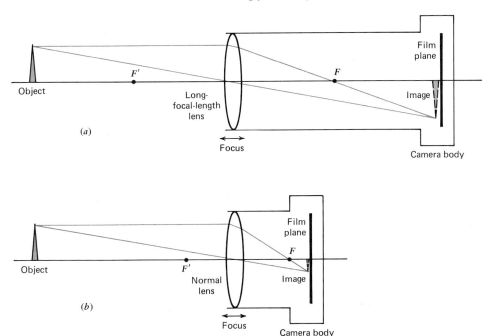

lens. Notice that the telephoto lens must be mounted at a greater distance from the film plane. You have probably seen photographers taking telephoto shots with long lenses protruding from their cameras.

The size of the image produced by a telephoto lens compared to the size produced by the normal lens is equal to the ratio of the focal lengths. Thus, a 200-mm lens will produce an image four times larger than that formed by a 50-mm lens. In this case we would say that the magnification is $4\times$.

The image size produced by a telephoto lens is conveniently described in terms of magnification. For a lens with a focal length shorter than that of the normal lens, it is more useful to state the *field of view* of the lens. Figure 8-20 shows how we do this. In comparing the two lenses, we adjust the object-to-lens distance until the image just fills the film frame. Then we measure the angle between the rays that enter the center of the lens from opposite ends of the object. Notice that the short-focal-length lens has a wider field of view than does the

Figure 8-20 *Schematic comparison of the fields of view of a normal camera lens and a short focal-length (or wide-angle) lens. Notice that the image fills the film frame in each case.*

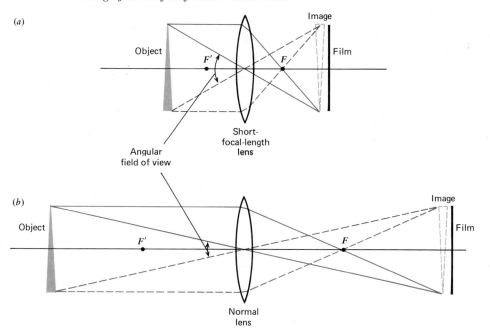

normal lens. Short-focal-length lenses are usually called *wide-angle* lenses.

In a typical 35-mm camera, the normal lens has a focal length of 50 mm and an angular field of view of about 46°. A 28-mm wide-angle lens has a coverage of about 75°, and a 200-mm telephoto lens accepts a narrow cone of light with an angle of about 12°. Super wide-angle (or *fisheye*) lens have fields of view up to 180°.

In Figs. 8-19 and 8-20 we see that any lens must be mounted at a distance from the film that exceeds the focal length of the lens. For telephoto lenses with very long focal lengths, this arrangement becomes quite awkward. However, if a diverging lens element is added, the physical length of the system can be made less than its effective focal length. A simple, two-element telephoto lens is shown schematically in Fig. 8-21. The diverging lens is located between the converging lens and the focal point. A single ray (parallel to the lens axis) from a distant object is traced through the system. Notice that if the diverging lens were not present, the focal point would be that of the converging lens alone, F_c. The effect of the diverging lens is to lengthen the focal distance from F_c to F_t. By tracing backward the final ray, we can find the position at which a single converging lens would have to be located in order to produce the same focal length as the combination. This position is shown in front of the converging lens. Thus, we see that the addition of the diverging lens has produced a shortening of the physical length of the system for a given focal length. Most telephoto lenses have this basic type of construction, although more lens elements are almost always used.

Figure 8-21 *Schematic of a simple, two-element telephoto lens. Notice how the addition of the diverging lens increases the focal length.*

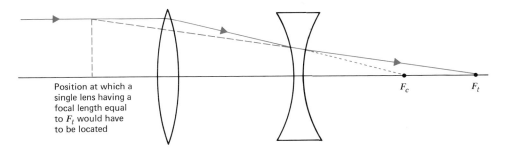

238   Light and Optical Instruments

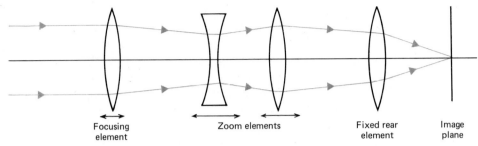

Figure 8-22 *Schematic of one type of zoom lens. Movement of the middle two-elements produces a change in focal length for the system.*

Probably the most spectacular type of camera lens is the variable-focal-length or *zoom* lens, widely used in movie and television cameras. There are many different designs capable of producing the same effect. (Between 1962 and 1966, there were at least 16 U.S. patents issued for zoom lens systems.) One of the designs is illustrated in Fig. 8-22. This zoom lens consists of four elements. The front element is the focusing element and its position is adjusted to give a sharp focus for the particular object to be viewed or photographed. The rear element remains in a fixed position relative to the image plane. The zoom effect is provided by the movement of the two middle elements. As the zoom control is changed, each of these lenses follows a path defined by a system of cams and guides. In many zoom lenses, the focal length can be changed over a range of 8-to-1 or 10-to-1. Some zoom lenses must be refocused when the focal length is changed, whereas others remain in focus (or nearly so) over the entire zoom range.

Lens Aberrations

Each of the lenses we have discussed so far has been pictured as a single piece of glass with each of its two faces having the same curvature. Actually, real camera lenses are not constructed in this way. The reason is that no lens is capable of producing an absolutely perfect image. All lenses, particularly those that consist of single pieces of glass, suffer from various types of aberrations and distortions. Two of these effects (there are several others in addition) are illustrated in Figs. 8-23 and 8-24. Figure 8-23a shows the focusing of rays parallel to the lens axis by a hypothetical "perfect" lens. These rays

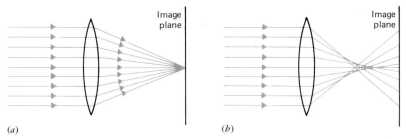

Figure 8-23 *The failure of a lens to bring a bundle of rays parallel to the axis to a focus at a single point is called* spherical aberration. (a) *"Perfect" lens.* (b) *Lens with spherical aberration.*

come together precisely at a point. A real lens, however, focuses parallel rays as shown in Fig. 8-23b. The rays close to the axis focus at one point, whereas the off-axis rays focus at other points. The result is that the total bundle of rays comes together, not at a single point, but in a small circle. The position at which this circle has minimum diameter is called the *circle of least confusion* and is the position at which the film should be located to produce the sharpest image. This effect is called *spherical aberration*. No lens or lens system is completely free of this type of defect, although by careful design of the lens curvature in multi-element lenses, the effect can be minimized.

Figure 8-24 shows what happens to a bundle of rays that make an angle with respect to the lens axis. These rays also come to focus at different points. This aberration is called *coma*. Notice that the extreme rays in the bundle come to a

Figure 8-24 *The failure of a lens to bring a bundle of rays that are at an angle to the axis to a focus at a single point is called* coma. (a) *"Perfect" lens.* (b) *Lens with coma.*

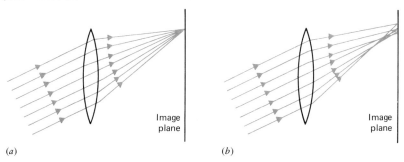

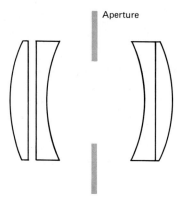

Figure 8-25 *The Zeiss* Tessar *lens, one of the early lenses that reduced aberrations by incorporating several elements into the lens.*

focus at a position on the image plane that is *higher* than the point of focus for the central rays. By changing the curvature of the lens, it is possible to make the extreme rays focus at a position *lower* than that for the central rays. Therefore, by using two lenses that have opposite coma effects, this type of aberration can be largely eliminated. Indeed, the way in which we deal with all lens defects is to construct lens systems with several elements chosen to cancel as nearly as possible the aberrations and distortions. One of the famous early lenses that successfully compensated various defects to a high degree is the Zeiss *Tessar* lens, shown in Fig. 8-25. This lens system consists of four elements—a separated doublet and a cemented doublet. Notice that the aperture is located between the two doublets in this system.

Chromatic Aberration

The lens aberrations we have mentioned so far are caused by the *geometrical* properties of lenses. There is an entirely different problem that is due to the fact that light consists of different *colors*. Figure 8-7 shows a light ray passing through a triangular-shaped piece of glass (which we call a *prism*). In the diagram we see a single ray emerging at point *B*. This behavior of a light ray passing through a prism is correct if the light has a single color. However, if we allow light from an ordinary light bulb (that is, *white* light) to be incident on a prism, we find the effect illustrated in Fig. 8-26. White light actually consists of a mixture of light with different colors. Red light has the longest wavelength (about 7.5×10^{-7} m) and violet light

Chromatic Aberration 241

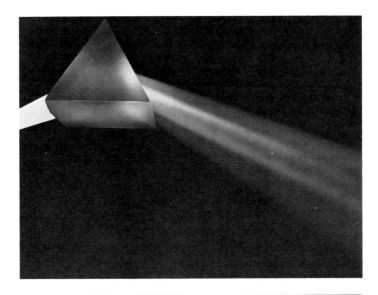

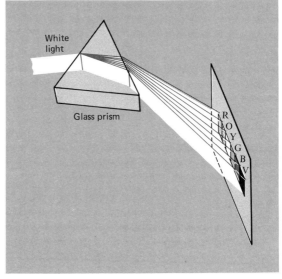

Figure 8-26 *When white light, consisting of all colors, passes through a prism, the light is dispersed into a spectrum of colors. Red light (R) is refracted least and appears on the left-hand side of the spectrum in this diagram, followed by orange (O), yellow (Y), green (G), blue (B), and violet (V).*

has the shortest wavelength (about 4×10^{-7} m). The range of wavelengths for *visible* light lies between these extremes. The amount of deviation that a light ray undergoes in passing through a prism (or a lens) depends on the index of refraction of the glass from which the prism (or lens) is made. All types of glass have the property that the index of refraction n de-

242 *Light and Optical Instruments*

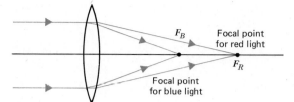

Figure 8-27 *Because the index of refraction of all glasses depends to some extent on the wavelength of the light, different colors come to a focus at different points. This is called* chromatic aberration. *The difference between F_B and F_R is exaggerated here.*

pends to a small degree on the wavelength of the light. The value of n tends to increase as we go from red light to blue light. As a result, when a ray of white light enters a prism, the various colors are dispersed into a spectrum of colors, as indicated in Fig. 8-26. (Tiny droplets of water in the air act in the same way and can disperse sunlight into a *rainbow*.)

When white light is focused by a lens, the variation with wavelength of the index of refraction of the lens causes the different colors to focus at slightly different points. This effect is shown in Fig. 8-27, where the difference between the blue focal point F_B and the red focal point F_R is exaggerated. This kind of lens defect is called *chromatic aberration*.

The only way to correct chromatic aberration is to make lenses from two (or more) pieces of glass with different indexes of refraction. Figure 8-28 shows how this is done. This lens consists of a diverging element made from light flint glass ($n = 1.55$) cemented to a converging element made from barium crown glass ($n = 1.61$). The red and blue rays pass through this doublet along slightly different paths. By choosing the curvatures of the faces properly, the red focal point and the blue focal point can be made to coincide. Such a lens is called an *achromatic* lens.

Figure 8-28 *By making a lens from two elements with different indexes of refraction, the red and blue rays can be brought to a focus at the same point. Such lenses are called* achromatic *lenses.*

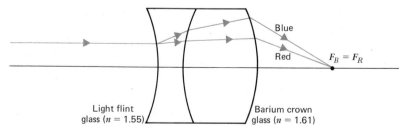

A two-element lens is capable of producing a common focal point only for two colors—the remaining colors still focus in slightly different positions. By adding a third element to the lens, *three* colors can be brought to the same focus. This more highly color-corrected lens is called an *apochromatic lens* (or *apochromat*). Most high-quality lenses are apochromats. The effect of chromatic aberration on the image is sufficiently small that the correction supplied by an apochromat is satisfactory for most photographic situations. More highly corrected lenses are generally not worth the increased expense.

Diffraction

There is one additional effect that limits the sharpness of the image that can be formed by a lens. This effect is called *diffraction* and is due to the wave property of light. Let us think first about what happens when water waves or sound waves strike a barrier in which there is a narrow slot, as shown in Fig. 8-29. The incident waves cause the water (or air) within the slot to be agitated with the same frequency as that of the wave. If the width of the slot is small compared to the wavelength of the waves, the slot acts essentially as a point source

Figure 8-29 *A wave is incident on a barrier containing a narrow slot. To the right of the barrier, the outgoing wave is circular because of diffraction at the slot.*

of the wave motion to the right of the barrier. Therefore, in this region, the outgoing waves are circular in shape. We say that the waves have been *diffracted*. Diffraction is a very common occurrence with water and sound waves. You are able to hear someone speaking in another room, even though you cannot see him, because the sound waves are diffracted in passing through the doorway. (Reflection from the walls adds to the effect.)

Light waves also undergo diffraction, but the effect is not as obvious as that for water or sound waves because the wavelength of light is so small. If light passes through a narrow slot, even a tiny hole, the size of the opening is still many times larger than the wavelength of the light wave. In this case, we can imagine that every point within the opening acts as a source of outgoing light waves. Let us first see what happens for *two* point sources of light. Figure 8-30 shows sources S_1

Figure 8-30 *Pattern of bright a dark lines due to the interference of the light from the sources S_1 and S_2.*

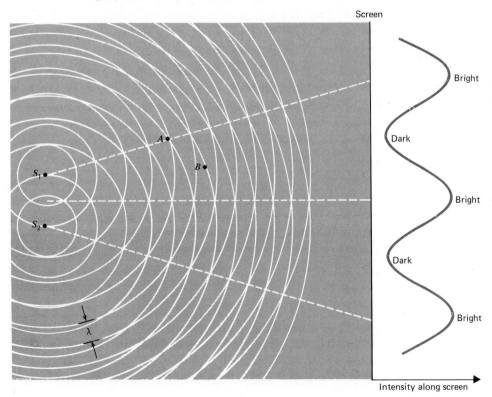

and S_2 each radiating waves with wavelength λ. Look at the pattern of outgoing circular waves. In some positions, such as point A, the wave crests (indicated by the circular arcs) of the two waves coincide. Therefore, at point A and all similar points there is *constructive* interference and the wave amplitudes add together. Point B, on the other hand, is located on a crest of the waves from S_2 and in the trough of the waves from S_1. Therefore, at point B and all similar points there is *destructive* interference and the wave amplitudes cancel.

Figure 8-30 shows a viewing screen located some distance away from the light sources. What will we see on this screen? Notice that we can connect with a series of straight lines all of the points (such as A) that correspond to positions of constructive interference. The places where these lines intersect the screen are regions of strong illumination. Between these bright areas are dark regions where there is destructive interference. Thus, the pattern of light on the screen from the two interfering sources is a series of alternating bright and dark lines.

Now, we return to our original problem. What do we see on a screen that is illuminated by light passing through a narrow slot or hole? Every point within the opening acts as a source. We can imagine choosing all possible pairs of points and plotting the light patterns as in Fig. 8-30. We would then add all of these together to obtain the complete pattern. Clearly, this is a tedious procedure, but there exist mathematical techniques for computing the result. Figure 8-31 shows what happens. The central bright region remains, but the other bright lines in the two-source pattern are much reduced in intensity. There is no sharp outline of the hole. Instead, the diffracted light produces a fuzzy spot on the screen.

A similar effect takes place when we focus light with a lens. Figure 8-32 shows a bundle of rays from a distant object located below the axis of the lens. These rays would come together at a point if the lens were "perfect" and if there were no diffraction effects. Because diffraction does occur, even a "perfect" lens will not produce a sharp, point-like image. Every point source of light will have a fuzzy image on the screen. Figure 8-32 also shows the outline of another bundle of rays from a source located on the axis of the lens (dashed lines). These rays produce another diffraction-broadened spot at the center of the screen. If the two light sources (both con-

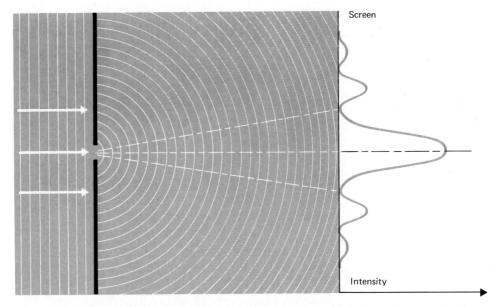

Figure 8-31 *The diffraction of light by a small hole produces a distribution of light intensity across a screen.*

Figure 8-32 *Because of diffraction, a lens does not focus the light from a distant source as a point. Instead, a fuzzy spot is formed. If two sources are close together, the diffracted images will merge.*

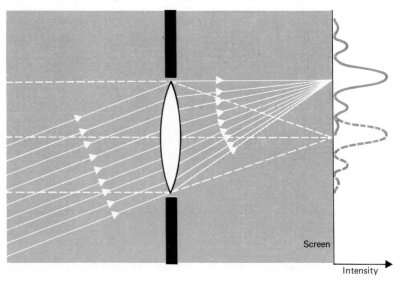

sidered to be points) are close together, the diffracted images will merge together and it will not be possible to distinguish the sources. Thus, it is never possible to obtain a perfectly sharp image of an object. In all kinds of lens systems, diffraction effects limit the ultimate detail that can be observed.

Black-and-White Film

The object of photography is to produce *pictures*. All of the optical apparatus we have been discussing serves merely to bring the light to the crucial element in the system, namely, the *film*. How does light affect film, and how do we produce a final picture of the subject photographed?

First, we consider the simplest type of black-and-white film. Such films consist of a thin layer of inert gelatin in which are dispersed tiny grains of silver bromide (AgBr). These grains have sizes about equal to the wavelength of visible light. The silver bromide is in the form of crystals in which the silver and the bromine exist as *ions*. That is, every silver atom donates one of its atomic electrons to a bromine atom. In this way, the silver atoms become positively charged ions (Ag^+) and the bromine atoms become negatively charged ions (Br^-). Each crystal grain contains about a billion (10^9) silver ions and an equal number of bromine ions.

When silver bromide is exposed briefly to light, the light causes some of the bromine ions to give up their excess electron. These electrons combine with silver ions and produce normal atoms of silver. The number of silver atoms formed depends on the amount of light to which the film was exposed. Tiny clusters of from 10 to 500 silver atoms in the grains constitute the *latent image* in the film.

The exposed grains of silver bromide appear no different from unexposed grains, and so the latent image is not visible until the film is *developed*. This process, which must be carried out in darkness to prevent further exposure of the still-sensitive grains, involves placing the film in a chemical bath. The chemical reactions that take place in the developing solution tend to convert the silver ions to silver atoms. But the silver atoms that were formed because of the exposure to light have the effect of speeding up the reactions in their vicinity. Therefore, the grains that constitute the latent image are converted to silver much more rapidly than are the unexposed grains. The developing action must be stopped after a time, for otherwise all of the grains would be reduced to silver and the film would be *overdeveloped.*

After developing, the film is placed in a *fixing* solution where the unexposed silver bromide grains are dissolved and washed away. The film now contains silver atoms only in those areas that were exposed to light; the greater the light exposure, the more concentrated will be the silver atoms. Therefore, the exposed areas of the film are dark with silver, whereas the unexposed areas are clear. Thus, the film has the reverse appearance of the optical image. The film is a *negative.*

Figure 8-33 shows in a schematic way the formation of the latent image in the film and the appearance of the film after developing and fixing.

In order to produce a positive print of the scene on the neg-

Figure 8-33 *Schematic representation of the formation of an image in a film.* (a) *The number of silver bromide grains that are rendered developable is proportional to the amount of light to which they are exposed.* (b) *The developed film is dark in the region of strong illumination and clear in the region of no illumination.*

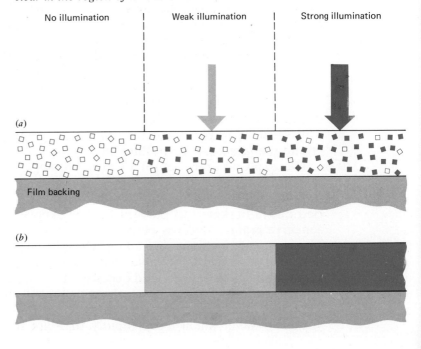

ative, the negative is placed in contact with a piece of printing paper and exposed to light. The latent image in the paper is now the reverse of the negative image: The dark areas of the negative prevent exposure of the paper and the clear areas of the negative allow maximum exposure of the paper. The paper is now developed and fixed to produce the final print. Some films are prepared and processed in special ways so that a positive print is obtained directly from the film. These special films are used in pictures-in-a-minute cameras.

Photographic films are manufactured in a variety of *speeds,* that is, with different sensitivities to light. In *fast* films, the latent image is produced with only a small amount of light, whereas *slow* films require much more light to produce a comparable image. The speed of a film is determined by the size of the silver bromide grains. Large grains are more sensitive to light than small grains. But this increased sensitivity of fast films is accompanied by a disadvantage. The developed large grains in such films degrade the sharpness of the image. We say that fast films are *grainy.* For maximum detail, a fine-grained (that is, a *slow*) film must be used.

Film speeds are rated by a numerical index. In the United States, the so-called ASA scale is used. Slow films have speeds of about ASA 20 and very fast films have speeds up to ASA 1000. With special processing, these films can be pushed to ASA 2000 or even higher. Fast color films are more limited and generally have ratings of no more than about ASA 200, although some new films have ratings of ASA 500.

The ASA number of a film is directly proportional to its sensitivity. If a certain scene requires $f2$ at 1/50 s for ASA 100 film, the same scene can be photographed at $f2$, 1/100 s with ASA 200 film. That is, if the film speed is increased by a factor of 2, the shutter speed (or aperture) must be adjusted to admit only half the amount of light.

Color Photography

Color photography is more complex than taking black-and-white pictures. We know that light waves with different wavelengths are seen by the eye as different colors of light. But the eye does not have a special sensor for each wavelength or color of light. Instead, there are only three types of color receptors in the eye. One is sensitive to a band of wavelengths in the red part of the spectrum, one is sensitive in the green part of the spectrum, and one is sensitive in the blue part of the spectrum. We sense all colors through the stimulation of these receptors in varying degrees by the incident light. For example, the sensation of the color *yellow* is produced by the stimulation of the red and green receptors in approximately equal amounts. If the red receptor were more strongly stimulated, we would sense the color *orange*.

We call the colors red, green, and blue the *primary* colors because we can produce all of the visual colors by various mixtures of these three colors. If you project onto a screen overlapping beams of red, green, and blue light, you will see the various colors indicated in Fig. 8-34. Notice that the combination of all three primary colors produces *white* light. Different relative intensities of the primary colors will produce other shades and hues.

We take advantage of the nature of color vision in the preparation of color film. Instead of having many layers of film, each sensitive to a different color or wavelength of light, we can use only three layers, each sensitive to a different color. Each layer contains a dye of one of the primary colors together with light-sensitive silver bromide grains. The developing procedure is different from that used for black-and-white films. When the processing is complete, the silver and

Figure 8-34 *A variety of colors is produced by the additive mixing the primary colors, red, green, and blue.*

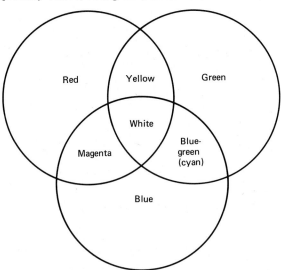

silver bromide are washed away and only the dye remains. By projecting white light through the transparent film, the image can be seen in color on a screen.

In addition to color transparency film, color negative film is also available. This film forms the image in *complementary* colors and is used to make color prints.

Questions and Exercises

1. Refer to Fig. 8-2. Make a similar diagram and draw several rays from A to B reflecting from the mirror at different points. Measure the length of each ray and show that the angle of incidence is equal to the angle of reflection only for the ray of minimum length.

2. A squad of soldiers are marching shoulder-to-shoulder across dry ground in a certain direction. They encounter a muddy area, the boundary of which makes an angle with their direction of march. The soldiers march more slowly in the mud than on dry ground. Sketch what happens to their direction of march. To what light phenomenon is this analogous? Carry the analogy further and make the muddy area triangular in shape. What effect can you now duplicate?

3. What is the speed of light in glass that has $n = 1.5$? [Ans. 200 000 000 m/s.]

4. Suppose that you are under water in a swimming pool and look at a person standing at the edge of the pool. Make a sketch to show the path of a light ray from the person's head to your eye.

5. How is it possible to start a fire with a magnifying glass?

6. A diverging lens has a focal length of 1 unit. If an object is located at a distance of 3 units from this lens, make a sketch to show the location of the image. Is this image real or virtual, erect or inverted, magnified or reduced?

7. Use a simple ray diagram to determine the position of the image in each of the following cases for a *convex* lens with focal points located at equal distances f from the lens. (a) The object is at a distance $2f$ from the lens. (b) The object is at a distance $\frac{1}{2}f$ from the lens. In (a), how large is the image compared to the object?

8. Repeat Exercise 7 for the case of a *concave* lens.

9. Can a diverging lens be used as a magnifying glass? Explain.

10. With a particular film, a certain scene requires an exposure of 1/1000 s at $f2$. What aperture would you use if the shutter speed were changed to 1/125 s? [Ans. $f5.6$.]

11. How much more light is admitted during an exposure at $f2$ than at $f22$ for the same shutter speed? [Ans. about 125 times.]

12. Extension bellows are sometimes used to produce large close-up images of small objects. The lens is removed from the camera and the extension bellows are inserted between the lens and the camera body. To see how this system works, refer to Fig. 8-19b. Move the object closer to the lens. (Make the distance one-third that in the diagram.) Sketch the ray diagram and show that the lens-to-film distance must be increased by an amount greater than that normally allowed for focusing. The bellows make possible this increased lens-to-film distance. By how much has the image size increased? [Ans. 9 times.]

13. Suppose that you connect a tone generator to a pair of loudspeakers that are separated by a certain distance along the wall of a room. As you walk around the room listening to the tone, what changes in sound intensity will you hear? If you now change the tone generator to a different frequency, will the effect of changing your position be exactly the same? Explain.

14. If you can photograph a certain dimly lit scene with an $f2$ lens using a film with a speed of ASA 200, why would you consider changing to an $f1.4$ lens and a film with a speed of ASA 100?

Additional Details for Further Study

The Lens Formula

The relationship connecting the focal length of a lens and the object and image distances is easy to obtain from the ray diagram shown in Fig. 8-35. The symbols have the following meanings:

x_o = distance from lens to object
x_i = distance from lens to image
f = focal length of lens
h_o = height of object
h_i = height of image

The Lens Formula   253

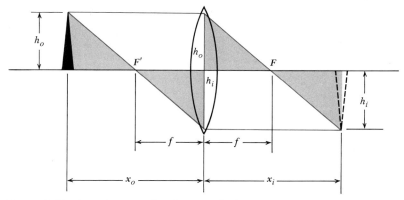

Figure 8-35 *Ray diagram for deriving the lens equation.*

Look at the two shaded triangles to the left of the lens. These triangles are *similar,* that is, they have equal angles. Therefore, the lengths of the sides of the triangles are proportional, and we can write

$$\frac{h_o}{h_i} = \frac{x_o - f}{f}$$

Similarly, from the two shaded triangles to the right of the lens, we can write

$$\frac{h_o}{h_i} = \frac{f}{x_i - f}$$

These two expressions for h_o/h_i can be equated and simplified:

$$\frac{x_o - f}{f} = \frac{f}{x_i - f}$$
$$(x_o - f)(x_i - f) = f^2$$
$$x_o x_i - f x_i - f x_o + f^2 = f^2$$

Cancelling the f^2 terms and rearranging, we have

$$f x_i + f x_o = x_o x_i$$

Finally, dividing this equation by $f x_o x_i$, we obtain

$$\frac{1}{x_o} + \frac{1}{x_i} = \frac{1}{f}$$

This expression is called the *simple lens formula.* If we know the focal length of a lens, then we can calculate the position of the image for any location of the object.

Light and Optical Instruments

Suppose that we have a lens with a focal length of 8 cm. If we place an object at a distance of 24 cm from the lens, where will the image be located? Rewriting the lens formula,

$$\frac{1}{x_i} = \frac{1}{f} - \frac{1}{x_o} = \frac{1}{8} - \frac{1}{24}$$

$$= \frac{3}{24} - \frac{1}{24} = \frac{2}{24} = \frac{1}{12}$$

Therefore,

$$x_i = 12 \text{ cm}$$

The *size* of the image formed by a lens is also easy to obtain by appealing to the geometry of the ray diagram. Look at the shaded triangles in Fig. 8-36. Again, these are similar triangles, and we can write

$$\frac{h_i}{h_o} = \frac{x_i}{x_o} = \text{magnification}$$

Thus, the ratio of the image height to the object height is equal to the ratio of the image distance to the object distance. The ratio h_i/h_o is just the magnification of the lens.

For the lens and object discussed above, the magnification is given by

$$\text{magnification} = \frac{x_i}{x_o} = \frac{12}{8} = 1.5$$

That is, the height of the image is 1.5 times the height of the object.

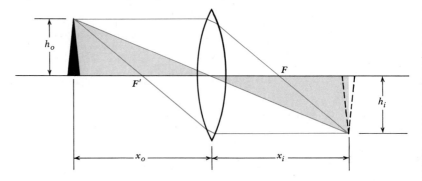

Figure 8-36 *Ray diagram for obtaining the magnification of a lens.*

Catadioptric (Mixed Reflection-Refraction) Systems

The limitation imposed on the image quality of a lens due to chromatic aberration was known to Isaac Newton in the seventeenth century. Newton also realized that there are no chromatic problems associated with the reflection of light from mirrors. Understanding these facts, Newton proceeded to invent a new type of telescope based on *reflection* instead of *refraction*. Reflecting telescopes (called *reflectors*) are now used almost exclusively for astronomical observations because they are free from chromatic aberration and because refractors of large size are subject to cracking and other damage.

Figure 8-37 shows how parallel rays of light can be brought to a focus by reflection from a curved surface. If the surface is spherical in shape, the rays will be brought to an approximate but not a perfect focus. In order for the focus to be exact, the shape of the mirror surface must be that of a *parabola*. (If you throw a ball across the room, the path of the ball is parabolic. A shallow parabolic arc is very close to a circular arc. For this reason, a spherical mirror of large radius will produce a focus that is satisfactory for many purposes.)

A Newtonian-type reflecting telescope is illustrated in Fig. 8-38. Notice that the converging rays are intercepted by a flat mirror before they come to a focus. This mirror diverts the rays to the side of the telescope for photographing or (with the addition of an eyepiece) for visual observations.

The grinding and polishing of parabolic surfaces is tedious and expensive. But a less expensive spherical mirror will not produce a sufficiently accurate focus for high-precision work. It is possible, however, to produce a high-quality image by

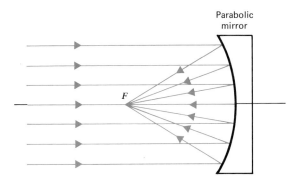

Figure 8-37 *Parallel rays of light are brought to a perfect focus by a parabolic mirror. A spherical mirror will produce an approximate focus.*

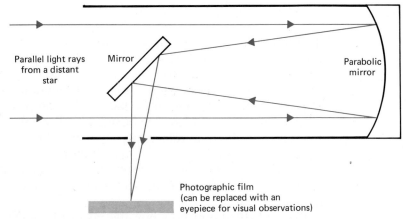

Figure 8-38 *Schematic of a Newtonian reflecting telescope. The deflecting mirror in the middle of the telescope blocks out some of the incident rays but this loss is not serious.*

using a spherical mirror in conjunction with a special lens that corrects for the defocusing property of the mirror. Such systems are called *catadioptric systems* (which means that they combine both refractive and reflective elements).

The first successful catadioptric system to be constructed used an aspheric (that is, not spherical) correcting lens in front of the spherical mirror. Although the aspheric lens had to be prepared very carefully, this disadvantage was more than offset by the relative ease of constructing the spherical mirror. (Refer to Exercise 6 at the end of this section.) This development led to the famous Schmidt telescopes, which are now widely used in astronomical research. The great advantages of these telescopes are their wide angles of view and their high speeds (about $f1$).

More recently, catadioptric systems have been designed that consist entirely of spherical surfaces. One such design is shown in Fig. 8-39. Here, there are two correcting lenses—the primary correcting lens and the field-flattening lens—both of which have only spherical surfaces. Notice also that the deflecting mirror is curved (spherical) and directs the converging rays through a hole in the main mirror.

Catadioptric systems (with aspheric or all-spherical correcting lenses) are now coming into wider usage. Some long-focal-length telephoto lenses of different catadioptric designs are now available for use with ordinary cameras.

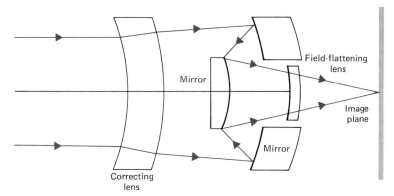

Figure 8-39 *Schematic of an all-spherical catadioptric system for long focal-length photography.*

Additional Exercises

1. An object is located at a distance of 30 cm to the left of a converging lens that has a focal length of 10 cm. Where is the image located? [Ans. 15 cm to the right of the lens.]

2. An object is located at a distance of 5 cm to the left of a converging lens that has a focal length of 10 cm. Where is the image located? (Your result will now carry a *negative* sign. What does this mean? You can easily see the answer if you make a sketch of the situation.) [Ans. 10 cm to the left of the lens.]

3. What is the size of the image compared to that of the object in each of the examples above? [Ans. (1) $\frac{1}{2}$, (2) 2.]

4. When parallel rays are incident on a converging lens from the left, the rays are brought to a focus at a point to the *right* of the lens (see Fig. 8-9). When parallel rays are incident on a diverging lens from the left, however, the rays appear to diverge from a point to the *left* of the lens (see Fig. 8-13). For this reason, we say that the focal length of a diverging lens is *negative*. That is, we can use the simple lens formula for the case of a diverging lens by inserting a negative number for the focal length. Then, you will always obtain a negative value for the image distance. (Can you see why?) What does a negative value for x_i mean? Suppose that you have a diverging lens with $f = -8$ cm. Where will the image be located if $x_o = 24$ cm? [Ans. 6 cm from the lens on the same side as the object.]

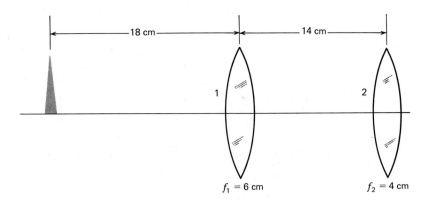

5. Find the position of the image for the two-lens system shown in the diagram. First, locate the image formed by lens 1 and then use this image as the object for lens 2. What is the magnification of the system? [Ans. 20 cm to the right of lens 2; 2.]

6. Why is it so much easier to grind and polish lenses and mirrors to *spherical* shapes than any other shapes?

Time Travel - Is it really possible?

9 RELATIVITY THEORY

In science fiction stories we read about all sorts of preposterous gadgets and impossible situations. For example, a space voyager leaves the Earth on a trip to a distant star. He makes the round trip in his high-speed rocketship and returns to find the Earth advanced by several centuries from the time of his departure. The space traveler has aged by only a few years, but the Earth has aged by hundreds of years. The space traveler has become a *time* traveler, leaping over centuries of Earth history into a new (and better?) era.

Fantastic, you say? Impossible? Strange and exciting, but not impossible. The laws of physics actually allow a voyage into the future such as that taken by our space traveler. But there *are* limitations. It is possible to go *forward* in time by taking a high-speed trip during which the traveler ages more slowly than his Earth-bound friends. However, it is never possible to go *backward* in time to visit the "good old days." Second, although we know that space travelers age more slowly during high-speed trips than do their stay-at-home friends, we have no idea at present how we might take advantage of this marvelous opportunity. We do not know how to produce the huge amount of energy that would be required to boost a spaceship to the necessary high speed.

In this chapter we will examine the physical principles that apply when motion takes place with very high speeds, speeds that are close to the speed of light. This is the realm of Einstein's *theory of relativity*. We will look first at the ideas that are basic to the theory and then we will discuss some of the interesting phenomena that are predicted by the theory (and are verified by experiment).

Relative Motion—Who Is Really Moving?

Suppose that you are traveling along a street in a bus and are seated next to a friend. You look out the window and see another friend who is standing on the street corner. What do your two friends conclude about your motion? Your friend in the bus sees you sitting motionless by his side; he says you are at rest. But your friend on the street corner sees you moving along the street with the bus; he says you are in motion. Who is right? Are you moving or are you at rest?

There is no "right" or "wrong" in this situation. Each of your friends is correctly describing the situation, even though one says that you are at rest and the other says you are in motion. The point is that whenever we wish to describe the motion of an object, we must first establish some coordinate system or *frame of reference* with respect to which the motion can be observed. Your friend on the street corner used the street itself as his reference frame and observed the bus and its occupants moving with respect to this frame. Your friend in the bus decided to use the bus as his reference frame and correctly concluded that you were at rest in this frame.

Usually, we imagine that every observer establishes a reference frame in which he is at rest for the purpose of describing motion or other events that take place. Thus, your friend on the street corner finds that you are moving in his reference frame. But it is also true that this friend, as well as the whole street, is moving in *your* frame which is attached to the bus! This appears to be rather artificial; surely, it is clear to all observers that the street is fixed and that the bus is *really* moving. But is this the case? An observer in space who is viewing the situation would conclude that the Earth, together with the street and your friend, is in motion. Or, have you ever been seated in a bus when another bus in the next lane started to move? For a few moments, were you uncertain *which* bus was moving? Whether an object is moving or is at rest depends on your point of view.

The Rise and Fall of the Ether Concept

Does there exist a reference frame somewhere in which we will always obtain a "correct" description of rest and motion? That is, can we find a reference frame that is at *absolute rest?* If so, we could refer all of our observations to this primary frame and thereby decide unambiguously who is moving and who is at rest.

Until about 70 years ago, most scientists believed that *space* itself constituted the primary reference frame at absolute rest. It was difficult to accept the complete void of space as defining a reference frame, so an hypothesis was formulated by which all space was imagined to be filled with a mysterious substance called the *ether*. Thus, the ether became the primary frame of reference.

The ether was the medium that allowed electrical and gravitational forces to act between bodies not in contact. And the ether was the medium that supported the wave motion necessary for the propagation of light. It was all very neat. Even though one could not *see* the ether—indeed, there was no way at all to detect the ether directly—it surely did exist because some sort of medium was necessary to support the many effects associated with electricity, light, and gravity. At least, this was the view of most nineteenth-century scientists.

The ether concept satisfied the intuitive need for something to fill the vacuum of space. For some reason, scientists felt easier about light propagating through the poorly understood ether than through completely empty space. But the ether turned out to be more trouble than it was worth. When the ether concept was used to interpret the results of various experiments involving the propagation of light, serious contradictions were discovered. Many efforts were made to modify the ether idea to bring it into agreement with experiment, but none was successful.

When Albert Einstein (1879–1955) became interested in the problems associated with light propagation in the ether, he decided that it would be unproductive to attempt once more to patch up the old theory. Instead, Einstein solved the problem by taking a new and radical view in which he abolished the concept of the ether!

During the latter part of the nineteenth century, when the development of the ether idea reached its peak, a light wave was considered to be a vibration of the ether. That is, a light wave was believed to propagate through the ether in much the same way that a sound wave propagates through air. The ether was regarded as a *real* substance, and a light wave was considered only as a distortion of the ether.

Einstein took a different view. To him, the physical reality was the electromagnetic field, not the ether. The ether proponents believed that electrical and magnetic effects (including light) were the result of distortions in the ether. Einstein realized that a light wave is a propagating disturbance of the electromagnetic field and that the field can exist in empty space. Thus, light can propagate through space without the necessity of invoking an ether or some other type of medium.

When Einstein discarded the ether concept, he simultaneously abolished the ideas of "absolute rest" and "absolute motion." If there is no ether to serve as the primary reference frame with respect to which all motion is measured, then no state of rest or motion can be considered to be "absolute." The only way in which the motion of an object can be described is with respect to some other object. That is, all motion is *relative*. It is for this reason that Einstein's theory is called the *theory of relativity*.

Whenever we describe a physical event or situation, we do so in terms of a reference frame that we imagine is attached to some observer or some object. (The ether is no longer available!) It does not matter how we construct this reference frame, as long as it is not undergoing acceleration. The laws of motion are the same in all reference frames that move uniformly (that is, without acceleration) with respect to one another.

This idea is easy to understand. If you play catch with a friend in the aisle of an airliner that is flying with constant velocity, you will see the ball move in exactly the same way as you do when you play catch on the ground. As long as you perform your experiments in a nonaccelerating reference frame, you will obtain exactly the same results as will every other nonaccelerating observer.

The fact that the laws of motion are the same in all nonaccelerating reference frames is not a new idea. Newton understood that if his famous equation of motion is valid in one reference frame, then it is also valid in any other frame that moves uniformly with respect to the first frame. Einstein extended this idea to include electromagnetic effects as well, and he made the following postulate basic to his theory:

> I. All physical laws are the same in all nonaccelerating frames of reference.

The Velocity of Light

Suppose that you are walking along the edge of a pond. A friend is standing beside the pond, and he touches the water with the end of a long pole to produce a series of circular water waves (Fig. 9-1). How would you and your friend, observers A and B in the diagram, describe these water waves? First, each observer recognizes that the waves propagate across the water in the pond. The water is a well-defined

264   Relativity Theory

Figure 9-1 *Two observers view water waves propagating across the surface of a pond. Observer* B *is in motion with respect to* A *and the pond.*

medium, and each observer agrees on the location of the pond. The waves have no existence independent of the water, so it is natural to describe the wave motion with respect to the water. Therefore, the simplest way for each observer to describe the situation would be to state that the waves propagate outward with a certain speed *with respect to the pond*.

In the case of water waves in a pond, both observers agree on what they see. Each observer sees a wave pattern radiating from a point that is at rest with respect to observer A. What will happen if we change the waves from water waves to light waves?

Let us suppose now that observers A and B are located in spaceships, so that there is no other matter around to confuse the issue. Observer B is moving with a velocity **v** with respect to observer A, as shown in Fig. 9-2. (We could also say, with equal correctness, that A is moving with a velocity $-\mathbf{v}$ with respect to B.) At a certain instant, there is a flash and a burst of light waves move outward through space, as shown in the diagram. What do the two observers see? We must be very careful in formulating our answer to this question because it involves an essential point in relativity theory.

In the case illustrated in Fig. 9-2 we have no pond to serve as a common basis for a comparison of the two observations. Therefore, each observer uses his own reference frame to describe the situation. The object that produced the flash might have been at rest with respect to one of the observers, or it could have been in motion with respect to one or both of the observers. But no matter how the source was moving, this motion can have no influence on the way the light waves radiate outward as seen by the observers. To see why this is true,

The Velocity of Light 265

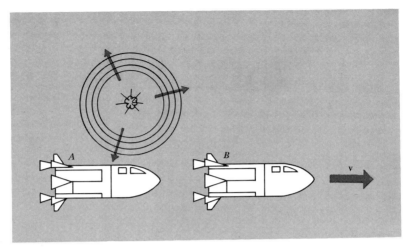

Figure 9-2 *Two observers who are moving with respect to one another view a burst of light waves radiating outward from a flash.*

let us return to the case of the water waves in the pond (Fig. 9-1). Suppose now that the pole is carried by some third party who moves the pole over the surface of the water and at some instant lets the pole strike the surface to initiate the wave motion. The pole then continues to move along its path. Does the fact that the source of the waves (namely, the pole) was in motion affect the wave pattern seen by the observers? Certainly not. The waves do not move more rapidly in the direction in which the pole was moving, nor do they move less rapidly in the opposite direction. The wave motion has a definite speed *with respect to the water* regardless of the motion of the source.

We must emphasize again that the observers in Fig. 9-2 have no "fixed" reference frame to which they can refer. Each observer expresses his observations in terms of his own reference frame. Observer *A* says that he sees light waves radiating outward from a particular point in his frame. Observer *B* makes the same statement! *B* says that he sees light waves radiating outward from a particular point in *B*'s own reference frame.

Can both observers be right? First of all, we have argued that the motion of the source with respect to the observers can make no difference in the observed pattern of waves. As far as an individual observer is concerned, the effect is the same as if the source were at rest in *his* reference frame. Thus, each observer sees a growing sphere of light centered in his own reference frame! This curious feature of the two observations is a result of the fact that there is no identifiable material medium that both observers can agree is the medium through which the light pulse propagates and of the fact that any motion of the source cannot influence the shape of the expanding sphere of light.

What are the consequences of the way in which two observers in relative motion view the propagation of light waves from a single source? Each observer sees the light waves moving away from the flash point with the same speed in all directions. (If the speed were different in different directions, the wave front would not be spherical in shape.) Upon measuring the propagation speed, each observer finds a value of 300 000 km/s (3×10^8 m/s). (We usually denote the speed of light by the symbol c.) That is, two observers who are in motion with respect to one another see the light waves from a single source and each measures the same value for the speed of propagation of the light waves!

The conclusion we have reached is really quite remarkable. The behavior of light waves is not at all the same as the behavior of material particles. For example, consider the two situations shown in Fig. 9-3. In each case we have an observer A who is moving toward observer B with a speed V. In Fig. 9-3a, A throws a ball toward B; in A's reference frame, the speed of the ball is v. What does B measure for the speed of the ball? Our experience tells us that B finds the ball's speed to be $V + v$ with respect to his own reference frame. (Actually, relativity theory shows that the speed is not exactly equal to $V + v$. For all speeds that do not approach the speed of light, however, $V + v$ is sufficiently close to the true value for all of our purposes.)

In Fig. 9-3b, observer A projects a burst of light toward B. What does B measure for the speed of this light pulse in his reference frame? It seems natural to use the same type of reasoning as we used for the case of the thrown ball. Then, B should measure a speed of $V + c$ for the light pulse. But we have just worked through a lengthy argument to the effect that two observers will measure the same value for the speed of light from a source even though the observers are in a state of relative motion. Thus, we must conclude that observer A in Fig. 9-3b measures a speed c for the light pulse in his refer-

The Velocity of Light 267

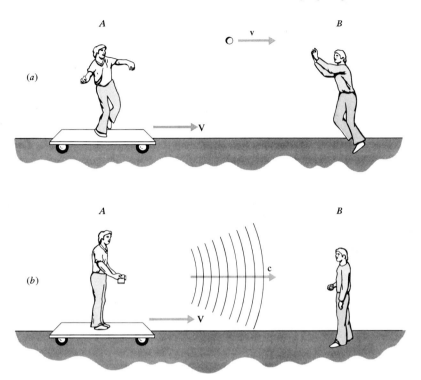

Figure 9-3 (a) *Observer B finds the speed of the ball to be* V + v. (b) *Observer B finds the speed of the light waves to be* c.

ence frame and that observer *B* measures the *same* value in *his* reference frame.

Notice that there is no essential difference between the situations illustrated in Fig. 9-2 and Fig. 9-3b. Even though the light source is clearly moving toward observer *B* in Fig. 9-3b, this does not influence the speed of the light as measured by either observer. (But there *is* a difference in some feature of the light as measured by the two observers. Can you see what this is? Refer to Exercise 3 at the end of the chapter.) We must not be misled by the existence of a "fixed" reference frame (the ground) in Fig. 9-3b. The propagation of light has nothing to do with the presence or absence of such a reference frame.

The argument we have just presented constitutes the second idea that is basic to the development of relativity theory. We can summarize all of our remarks by the following statement:

> II. The speed of light (in vacuum) is the same for all observers in nonaccelerating reference frames regardless of any relative motion between observer and light source.

The Theory of Relativity

Using as his basis only the two postulates that appear above in the boxes, Albert Einstein was able to construct his *theory of relativity*. This theory contained revolutionary new ideas about space and time, ideas that have completely restructured scientific thought. Although we now recognize Einstein's theory as a crucial landmark in the development of physics, the scientists of the day were not quick to adopt these new ideas. Einstein's first paper on relativity was published in 1905. In 1921 he received the ultimate recognition of his work—the Nobel Prize in physics. But his relativity theory was judged not to have sufficient merit for this award: Einstein's Nobel Prize was received for his work on the theory of the photoelectric effect! (This work was basic to the later development of quantum theory.)

In the prerelativity view of Nature, the concepts of space and time were considered to be completely separate. A series of events could be located in definite positions in space and could be specified in terms of precise times of occurrence. Moreover, every observer, regardless of his motion, would agree on the distances and the time intervals between the various events. Einstein showed that this view of Nature is not correct. The ideas of space and time are linked together. Distances and time intervals have different meanings for observers who are in motion with respect to one another. It is not possible to describe an event in terms of a location (that is, by using three space coordinates, x, y, and z) plus a separate specification of the time t. In the Einsteinian world of moving observers (the *real* world), the space coordinates and time values are dependent on one another. The four coordinates—x, y, z, and t—all have equal importance and effect. For this reason, relativity theory is said to deal with a *four*-dimensional world.

It is easy to understand how an observer makes measurements on an object at rest in his reference frame or measures a series of events that take place at a fixed location. The observer equips himself with a standard meter stick, a calibrated clock, and a standard kilogram mass. To measure the length of an object, for example, he merely places the meter stick alongside and reads the value of the length.

But now suppose that two observers measure the length of an object. The object is at rest in the frame of observer A, but it is moving in the frame of observer B. When A and B compare their length measurements, will they agree? They will not! The process of making a length measurement requires the observer to have information concerning the position of both ends of the object at the same instant. For an object at rest, the observer has no difficulty. But when the object is in motion, space and time values become interrelated and the measurement actually produces a different result. That is, two observers who are in relative motion will not agree on the length of an object. Nor will they agree on the time interval that takes place between a pair of events.

If these effects actually occur, how can we ever agree on distances and times? The occupants of moving automobiles and buses will have different views of space and time compared to persons on the street. True enough. But at the speeds with which we usually move, relativistic effects are exceedingly small. Thus, our everyday world is a nonrelativistic world. Only when speeds approaching the speed of light are involved do relativistic effects become apparent. Relativity plays an important role in describing the behavior of high-speed subatomic particles, and the theory is crucial in our attempts to understand the large-scale structure of the Universe. And, as we will see, relativity theory explains how we are able to obtain huge amounts of energy from the processes of nuclear fission and fusion.

Time Dilation

As we have pointed out, observers who are in relative motion do not agree on measurements of distances or time intervals. To see why this is so, let us examine the way in which two observers view the rate of ticking of a clock. We imagine that we construct a clock as shown in Fig. 9-4a. The main element of

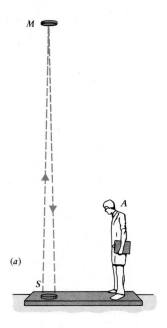
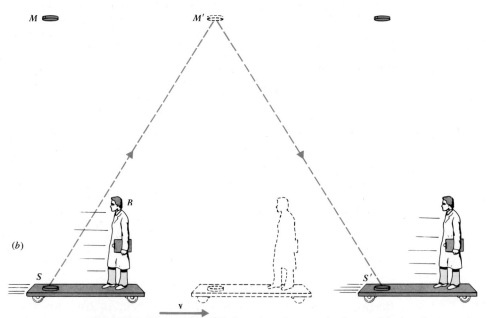

Figure 9-4 *The paths of light pulses in two clocks as seen by observer* A. *Observer* B *and his clock are moving at a constant velocity* v *with respect to* A. *Observer* A *concludes that* B's *clock runs more slowly than his own clock.*

the clock is a light pulse that travels at the known speed, $c = 3 \times 10^8$ m/s, between a light source S and a mirror M. When the light pulse reaches the mirror, it is reflected back to S where it is detected. One round trip, $S \to M \to S$, constitutes a "tick" of the clock.

Suppose that we have two identical light-pulse clocks: One is placed in the laboratory of observer A and one is placed in the laboratory of observer B. Now, let B's laboratory be in motion with a velocity **v** relative to A, as shown in Fig. 9-4. At the instant when B's clock passes A's clock, each observer begins a "tick" of his clock by flashing the source.

Let us imagine that we station ourselves with A and observe the "ticks" of the two clocks. We see, of course, the light pulse rise straight up to the mirror M and return to the source in A's own clock. But while this is happening, observer B and his clock are moving toward the right. We see the light pulse in B's clock rise to the mirror along a diagonal path; the return of the pulse is likewise along a diagonal path. That is, the path $S \to M' \to S'$ taken by the pulse in B's clock (Fig. 9-4b) is much longer than the corresponding path taken by the pulse in A's own clock. Now, we know that the speed of the light pulses in the two clocks, as we see them, are exactly the same. (The fact that B's light source is moving with respect to our reference frame does not change the speed of the light from this source.) We must conclude, then, that the time required for the completion of a "tick" in B's clock is longer than that for a "tick" of A's clock. That is, we (and observer A) find that B's clock runs more slowly than A's own clock.

Next, let us move from A's laboratory into B's laboratory and repeat the experiment. Now we see B's clock at rest and A's clock moving past with a velocity $-\mathbf{v}$. When the light sources are flashed, we see the pulse in B's clock move straight up and down, whereas the pulse in A's clock moves along diagonal paths. This is the same view we had before, only reversed. We now find that A's clock runs more slowly than B's clock!

In the first experiment, we concluded that B's clock runs more slowly than A's clock; in the second experiment, the conclusion was exactly the reverse. How can both results be correct? Notice that in both cases we found the *moving* clock to run more slowly than the *stationary* clock. In fact, we can state our conclusion in a precise way as follows:

Any nonaccelerating observer will always find that a clock that moves at constant velocity in his reference frame will run more slowly than an identical clock that is at rest in the observer's frame.

This behavior of moving clocks is called the *time dilation* effect. (To *dilate* means to enlarge or expand; the time interval is expanded when the clock runs more slowly.) Although we imagined the experiments performed with light-pulse clocks, any other kind of clock would have led to the same conclusion (but the analysis would have been more difficult).

The time dilation effect can be stated by means of an equation. The expression is easy to derive (see the section on *Additional Details* at the end of this chapter), but we give here only the result:

$$t = \frac{t_0}{\sqrt{1 - (v^2/c^2)}}$$

where the symbols have the following meaning:

t = duration of a "tick" of the moving clock according to the stationary observer

t_0 = duration of a "tick" of the stationary clock according to the stationary observer

v = speed of the moving clock with respect to the stationary clock and observer

c = speed of light

The speed v of the moving clock is always less than c. But when v is a large fraction of c, the denominator of the time dilation expression becomes very small. This means that t becomes large compared to t_0. For example, when v is 87 percent of the speed of light, t is twice as large as t_0. That is, the duration of a "tick" of the moving clock is twice as long as that of a "tick" of the observer's own clock. Thus, according to the stationary observer, the moving clock runs at a rate that is half the rate of the stationary clock.

Length Contraction

We have remarked that space and time are interconnected. We have just seen that observers who are in relative motion

Length Contraction   273

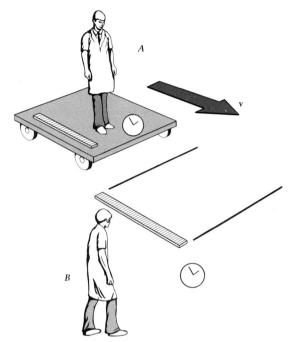

Figure 9-5 *Two observers use clocks and meter stick to measure their relative velocity.*

measure different time intervals. Next, we would like to know how their length measurements compare.

Consider the situation illustrated in Fig. 9-5. Observer *B* has a laboratory equipped with a meter stick and clock. Observer *A* has similar equipment, but his laboratory is located on a platform that moves with a certain constant velocity relative to *B*'s laboratory. The two observers decide that each will make a measurement of the relative velocity between the two reference frames. Observer *B* makes his measurement by recording the time required for the front of *A*'s platform to travel the distance between two lines on the laboratory floor; *B* spaces these lines exactly 1 m apart. Observer *A* makes his measurement by recording the time required for one of *B*'s lines to move a distance corresponding to the length of the meter stick that lies on *A*'s platform. (Remember, observer *A* sees *B* and *B*'s laboratory moving past *A*'s own laboratory.) Thus, each observer will measure, using his own clock, the time required for a movement of 1 m according to his own meter stick. The relative velocity *v* will then be equal to 1 m divided by the measured time interval.

When the experiment is completed, the observers compare their results and are satisfied to find that each has measured the same number of meters per second for the relative velocity. This really is not an unexpected result. The observers are measuring the *relative* velocity of their two reference frames. The only possibility is that A moves in B's frame with the same velocity that B moves in A's frame. How could it be otherwise?! But observer B, recalling the previous discussion about time dilation, says, "Something must be wrong. The clock used by A runs more slowly than my own clock. If A used a slow clock and yet still obtained the same value for the relative velocity that I measured, then A's distance measurement must have been in error. Observer A must have used a faulty meter stick! In fact, A's meter stick must be in error (too short) by exactly the right amount to cancel the time dilation effect and give the correct value for the relative velocity. Curious!"

While observer B is expressing his suspicion about A's meter stick, observer A is voicing his concern over B's length measurement. He arrives at the same conclusion: "B must have used a meter stick that is too short; otherwise, he would not have found the same velocity using that slow clock."

The two observers decide to repeat the experiment using a different relative velocity. Again, each observer finds that his measurements yield the same value for the relative velocity as do the measurements of the other observer. Each observer concludes that the moving meter stick becomes shortened by exactly the right amount to compensate for the slowly running clock. That is, each observer finds that as the duration of a "tick" of the moving clock becomes *larger* by the factor $1/\sqrt{1 - (v^2/c^2)}$, the length of the moving meter stick must become *shorter* by the factor $\sqrt{1 - (v^2/c^2)}$. We can summarize this statement by the expression

$$l = l_0 \sqrt{1 - \frac{v^2}{c^2}}$$

where

l = length of a moving object according to a stationary observer

l_0 = length of the same object when it is at rest with respect to the observer

The shortening of an object when there is relative motion between the object and the observer is called the *length contraction* effect.

It should be pointed out that the length contraction effect takes place only *along* the direction of relative motion; lengths that are *perpendicular* to the direction of motion are not affected. In Fig. 9-5, both meter sticks are oriented along the direction of **v**, so that each observer sees a contraction of the meter stick that is in the other laboratory. Notice that if B measures the *length* of A's platform, he will find a value *smaller* than A measures. But if B measures the *width* (or the *height*) of A's platform, he will find the *same* value that A measures. (In Fig. 9-4, we were careful to arrange that the light pulse travels from source to mirror in a direction perpendicular to the relative velocity **v**. Can you see why this was done?)

We have found that the time dilation and length contraction effects are not independent phenomena. Instead, they are intimately coupled together. In relativity theory, space and time cannot be separated—both have equal importance in the four-dimensional world in which we live.

Are Relativistic Effects REAL?

Is it really true that moving clocks run more slowly than stationary clocks and that moving meter sticks contract? If observers A and B are in relative motion, B will find that A's meter stick is shorter than his own and that A's clock runs more slowly than his own. But if we join A and observe his meter stick and clock, we see nothing unusual. The reason is that the length contraction effect and the time dilation effect are the result of using length and time standards in one's own reference frame to measure lengths and times in a moving frame. The observer who is at rest relative to a meter stick or clock finds no contraction or dilation effect at all. So, are these effects *real?*

It seems only reasonable to define *reality* in terms of *measurements* that we can make. Then, relativistic effects are certainly real. Every experiment that has ever been carried out involving the measurement of length and time (and, as we will see, also *mass*) in rapidly moving systems has verified the predictions of relativity theory. How can one refute the reality of direct measurements? Relativistic effects are not optical illusions nor are they some kind of scientific hocus pocus. Relativistic effects are *real*.

A specific example of an easily observed relativistic effect will serve to reinforce this idea. *Muons* are subatomic particles that are created in certain types of high-energy processes involving elementary particles. Experiments with special detectors reveal that muons are continually shooting through the air all around us. These muons are created high in the atmosphere as the result of processes initiated by energetic *cosmic rays* striking the atoms of the air. Muons are radioactive particles: On average, a muon will live about two millionths of a second (2×10^{-6} s) before it decays into an electron (plus two other particles called *neutrinos*). If a muon travels with a speed close to the speed of light, it will move only about 600 m during a time of 2×10^{-6} s. How, then, is it possible that muons created high in the atmosphere can reach ground level? Relativity theory provides the answer. The property of a muon that dictates its decay after a certain time is a kind of clock. If we travel along with a muon (that is, if we have a muon at rest in our reference frame), then the muon "clock" will tick away about 2×10^{-6} s between the instant of creation and the instant of decay. In fact, this is the way in which the muon lifetime has been measured. On the other hand, if we observe a rapidly moving muon from our Earth-based reference frame, we find that the muon clock runs more slowly than our own clock. Therefore, according to our clock, the muon will live much longer than 2×10^{-6} s. If the muon travels with a speed close to the speed of light, it will move a distance much greater than 600 m before decay. Thus, the fact that muons created high in the atmosphere can survive to reach the surface of the Earth is a verification of the relativistic time dilation effect.

Time Travel

If a muon can travel a distance greater than its lifetime would seem to allow, can people do the same? Would it be possible for space travelers to make a journey to a distant star within an average lifetime? Relativity theory tells us that this is so!

Consider a star that lies at a distance from the Earth such that light, traveling at 3×10^8 m/s, requires 1000 years to reach the Earth from the star. This distance, which is 10 billion billion meters, we call 1000 *light years* or 1000 L.Y. (By astronomical standards, 1000 L.Y. is not a large distance: The diameter of the local collection of stars, the Milky Way Galaxy, is about 100 000 L.Y.) Suppose that we wish to visit this star. Suppose also that, by some means yet unknown, we construct a spaceship capable of reaching a speed close to the speed of light. Then, the spaceship would require a time only a little longer than that required by light to make the trip. Thus, to an Earth observer, the spaceship would reach the star in just over 1000 years. It seems as though only the descendants of the original space travelers, many generations along, will arrive at the distant star.

But this conclusion neglects the effects of time dilation. To an Earth observer, the clocks in the spaceship run slowly—not only the mechanical or electronic clocks, but the biological clocks as well. Every activity in the spaceship takes place slowly according to the Earth observer. The space travelers move in a kind of super slow motion; they work and they age at this same slow pace. The passage of 1000 years according to an Earth observer (and his descendants) is equivalent to the passage of only a short time to the occupants of the spaceship.

How does the trip appear to the space travelers? To an Earth observer, the spaceship is moving with a very high speed toward the distant star. To the occupants of the spaceship, however, the star appears to be rushing toward them at the same high speed. According to the space travelers, the distance to the star, which an Earth observer measures to be 1000 L.Y., is contracted to very much less than 1000 L.Y. Therefore, because of the very high speed with which the star is approaching, the star will arrive at the spaceship (or the spaceship will arrive at the star!) in a time much less than 1000 years.

Table 9-1 lists the times of travel for a trip to a star 1000 L.Y. away for various speeds of the spaceship. Two columns of times are given—one according to the space travelers' clock and one according to the Earth observer's clock. Notice that doubling the speed from one-half the speed of light to just under the speed of light causes the travel time to decrease by one-half, according to the Earth observer. But, because of the

Table 9-1 Travel Times to a Star 1000 Light Years Away

Speed of the Spaceship as a Percentage of the Speed of Light	Travel Time (years)	
	According to Space Travelers' Clock	*According to Earth Observer's Clock*
10%	9950	10 000
50%	1732	2000
90%	484	1111
99%	142	1010
99.9%	45	1001
99.99%	14	1000.1
99.999%	4.5	1000.01
99.9999%	1.4	1000.001

time dilation effect, this same speed change causes a much greater reduction in travel time according to the space travelers.

It must be emphasized that the space travelers sense nothing unusual about the passage of time during their high-speed voyage. To them, events seem to happen at the normal rate. According to their clocks, they age in a perfectly natural way. It is only to an observer moving with a high speed relative to the spaceship (for example, an Earth observer) that the activity in the spaceship seems slowed down. Thus, the space traveler ages much more slowly than his Earth-bound friends.

If the spaceship makes the trip to a star 1000 L.Y. away at a speed of 99.999% of the speed of light, the one-way passage requires 4.5 years (see Table 9-1). If the return trip is made at the same speed, the space travelers will arrive back at Earth 9 years after their departure, according to their own clocks. But they will find an Earth that has aged by 2000 years! Relativity theory shows us the way to skip over hundreds or thousands of years, to become time travelers voyaging into the future. Such excursions, however, are one-way trips. There is no way to reverse the procedure and to travel backward in time. Time moves only forward, and relativistic time travel permits us to go only into the future, not into the past.

Perhaps by now you have uncovered a flaw in the argument about time travel. We have presented the situation as follows. The Earth observer sees the spaceship moving with a high speed and sees the clocks aboard running slowly. Then, when the spaceship returns to Earth, the net passage of time according to the space travelers is less than the time recorded on Earth. Can this really be correct? Previously, we have emphasized that motion is relative. Therefore, we should be able to view the space voyage as one in which the Earth moves away from the spaceship, travels to a distance of 1000 L.Y., and then returns to the spaceship. According to the occupants of the spaceship, the Earth clocks should run more slowly than their own clocks. Therefore, when the Earth returns from its trip into space, the Earth inhabitants should be younger than the occupants of the spaceship! This is exactly the opposite of the conclusion we reached before.

Sometimes this situation is described in terms of the aging of twins. One twin stays on Earth and the other twin is the space traveler. According to the first view, the space-traveling twin should remain young while the Earth twin grows old. According to the second view, the Earth twin should be the one to remain young. For this reason, the problem is often called the *twin paradox*.

Actually, there is no paradox at all. The reasoning that leads to the paradox assumes that there is no difference between viewing the motion of the spaceship from the Earth and viewing the motion of the Earth from the spaceship. But there *is* a difference in this case. When the spaceship leaves the Earth, it *accelerates* to a high speed and then drifts toward the star maintaining this constant speed. At the star, the spaceship must *accelerate* again to turn around and head back toward Earth. Finally, upon reaching the Earth, the spaceship must *accelerate* once more to land on Earth. (Remember, *acceleration* means any change in velocity, whether speeding up, slowing down, or changing direction.) Thus, the spaceship undergoes several accelerations during the voyage, whereas the Earth experiences no acceleration. Therefore, when the space travelers view the Earth moving away and then returning, it is the spaceship and not the Earth that is accelerating. The space travelers are *not* nonaccelerating observers and they cannot apply the simple relativistic rules we have been using. Only the Earth observer is a nonaccelerating observer, and only his conclusion about the aging of the space travelers is correct. By using a more complicated analysis that applies for the case of accelerating observers, the same conclusion is reached. Thus, there is no paradox and the space-traveling twin does in fact age more slowly than his stay-at-home brother. Time travel *is* possible!

Do not plan your trip into the future just yet, however. Even though relativity theory shows us how time travel is possible *in principle,* we now have no idea how we might accomplish a high-speed space voyage. In order for the time dilation effect to be useful for long-distance space trips, the speed of the spaceship must be very close to the speed of light. Table 9-1 shows that a visit to a star 1000 L.Y. away becomes feasible only if a speed of at least 99.99% of the speed of light is reached. At the present time we have absolutely no idea how to propel a spaceship to such a high speed. The best that we have been able to do so far is to accelerate *atomic particles* to these speeds. We do not know how to project a grain of sand, much less a spaceship, to a speed of 99.99% of the speed of light! (For further comments on this problem, see Chapter 3.)

The Increase of Mass with Speed

Measurements of both length and time in moving systems are subject to relativistic effects. What about the third fundamental quantity—mass? In Chapter 5 we learned that a charged particle moving in a magnetic field is subject to a force. If we know the charge and the velocity of the particle and the strength of the magnetic field, we can calculate the force that acts on the particle. Then, if we know the mass of the particle, we can compute the path that the particle will follow in the magnetic field.

Suppose that we design an experiment to check the results of our calculation of the path followed by a charged particle in a magnetic field. Let us select an *electron* for our particle because we know both the charge and the mass of an electron and because we know how to propel electrons at high speeds. We accelerate the electron to a speed of, say, 1000 m/s and project it into a magnetic field whose strength is known. We find that the electron's path is as close to the predicted path as we can measure. If we increase the speed to $10\,000\,(10^4)$ m/s or to $100\,000\,(10^5)$ m/s, we also find agreement with the prediction. Next, we check the prediction at the much higher speed of 10^8 m/s. (Remember, the speed of light is 3×10^8 m/s.) This time, the electron does not follow the predicted path. The high-speed electron is more difficult to bend than we expect—its reaction to the field is more sluggish than at the lower speeds. The electron behaves as if its mass is greater than the value we measure when the electron is at rest or when it moves slowly. If we increase the speed still further, we find an even greater mass. If the electron's speed approaches the speed of light, its mass becomes enormous.

The bending of a charged particle in a magnetic field is one of the most precise methods available for determining the mass of a particle. Measurements of this type have shown that the mass of a particle increases with speed according to the expression

$$m = \frac{m_0}{\sqrt{1 - (v^2/c^2)}}$$

where

m = mass of a moving object according to a stationary observer

m_0 = mass of the same object when it is at rest with respect to the observer

(To see how this expression for the mass arises, see the *Additional Details* at the end of this chapter.)

When an observer measures the mass of an object at rest in his reference frame, he obtains the value m_0, which we call the *rest mass* of the object. If this object is in motion with respect to the observer, he will always measure the mass, by whatever method, to be greater than the rest mass. When we state *the* mass of an object, with no reference to speed or relativistic effects, we always mean the *rest mass* of the object.

Table 9-2 Increase of Mass with Speed

Speed of an Object as a Percentage of the Speed of Light	Mass of the Object in Units of its Rest Mass
0	1
10%	1.005
50%	1.16
90%	2.29
99%	7.09
99.9%	22.4
99.99%	70.7
99.999%	224
99.9999%	707

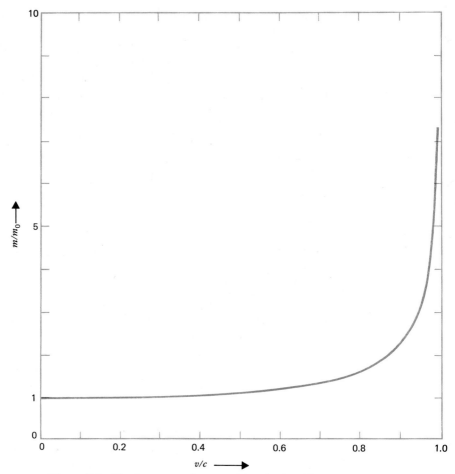

Figure 9-6 *The increase of mass with velocity. The mass* m *is given in units of the rest mass* m_0 *and the velocity is given as a fraction of the velocity of light.* m/m_0 *becomes indefinitely large as* v *approaches* c (*but* v *can never equal* c).

Table 9-2 shows how the mass of an object increases with speed; this same information is presented graphically in Fig. 9-6. For speeds less than a few percent of the speed of light, the mass increase is negligible. At everyday speeds, the increase cannot even be detected with our most sensitive instruments. For example, a 100 000-kg jetliner moving with a speed of 300 m/s (670 mi/h) will have a mass that is 0.00005 grams greater than its mass when at rest.

The most rapidly moving particles that have ever been produced in the laboratory are the electrons that emerge from the machine at the Stanford Linear Accelerator Laboratory (SLAC). These electrons have a speed that is 99.999999995% of the velocity of light! If these electrons raced a light pulse to the Moon, they would lose by a distance of only 2 cm (less than an inch). At such a tremendous speed, the mass of an electron, as measured by a laboratory observer is 140 000 times its rest mass. The electrons from the SLAC machine have a mass greater than that of an iron atom.

The Ultimate Speed

We have written down the relativistic expressions for the three fundamental physical quantities—length, time, and mass. In each of these expressions we find the factor $\sqrt{1 - (v^2/c^2)}$. When the speed v of relative motion becomes almost as large as c, this factor becomes very small. If v were equal to the speed of light c, then the factor would be zero. This means that lengths would contract to zero, that clocks would cease to run, and that masses would become infinitely large. These effects are completely without physical meaning. We must conclude, therefore, that the speed v can never equal (or exceed) the speed of light. That is, no observer can ever measure the speed of a material particle to be equal to or greater than the speed of light. The speed of light is the universal speed limit, the ultimate speed.

Sometimes we hear the following kind of question: "I understand that, according to relativity theory, no object can ever exceed the speed of light. But what would happen *if* we could make an object travel with such a speed?" The only way we can answer a question in physics is by the conclusive test of reality, namely, by *measurements* that we can make. If there is no conceivable way of making a measurement to answer a question about the behavior of Nature, then the question makes no sense. According to our present knowledge, there is no method by which a material object can ever be made to equal or exceed the speed of light. Therefore, until some measurement provides us with the essential clue and shows that the present theory is incorrect, we can give no answer to "what if" questions about speeds greater than the speed of light.

Mass and Energy

If we supply energy to an object by doing work on it, then we can increase the speed of the object. The more energy we supply, the greater will be the speed. We have just learned that the speed of an object can never be made equal to or greater than the speed of light. On the other hand, there is no conceivable physical limit to the amount of *energy* we can give to an object. If the speed of an object can never be pushed to the speed of light, what is the result of imparting more and more energy to the object?

The motional or kinetic energy of an object depends on its *mass* and on its *speed* (see Chapter 4). Because there is a limit to the speed of an object, increasing the *energy* of the object to a very high value must result in increasing the *mass* of the object. This connection between mass and energy is one of the fundamental and important ideas in relativity theory. Einstein succeeded in showing that the relationship uniting the concepts of mass and energy involves the speed of light. We can state Einstein's result in the following way:

(change in energy of an object)
$$= \text{(change in mass of the object)} \times c^2$$

This is the famous Einstein mass-energy relation, which we usually hear expressed as the familiar equation,

$$\boxed{E = mc^2}$$

The Einstein equation does *not* say that mass and energy are the same. The equation does say that mass and energy are related: Mass has equivalent energy and energy has equivalent mass. Because mass changes and energy changes are connected, we must now restate the conservation laws for mass and energy. Prior to the development of relativity theory, it was believed that there were two distinct conservation principles: *Mass is conserved* and *energy is conserved*. But Einstein showed that mass can be converted into an equivalent amount of energy and that energy can be converted into an equivalent amount of mass. Therefore, we really have a single conservation law, which states that *mass-energy is conserved*.

If we put a value for the mass into the equation $E = mc^2$, we find that even a small amount of mass has a tremendous equivalent energy. For example, if 1 kg of matter could be converted entirely into energy, this process would release about 30 billion kilowatt hours of energy! This is approximately one-half of the total amount of energy consumed in the United States per day.

Unfortunately, the entire mass of an object cannot be converted into useful energy. We cannot destroy the fundamental units of matter, protons and neutrons. However, we *can* rearrange protons and neutrons into forms that have different masses. Then, the mass change can be released as energy. This is the type of change that takes place in the process of *fission* (which we will discuss in Chapter 13). When a heavy nucleus undergoes fission, it splits into two lighter nuclei whose combined mass is less than the initial mass. The mass difference is converted into the kinetic energy of the light fragments. In this process the mass does change, even though the number of protons and neutrons remains the same. The fission of a heavy nucleus such as uranium-235 results in a mass change of about 0.1 percent. Thus, about 2 kg of uranium per day must undergo fission to run a modern power generating plant with an output of 1000 MW.

Nuclear rearrangements can involve mass changes that release substantial amounts of energy. We also obtain energy from chemical processes such as combustion or burning. Are there mass changes involved in these energy releases? Indeed there are. But chemical reactions are the result of *atomic* changes, not *nuclear* changes. In nuclear fission, the nuclear protons and neutrons are rearranged; in a combustion reaction, the atomic electrons are rearranged. The forces that bind protons and neutrons in nuclei are much stronger than the forces that bind electrons in atoms. Consequently, chemical processes involve mass changes and energy releases that are much smaller than those in nuclear processes. When a kilogram of uranium undergoes fission, the mass change is about 10 million times greater than when a kilogram of gasoline is burned. Thus, the extraction of energy from heavy nuclei by fission is vastly more efficient than the extraction of energy from chemical fuels by combustion.

Not only can mass changes produce energy, but energy changes can result in the creation of mass. The first direct proof that mass can be created from energy was obtained in 1932 by the American physicist, Carl D. Anderson. While studying the interaction of cosmic rays with matter, Anderson found that some of the energetic gamma rays that entered his apparatus produced pairs of electrons. A gamma ray would simply disappear and in its place would be two electrons. The energy of the gamma ray (that is, electromagnetic energy) was converted directly into the mass of the electrons. Anderson showed that one of the electrons of each pair produced in this way is an ordinary electron whereas the other electron has all of the properties of an ordinary electron except that it carries a positive charge. This new particle that Anderson discovered is called a *positron*. The positron is the *antiparticle* of the electron. (Notice that the creation of an electron-positron pair does not violate the principle of conservation of electrical charge. The particles created carry equal charges of opposite sign so that the total charge before creation of the pair and after creation is zero.)

Positrons are not a part of the matter of our world. These unusual particles have no place in the atoms of ordinary matter. After a positron is created, it will wander through matter for a brief time and then it will encounter an electron. The positron and electron will then combine and disappear, leaving behind only electromagnetic radiation. That is, a particle (the electron) and its antiparticle (the positron) annihilate one another, producing gamma radiation. Energetic gamma rays create electron-positron pairs and these pairs eventually annihilate, producing gamma rays again. Energy is converted into mass, and then mass is converted into energy—a striking confirmation of Einstein's theory.

The General Theory of Relativity

Relativity theory, as we have discussed it so far, deals exclusively with nonaccelerated systems and observers. This form of the theory, which applies only to cases involving uniform velocity, is called the *special* theory of relativity. In 1915 Einstein enlarged the scope of his theory to include accelerating reference frames. This broader theory is called the *general* theory of relativity.

We know that mass is associated with gravitational forces and that such forces produce accelerations. For this reason, gravitational effects fall outside the domain of the special theory. Einstein developed the more complete theory to deal specifically with mass and gravitational forces. Indeed, the general theory is a theory of gravitation.

In Einstein's four-dimensional, space-time world, mass is

given a geometrical interpretation. The presence of mass results in the curvature or warping of the space-time fabric; and, by the same token, a curvature of space-time reveals itself as a mass. In the general theory, not only are space and time woven together into a single concept, but mass is associated with distortions of the space-time surface.

We can see the effect of space-time curvature in one of the famous tests of the general theory. We usually think of light as traveling in straight lines. In fact, we can *define* a straight line to be the path followed by a light ray. According to the general theory, a light ray is affected by the presence of mass because mass represents a distortion of space-time. The effect of mass on a light ray is small, however, so that the distortion becomes observable only for cases in which a very large mass, such as the Sun, is involved.

Why should the Sun affect a light ray that passes near it? Einstein's mass-energy relation, $E = mc^2$, tells us that energy has equivalent mass. Therefore, a certain amount of mass is associated with a bundle of light energy. (A bundle of light energy is called a *photon*.) Thus, the photon "mass" is attracted by the Sun's gravity and follows a path that is curved. In the general theory, however, a light ray is always considered to move along a straight line. What we observe as a curved path for the light ray is actually a curvature of space. In the distorted region of space near the Sun, the light ray still follows a straight-line path as it is defined in the general theory.

Einstein suggested a way to detect the distortion effect. Look at Fig. 9-7. The star S is located on the opposite side of the Sun from the observer. The light from the star passes near the surface of the Sun and is bent slightly toward the Sun. When this light reaches the observer, if he considers the light to have traveled along a straight line (in an undistorted space), he places the star's position at S' instead of the real position S. Because the star's light is completely obscured by the intense light from the Sun, this observation must be made when the Sun's light is blotted out by the Moon. Therefore, the procedure is to photograph the stars near the Sun at the time of a solar eclipse. Another photograph is taken six months later when the Sun is in another part of the sky and the stars are visible at night. The two photographs are compared, and those stars whose light passed near the Sun during the eclipse are seen to occupy slightly different positions in the nighttime photograph.

When this experiment was first carried out during the solar

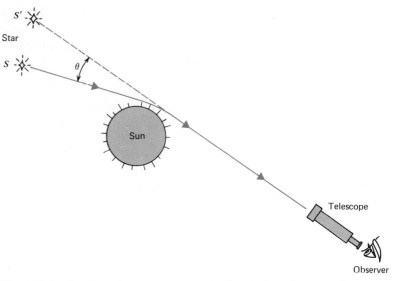

Figure 9-7 *Starlight that passes near the Sun is bent by a small amount due to the Sun's mass. The observer sees the starlight as if it had originated at S′ instead of S. Thus, the apparent position of the star is shifted by the amount θ due to the presence of the Sun.*

eclipse of 1919, Einstein's prediction was found to be correct. The observations and measurements are very difficult, however, and the best that has been done is to confirm the theoretical prediction to an accuracy of about 10 percent.

Because gravity is such a weak force—it is by far the weakest of the fundamental forces of Nature—all of the new effects predicted by the general theory have small magnitudes and are difficult to detect. For this reason, exacting tests of the theory have not yet been made. If all general relativistic effects are so small and so difficult to observe, why is the theory of any importance? First of all, the general theory of relativity has shown us how the ideas of space and time are united into a single fundamental concept, thereby providing us with a new and profoundly different way to view the behavior of Nature. Second, even though general relativistic effects are small for bodies such as the Earth and the Sun, the Universe as a whole contains an enormous amount of mass. General relativity therefore plays a crucial role in our attempts to understand how the Universe behaves, how it evolved, and what its destiny will be.

Questions and Exercises

1. How would the relativistic effects described in this chapter be changed if the speed of light were infinitely large?

2. Suppose that the speed of light were only 60 mi/h. Describe some of the effects this would have on everyday observations. For example, if a bicyclist were to pass you on the street, how would the rider and bicycle appear to you? Would you appear to the rider to be your usual self?

3. Consider two identical light sources. One source is at rest with respect to an observer and the other source is moving toward the observer at a certain speed. Does the observer see any difference in the light from the sources? Is the speed of the light from each source the same? What about the frequency of the light?

4. Suppose that a spaceship is near the Earth and is traveling with a speed of $0.5c$ toward a certain star that is 1 L.Y. from the Earth. The spaceship beams a radio signal toward the star. How long will be required, according to an observer on the Earth, for the signal to travel to the star?

5. According to what you have learned about relativity in this chapter, is there any truth in the following famous limerick?

    There was a young lady named Bright,
    Who could travel much faster than light.
      She departed one day,
        In a relative way,
    And returned on the previous night.

6. A square billboard is located parallel to a highway. If you travel along the highway with a speed that is an appreciable fraction of the speed of light, describe the appearance of the billboard as you pass it.

7. In this chapter we discussed the flight of a muon from a point high in the atmosphere to the surface of the Earth from the viewpoint of an observer on the Earth. How would the same situation appear to an observer moving along with the muon?

8. Will an observer measure the *density* of an object that moves relative to him to be the same as or different from that of an identical object at rest in his reference frame? Explain.

290 *Relativity Theory*

9. If you apply a constant force to an object, why will the acceleration of the object decrease as time goes on?

10. An observer on the Earth views a spaceship traveling with a speed equal to 0.8 of the speed of light, as indicated in the diagram. In the spaceship are two 6-ft astronauts, one "standing up" and one "lying down." What does the observer measure for the heights of the two astronauts? [Ans. 6 ft for A; 3.6 ft for B.]

11. Why is more energy required to boost the speed of an electron from $0.8c$ to $0.9c$ than is necessary to increase the speed from 0 to $0.1c$?

Surface of the Earth

Additional Details for Further Study

Time Dilation

The expression for the time dilation factor can be derived easily by referring to the geometry of the situation illustrated in Fig. 9-4. We repeat this diagram in Fig. 9-8, where we add the important distances.

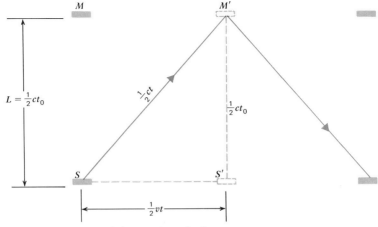

Figure 9-8 *Geometry of the moving clock.*

The distance between the source S and the mirror M is equal to L. This distance is one-half of the distance traveled by the light pulse during one "tick." Therefore, according to the stationary observer,

$L = \tfrac{1}{2} c t_0$

When the light pulse reaches the mirror M' in the moving clock, this represents one-half of a tick in this clock, that is, the time $\tfrac{1}{2} t$. Again, according to the stationary observer, the light pulse traveled from S to M', a distance of $\tfrac{1}{2} ct$, during this half-tick, and the source moved from S to S', a distance of $\tfrac{1}{2} vt$, during this same interval.

The triangle $SM'S'$ is a right triangle for which we know the length of each side. Using the Pythagorean theorem, we can write

$(\tfrac{1}{2} ct)^2 = (\tfrac{1}{2} vt)^2 + (\tfrac{1}{2} c t_0)^2$

Simplifying and rearranging,

$c^2 t^2 - v^2 t^2 = c^2 t_0^2$

$t^2 (c^2 - v^2) = c^2 t_0^2$

Dividing by c^2,

$t^2 \left(1 - \dfrac{v^2}{c^2} \right) = t_0^2$

so that

$$t^2 = \frac{t_0^2}{1 - (v^2/c^2)}$$

Finally, taking the square root, we have

$$t = \frac{t_0}{\sqrt{1 - (v^2/c^2)}}$$

The Increase of Mass with Speed

The first postulate of relativity theory is that all physical laws must be the same in all nonaccelerating reference frames. One of these laws is the conservation of momentum, and we can make use of this principle to obtain the expression for the increase of mass with speed.

Consider two observers, A and B, who are stationed in nonaccelerating laboratories that are equipped with clocks and meter sticks, as in Fig. 9-9a. We will assume that A is the

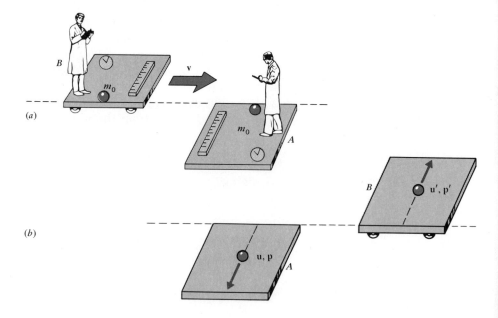

Figure 9-9 *Testing the conservation of momentum in moving reference frames.*

"stationary" observer and that B moves relative to A with a velocity **v**. In each laboratory there is an object with a rest mass m_0 situated at the edge of the platform. (The masses of the two objects were compared carefully while at rest with respect to one another.) When the two laboratories pass one another, a grazing collision between the objects takes place. That is, each object receives a small velocity at right angles to the velocity vector **v**. (In such a collision neither object will acquire any appreciable velocity in its laboratory along the direction of relative motion.) After the collision, the situation is that shown in Fig. 9-9b. The object in A has a velocity **u** and a momentum **p**; the object in B has a velocity **u'** and a momentum **p'**.

Each observer uses his clock and meter stick to determine the velocity of his object. Using this result together with the known mass of the object, each observer computes the momentum of his object. The observers compare their results and are happy to find that each has obtained the same value for the momentum of his object. They congratulate themselves on having verified the conservation of momentum in the collison.

Upon further reflection, each observer concludes that something must be wrong with the experiment. Observer A says to B: "I know that your length measurement was correct because your meter stick was oriented perpendicular to our relative velocity and therefore did not suffer any length contraction. But your clock ran slow. It actually required a *longer* time interval for your object to travel 1 m than you recorded. Therefore, the value you obtained for the velocity of your object was too *large*—in fact, too large by the time dilation factor $1/\sqrt{1 - (v^2/c^2)}$. Because you found the same value for the momentum of your object, you must have used a value for the mass that is too *small*—in fact, too small by the factor $\sqrt{1 - (v^2/c^2)}$." Observer A therefore concludes that the mass of B's object must be larger than m_0; according to A, the mass of the object in B's laboratory is

$$m = \frac{m_0}{\sqrt{1 - (v^2/c^2)}}$$

Observer B makes exactly the same argument; he finds that the mass of A's object is given by the same expression. The mass of a moving object is always greater than the mass of an identical object that is at rest with respect to the observer.

Additional Exercises

1. The average lifetime of a muon is 2×10^{-6} s. If a muon travels with a speed of $0.9c$ with respect to an observer, what lifetime for the muon will the observer measure? [Ans. 4.6×10^{-6} s.]
2. If the relative velocity is $v = 0.8c$ in Fig. 9-9, what is the time dilation factor and what is the mass increase factor? [Ans. 0.6; $1/0.6 = 1.67$.]
3. A meter stick is in motion in a direction along its length with a speed sufficient to contract its length to $\frac{1}{3}$ m. What is the mass of the meter stick in terms of its rest mass? [Ans. $3m_0$.]
4. What is the velocity of a particle whose mass is equal to twice its rest mass? [Ans. $v = (\sqrt{3}/2)c = 0.866c = 2.6 \times 10^8$ m/s.]

How do lasers work?

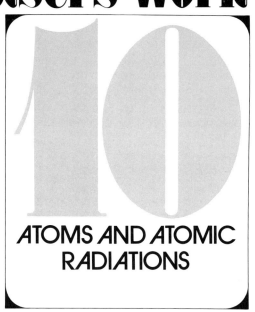

10 ATOMS AND ATOMIC RADIATIONS

Within the last few years, we have seen the discovery of the *laser* principle and the development of laser devices into a wide range of tools that are being used in many different areas. Lasers have proved useful in a variety of scientific studies that require light with high intensity and pure color; they have found industrial applications in microwelding, boring, and etching, as well as in precision surveying; they are used in medicine for delicate surgical procedures; they have been applied to environmental investigations of air pollutants; and they have been used by artists in developing "light pictures." In the future we will certainly see lasers used in the communications and entertainment industries. Indeed, lasers are already so reliable and so inexpensive that laser beams are sometimes used by lecturers as pointers!

In this chapter we will discuss the physics of lasers and how they are used in several situations. To understand how lasers work, we will need to explore the interiors of atoms. We will see how the structures of atoms are related to the radiations that they emit. We will see why atoms radiate light only with certain definite wavelengths. In Chapter 8 we mentioned some of the *wave* properties of light. Now, we will look more closely at the nature of light and we will find that light also has an interesting *particle*-like character.

The Photoelectric Effect and Quanta

During the early years of the nineteenth century, various experiments were performed that demonstrated that light exhibits *interference* effects. These observations were consistent with the idea that light is a wave phenomenon. For almost a hundred years this idea was unchallenged. Near the end of the nineteenth century, however, other experiments indicated that light is something more than an ordinary wave.

The developments that led to a new way of thinking about the nature of light began with a simple observation. A clean surface of a piece of zinc metal was prepared by polishing away the oxide layer that normally forms on zinc and similar metals. The piece of zinc was supported on an insulating material such as glass or wax. Next, the surface was exposed to a source of ultraviolet (UV) light. (Sunlight contains suffi-

The Photoelectric Effect and Quanta 297

cient UV light for this experiment.) Measurements with an instrument sensitive to electrical charge showed that the zinc acquired a positive charge upon exposure to the UV light. This phenomenon is called the *photoelectric effect*.

Was there anything extraordinary in this observation? Not really. It was known that light is a wave and that waves carry energy. When the UV light was incident on the zinc, energy was transferred to the material and this energy caused negative charge to be ejected from the metal; the removal of negative charge left the zinc with a positive charge. We now know that the basic unit of negative charge is the electron and that all atoms contain electrons. Therefore, in the photoelectric effect, electrons are literally knocked off the zinc (Fig. 10-1). The ejected particles are called *photoelectrons*. Interpreted in these terms, the observation of the photoelectric effect in no way contradicted the idea that light is a wave phenomenon. However, further experiments were to lead to a different conclusion.

First, there was the matter of the *frequency* of the incident light. Although UV light could eject electrons from zinc, it was found that red or yellow or green light could not. The only difference between light of different colors is the frequency (or the wavelength) of the light waves. Red light has the lowest frequency, yellow is next, then green, and then blue. Ultraviolet light, which lies just outside the range of visible light, has an even higher frequency. (Above UV light, in terms

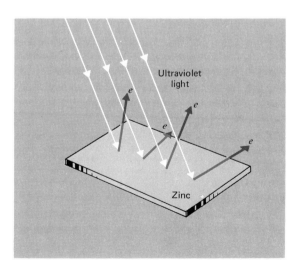

Figure 10-1 *The photoelectric effect. Ultraviolet light incident on a metal plate, such as zinc, ejects electrons and the plate acquires a positive charge.*

of frequency, we find X rays and gamma radiation.) Why is it that high-frequency light can eject photoelectrons from zinc whereas low-frequency light cannot?

Next, different materials were found to exhibit the photoelectric effect for different frequencies of light. For example, green light will produce photoelectrons when incident on sodium and potassium, and yellow light will eject photoelectrons from cesium. But UV or higher-frequency radiations are required for zinc and most other metals. Why do different elements respond to light in different ways?

As the experiments became more sophisticated, the energies of the photoelectrons were measured. Again, a surprising result was found. If an experiment were begun by irradiating a particular material with low-frequency light, no photoelectrons were observed, no matter how intense the light. As the frequency of the light was increased, there was a definite frequency at which photoelectrons were first ejected. This frequency is called the *threshold frequency,* and each element has its own particular value. As the light frequency is increased above the threshold frequency (for any material), the energy of the photoelectrons increases in direct proportion to the frequency. Moreover, the photoelectron energy depends *only* on the frequency of the incident light and *not* on its intensity.

The wave theory of light and the conventional way of thinking about the interaction of waves with matter was completely unable to explain the various experimental findings relating to the photoelectric effect. It was clear that some radically new interpretation was required.

The crucial idea was supplied by Albert Einstein, who, in 1905, developed a simple but complete and accurate theory of the photoelectric effect. (1905 was also the year that Einstein published the paper that developed the foundations of relativity theory!) Einstein realized that both the interference properties of light (the wave properties) and the properties revealed in the photoelectric experiments could be accounted for if all light exists as tiny oscillating bundles of electromagnetic energy (Fig. 10-2). These bundles of light are called *quanta* or *photons*. If a large number of photons act together, the individual character of the photons is lost and only the oscillatory property is evident. Thus, a collection of photons appears as a light *wave* and the frequency of the wave is the frequency of oscillation of the individual photons. The quantum hypothesis therefore does not affect the wave-theory interpre-

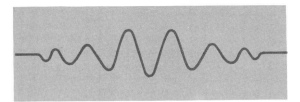

Figure 10-2 *Schematic representation of a photon. The electromagnetic energy of a photon is concentrated in a tiny bundle. The wavelength of light is so small that a photon contains a hundred thousand or so oscillations, instead of the half dozen shown here. It is for this reason that light photons retain their wave properties even though they can be localized in space.*

tation of the many interference experiments that have been performed.

Einstein's new quantum idea left intact the interpretation of a century of light-wave experiments while providing the basis for understanding the photoelectric effect. When UV light is incident on a piece of zinc, each photoelectron that is ejected is the direct result of the action of an individual photon. A photon does not interact with the piece of metal as a whole. Instead, a photon interacts with a single electron; the photon is absorbed and its energy is transferred to the electron, which can then escape from the material. In the photoelectric effect, light acts more like a *particle* than a *wave*. Indeed, this sometimes-wave, sometimes-particle aspect of radiation and matter is a basic feature of the modern *quantum theory* that evolved from the early work of Einstein and others.

How does the quantum idea explain the details of the photoelectric experiments? According to Einstein's theory, the energy of every photon is directly proportional to its frequency. That is,

(photon energy) = (constant) × (photon frequency)

In symbols, we can write

$E = hf$

where the constant of proportionality is given the symbol h and is called *Planck's constant* in honor of the German physicist Max Planck (1858–1947), who first suggested the discrete nature of radiation. (Planck's highly original idea was not fully developed—nor was it widely accepted by the scientific community—and it remained for Einstein to make this concept respectable by incorporating it into his successful photoelectric theory.)

When a photon, with a definite frequency and a definite energy, is incident on an atom, the photon is absorbed and its energy is transferred to one of the atomic electrons. The electrons in an atom are bound to that atom by electrical forces. To remove an electron from an atom requires the expenditure of a certain amount of energy, called the electron *binding energy*. Therefore, in the photoelectric effect, a portion of the photon energy is used to free an electron from the atom and the remainder appears as the kinetic energy of the ejected electron. In this process, the photon disappears. We can express these statements in terms of an energy equation:

(electron kinetic energy) = (photon energy)
− (electron binding energy)

or, using symbols,

K.E. = $hf - \phi$

where $E = hf$ is the photon energy and where the symbol ϕ (Greek letter *phi*) denotes the energy required to release the electron from the atom. As you can see, this equation is simply an expression of energy conservation in the photoelectric process.

Einstein's photoelectric equation shows immediately why there is a threshold frequency for the effect and why the electron kinetic energy depends on the frequency of the incident light. No photoelectrons can be ejected from a material until the photon energy equals or exceeds the electron binding energy. The threshold frequency f_0 is therefore determined by $hf_0 = \phi$. Exactly at the threshold frequency, the photoelectrons have zero kinetic energy. As the frequency of the incident light is increased above f_0, the kinetic energy of the photoelectrons increases in direct proportion.

Finally, the existence of different threshold frequencies for different elements is due to the fact that no two elements have exactly the same atomic electron structures. The electrons that are ejected in the photoelectric effect are bound with different energies in different elements. More energy is required to remove an electron from an atom of zinc, for example, than from an atom of cesium. Accordingly, higher-energy photons are required for the photoelectric effect to occur for zinc than are required for cesium. Low-frequency yellow photons will release photoelectrons from cesium, but high-frequency UV photons are necessary to release photoelectrons from zinc.

Einstein's quantum hypothesis and his simple but elegant theory of the photoelectric effect provided a complete and consistent explanation of the various results that had been obtained in a wide range of experiments with light. But it did more than that. The quantum idea—the idea that light and other electromagnetic radiations occur as discrete bundles of energy—profoundly affected scientific thinking and led eventually to a comprehensive theory of atomic structure and atomic radiations.

Rutherford and the Nuclear Atom

In the late 1890s, the French physicist Henri Becquerel (1852–1908) made one of the great accidental discoveries in the history of science. While performing some experiments in which photographic plates were used, Becquerel found that a rock containing a uranium mineral caused the plates to become fogged. He determined that light was not responsible for the exposure of the plates and concluded that some previously unknown radiation from the uranium mineral had caused the foggy images in the plates. Becquerel had discovered *radioactivity*. Many scientists quickly began investigations of uranium and other radioactive substances in order to determine the nature of the new radiations.

In Chapter 12 we will be concerned with more details of radioactivity. For the present, however, we need to know only that in one type of radioactive process, an atom spontaneously ejects a heavy, high-speed particle, called an *alpha particle* (α particle). We now know that an α particle is identical to an atom of the element helium from which two electrons have been removed. A few years after the discovery of α radioactivity, Ernest Rutherford (1871–1937), a New Zealander then working in England, decided that the tiny, high-speed α particles were ideal tools to study the interiors of atoms. Rutherford began a series of experiments in which he used α particles from radioactive substances to probe the atoms of various elements.

Rutherford used α particles to learn about atomic interiors in much the same way that we might use a long needle to examine the interior of a peach. By sticking the needle into

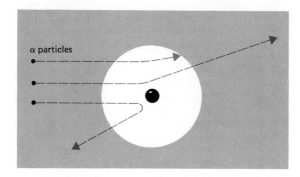

Figure 10-3 *By studying the way in which fast particles are deflected by atoms, Rutherford was able to conclude that most of the mass of an atom is concentrated in a tiny nuclear core.*

the peach at various positions and noting the resistance that it meets, we can deduce that the peach has a hard core of a certain size. We can discover the existence of the peach pit and can measure its size without ever actually seeing it. Rutherford studied the way in which rapidly moving α particles are deflected when they penetrate atoms (Fig. 10-3). He was able to conclude that most of the mass and all of the positive charge in an atom is concentrated in a tiny region in the center of the atom. Rutherford had discovered the atomic *nucleus*. Other experiments had shown that atoms have sizes of about 10^{-10} m (one ten-billionth of a meter); Rutherford's data showed that atomic nuclei are 10 000 times smaller and have sizes of about 10^{-14} m! (See Fig. 10-4.)

According to Rutherford's picture of the nuclear atom, essentially the entire mass of an atom is contained in the incredibly small nucleus, with the equally small atomic electrons

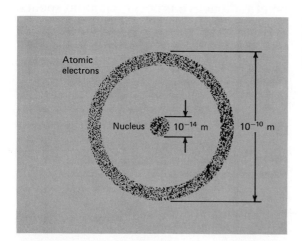

Figure 10-4 *Schematic comparison of atomic and nuclear sizes. Atoms are about 10 000 times larger than nuclei.*

distributed throughout the remainder of the atomic volume. Atoms are therefore mostly empty space!

Atomic Spectra

Rutherford's discovery, made in 1911, was quickly put to use by a young Dane named Niels Bohr (1885–1962), who developed the first model of atomic structure that was capable of explaining some of the features of the light emitted by atoms. Before we describe Bohr's atomic model, let us review some of the characteristics of light sources.

Everyone knows that different types of light sources produce light with different colors. Lighting fixtures that are used in advertising displays produce light by passing an electrical current through gas-filled tubes. These glowing gases emit red light, or yellow light, or blue light, or light with other colors. Lasers also produce light with a variety of colors; perhaps you have seen laser beams that are red or green. Why do these various light sources have distinctive colors? (Ordinary household lamps produce light by the heating of a filament. This is a different process, and the emitted light consists of radiations with all colors; we see this mixture as *white* light.)

One way to examine the light from a source is to *disperse* the light by passing it through a glass prism (see Fig. 8-26). Because the index of refraction of glass is different for light with different frequencies, the incident light is spread out into a spectrum of colors. If we study the light from various gaseous elements or from the vapors of solid elements, we find an interesting result. None of the spectra consists of light with a broad range of frequencies; we do not find each color blending smoothly into the next as we do for white light (Fig. 8-26). Instead, each spectrum consists of light with a few definite frequencies (or wavelengths). If we project the dispersed light onto a screen or if we use it to expose a photographic plate, we see that each spectrum contains a series of bright lines. Every bright line corresponds to light with a single well-defined wavelength, and each element has a spectrum with its own characteristic sequence of bright lines. The spectrum of light from neon gas, for example, consists of a few red lines and very little else. Sodium vapor produces a spectrum consisting primarily of two yellow lines. Some elements (mercury, for example) yield spectra that are incredibly rich in lines; the mercury spectrum contains thousands of bright

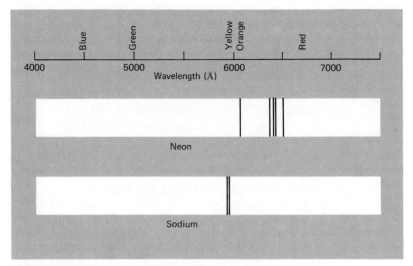

Figure 10-5 *Colors and wavelengths of the visible light spectrum. (1 angstrom = 1 Å = 10^{-10} m.) The two bright-line spectra show the principle lines from neon and sodium.*

lines. The brightest of the spectral lines of neon and sodium are shown in Fig. 10-5, along with an indication of the wavelengths and colors. Notice that in this diagram wavelengths are given in units of *angstroms*. One angstrom (1 Å) is equal to one ten-billionth of a meter (10^{-10} m). Thus, an angstrom is a length unit about equal to the size of an atom.

At the time of Rutherford's discovery of the nucleus, atomic spectra had been studied extensively for about 50 years. Detailed compilations had been made of the wavelengths of the bright lines in the spectra of many elements. But no one knew the significance of these mysterious lines.

The spectrum of the least massive element—hydrogen— was found to be particularly simple. The hydrogen spectrum, shown in Fig. 10-6, contains a series of lines that begins with a red line at a wavelength of 6562 Å and extends into the ultraviolet, terminating at 3646 Å. As the shorter wavelengths are approached, the spacing between the lines becomes smaller and smaller, until they seem to merge together at the short-wavelength end. Such an interesting spectrum must be due to some prominent and regular feature of the structure of the hydrogen atom. But what could this feature be? No one knew.

The first clue to understanding the hydrogen spectrum was

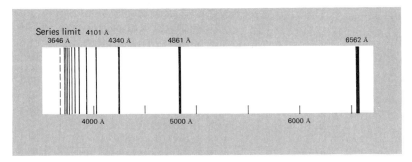

Figure 10-6 *The Balmer series in the spectrum of hydrogen.*

discovered in 1885 by Johann Balmer (1825–1898), an obscure Swiss music teacher. Balmer found that it was possible to write down a simple formula that expressed the wavelengths of the hydrogen spectral lines. If we rewrite Balmer's formula in terms of the frequency instead of the wavelength, we have

$$f = (\text{constant}) \times \left(\frac{1}{2^2} - \frac{1}{n^2}\right) \quad n = 3, 4, 5, \cdots$$

The substitution of the series of numbers 3, 4, 5, · · · for n in this expression permits the calculation of the frequencies of the various spectral lines. (Balmer found the value of the constant that gave the best agreement with the experimental values.) The line at 6562 Å (see Fig. 10-6) corresponds to $n = 3$; the line at 4861 Å corresponds to $n = 4$; the line at 4340 Å corresponds to $n = 5$; and so forth.

Whenever a large amount of experimental data can be accounted for by a simple formula such as Balmer's, it is quite likely that the formula has real validity. In such a case, it should be possible to derive the formula from a properly constructed theory. This, Niels Bohr set out to do, using Rutherford's nuclear model of the atom as his starting point.

Bohr's Theory of the Hydrogen Atom

After Rutherford had discovered that most of the mass and all of the positive charge of an atom resides in its nucleus, there was an obvious question to be answered: How do the electrons in an atom remain in a stable condition? Electrons were known to be negatively charged particles; as such, they must experience an attractive electrical force due to the positively charged nucleus. Why, then, do the electrons in an atom not fall into the nucleus? Bohr sought an answer to these questions by imagining an atom to be constructed in much the same way as the solar system. The Earth and the other planets are attracted toward the Sun by the gravitational force. But the planets do not fall into the Sun because their motions counterbalance the effect of the Sun's gravitational pull. The planets fall *around* the Sun, not *into* it (see Chapter 3).

The hydrogen atom, Bohr thought, should look very much like a one-planet solar system. The single atomic electron of hydrogen (the "planet") should move in a stable orbit around the nuclear "sun" (Fig. 10-7). In the solar system, the distance at which a planet revolves around the Sun is related to the planet's orbital speed. If the distance and speed are in the proper relationship, the planet can move in an orbit of any size, and for each orbit there is a particular value of the planet's energy.

The gravitational force between a planet and the Sun is very similar to the electrical force between an electron and a nucleus: Both forces depend inversely on the square of the separation distance, that is, on $1/r^2$. The dynamics of both systems should therefore be the same. But can an electron actually orbit around a nucleus at *any* distance? Bohr realized that this could not be the case. The reason is to be found in the occurrence of bright-line spectra. When a bright line appears in a

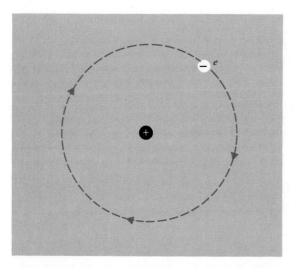

Figure 10-7 *Bohr's simple model of the hydrogen atom.*

spectrum, it means that the atoms in the sample are emitting light with a definite frequency; that is, each emitted photon has the same energy. Before an atom emits a quantum of light, it has a certain energy; after the emission, it has a smaller energy. Now, energy must be conserved during this process, so the energy decrease experienced by the atom must exactly equal the energy of the emitted photon. That is,

(photon energy) = (initial energy of atom)
$$- \text{(final energy of atom)}$$

or

$$E = hf = E_i - E_f$$

We say that the atom begins in a certain *energy state* characterized by E_i and makes a *transition* to a lower energy state characterized by E_f. This transition is accompanied by the emission of a photon with an energy $E = E_i - E_f$.

The energy of an atom depends on the size of the electron orbit: The larger the orbit radius, the greater will be the energy. Therefore, according to Bohr's model, the emission of a photon by an atom means that the atomic electron changes from some initial orbit to a final orbit that has a smaller radius (Fig. 10-8). This process takes place *spontaneously;* any atom that has been given an excess of energy will spontaneously return to its normal condition by emitting a photon.

The occurrence of bright lines in the spectrum of hydrogen must therefore mean that the atomic electron is restricted to move only in certain definite orbits. When a transition between two definite orbits takes place, a photon with a defi-

Figure 10-8 *When an atom makes a transition from one energy state to a lower energy state, the electron orbit changes and a photon is emitted.*

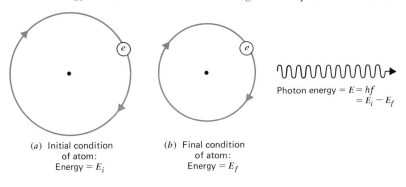

nite frequency is emitted. If every conceivable orbit were allowed, then light with all frequencies would be emitted and the spectrum would show a continuous range of colors instead of the bright lines that are actually observed.

In this way, Bohr reasoned that atomic energy states must be *quantized* and that the electrons in atoms are allowed to move only in certain orbits. Bohr did not know *why* this should be so, but he could see no other way to explain the occurrence of bright-line spectra. Using this quantization hypothesis, Bohr succeeded in showing that the radii of the orbits in the hydrogen atom can be characterized by a single index that takes on only integer (whole-number) values: $n = 1, 2, 3, 4, \cdots$. This index n is called the *principal quantum number*. Bohr found that the orbit radii depend on n^2 and that the energies of the corresponding states are proportional to $1/n^2$.

Bohr was now able to account for the Balmer series of lines in the hydrogen spectrum. If a hydrogen atom is initially in a state characterized by the quantum number n_i and then makes a transition to a state with quantum number n_f, the energy of the emitted photon will be

$$E = E_i - E_f$$
$$= (\text{constant}) \times \left(\frac{1}{n_f^2} - \frac{1}{n_i^2}\right)$$

The Balmer lines correspond to the set of transitions in which the final state always has $n_f = 2$ and $n_i = 3, 4, 5, \cdots$. Then, Bohr's formula becomes the same as Balmer's formula given above.

When an atom absorbs energy from some source, it makes a transition to an excited state with an energy greater than that of the original state. When an atom emits energy (in the form of a photon), it makes a transition to a state with a smaller energy. There is a balance of energy in all such processes.

Figure 10-9 shows some of the orbits and transitions in the hydrogen atom. Only the first five orbits are shown ($n = 1, 2, 3, 4, 5$). Each orbit corresponds to an energy state of the atom. The state with $n = 1$ is the state with the lowest possible energy; this is the normal or *ground state* of the atom. If a normal atom is excited by some means to any higher state, it will make a transition to a lower energy state by emitting a photon. After one or more transitions, the atom will again be

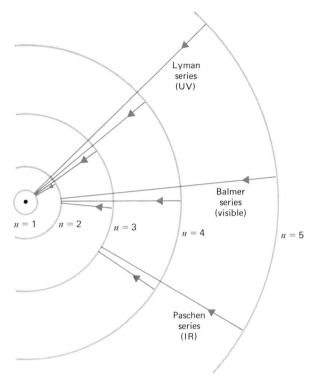

Figure 10-9 *The first five allowed orbits of the hydrogen atom and the transitions connecting them. Notice that the transitions in the Balmer series all terminate on the* n = 2 *orbit.*

in its ground state. Notice, in Fig. 10-9, that there is a series of transitions for each value of n. The Balmer series of spectral lines corresponds to transitions in which the final state has $n = 2$. Those transitions that terminate on the $n = 1$ orbit (the Lyman series) have higher energies than the Balmer transitions and these lines lie in the ultraviolet part of the spectrum. The Paschen series of lines are low-energy (infrared, IR) transitions that terminate at $n = 3$.

Later developments showed that Bohr's original planetary model of atomic structure is too crude a picture to explain the fine details of even the simplest atom, hydrogen. However, Bohr's idea of quantized energy states remains in the modern quantum theory of atoms, and it must be recognized that this theory is a direct descendant of Bohr's early work.

In the modern theory (which is described briefly in the sec-

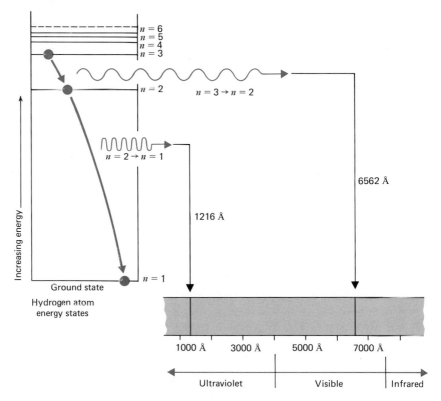

Figure 10-10 *Energy states of the hydrogen atom. Also shown are two transitions and the corresponding spectral lines.*

tion on *Additional Details* at the end of this chapter), we no longer refer to the "orbits" of atomic electrons. Instead, we phrase the description of atomic structure and processes entirely in terms of energy states. Thus, the diagram in Fig. 10-9 is not a realistic picture of the hydrogen atom. We can draw an energy diagram, however, that retains its validity in the modern theory. Such a diagram is shown in Fig. 10-10. The ground state (the state with $n = 1$) is the lowest possible energy state and is shown at the bottom of the series of horizontal lines that indicate the various energy states. All of the states with $n = 7$ and greater are crowded into the small space between the line for $n = 6$ and the dashed line (which corresponds to infinitely large n). If an electron has an energy greater than that indicated by the dotted line, the electron is no longer bound to the nucleus. Thus, the dashed line separates the energies that correspond to *atoms* from those that correspond to *ions*.

Also in Fig. 10-10 we show in a schematic way two atomic transitions. If an atom is initially in the $n = 3$ state, it can make a transition to the $n = 2$ state. This corresponds to the longest wavelength transition in the Balmer series and produces a spectral line with a wavelength of 6562 Å. Subsequently, the atom will make a transition to the ground state ($n = 2 \rightarrow n = 1$), and a high-energy (short-wavelength) photon will be emitted. This transition is in the Lyman series (see Fig. 10-9) and has a wavelength of 1216 Å, which is in the ultraviolet part of the spectrum.

Electron Waves

Although Bohr's atomic model was unable to explain some of the detailed features of atomic spectra and was later replaced by quantum theory, the model did score a remarkable success in accounting for the various series of lines in the hydrogen spectrum. The key ingredient of this theory was Bohr's idea that atomic energy states are quantized and that electrons are allowed to move only in certain definite orbits. This bold hypothesis led to a formula that reproduced the experimental values for the wavelengths of hydrogen spectral lines. But why should atomic energies and orbit radii be quantized?

In our study of the photoelectric effect, we found that light not only has *wave* properties but has *particle* properties as well. In 1924, a young Frenchman, Louis Victor de Broglie (pronounced *de Broy;* 1892–) asked an astonishing question: If light can have *particle* properties, can electrons have *wave* properties? No one had ever thought of things that we know as "particles" as having any of the features of waves, but a series of experiments was soon to show that de Broglie's idea was indeed correct.

De Broglie argued as follows. Light carries energy and momentum. In particular, a light photon of a certain frequency has a definite wavelength and carries a definite amount of momentum. The relationship connecting the wavelength and the momentum of a light photon is easy to derive (see the *Additional Details* at the end of this chapter); de Broglie found

$$\text{wavelength of photon} = \frac{\text{Planck's constant, } h}{\text{photon momentum}}$$

Now, *momentum* is a quantity that we also associate with particles. It was de Broglie's idea that if a particle can exhibit wave properties, the particle should have a wavelength given by the same expression as that for photons:

$$\text{wavelength of particle} = \frac{\text{Planck's constant, } h}{\text{particle momentum}}$$

We know that light waves exhibit *interference*. If we direct a beam of light with a certain wavelength through a narrow slit, we know that we will observe on a screen a broad interference pattern (see Fig. 8-31). The size and shape of the pattern depends on the wavelength of the light and on the width of the slit. Suppose that we accelerate a beam of electrons to a velocity such that the de Broglie wavelength of the electrons (Planck's constant divided by the electron momentum) is exactly equal to the wavelength of the light used in the interference experiment. Then, we direct the electron beam through the same slit used in the light experiment. What will we observe on the screen? If we record the impacts of the individual electrons on the screen (we could use a photographic plate), we would find exactly the same intensity distribution as we observed for light! (See Fig. 10-11.) The electrons exhibit interference effects in exactly the same way as light with the same wavelength!

How can this be so? How can "waves" be *particles* and "particles" be *waves?* The point is this. When we perform experiments, we arrange for photons and electrons to interact with our apparatus in a certain way. This is the only way we can "observe" these entities. Photons and electrons are a part of the microscopic world. Actually, they are *neither* particles

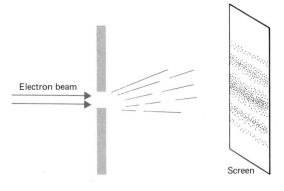

Figure 10-11 *A beam of electrons is directed through a narrow slit and the impacts on a screen some distance away are recorded. The distribution of electron impacts is exactly the same as the light intensity on the screen if the wavelengths of the electrons and the light are the same. Compare Fig. 8-31 for the distribution of light intensity.*

nor waves—they are *quantum* things. But when we allow photons and electrons to interact with our apparatus, we force an interpretation of the results in terms of our large-scale experience, namely, in terms of particles or waves. When we set up a photoelectric experiment and detect the individual photoelectrons that are ejected from the surface, we are forcing a *particle* interpretation on the incident photons. Or, when we send electrons through a narrow slit and detect the distribution on a screen, we are forcing a *wave* interpretation on the electrons. Waves or particles? It depends on how we observe them through interactions with our apparatus.

Now we can return to our original question and examine the connection between the occurrence of definite orbits for electrons in atoms and the fact that electrons can exhibit wave properties. Within the framework of his theory, Bohr was able to calculate the radii of the allowed electron orbits in the hydrogen atom and the electron velocity in each of those orbits. If electrons have wave properties, then the electron waves must fit into these allowed orbits in a precise way. As we proceed around a particular electron orbit, if the electron wave joins smoothly onto itself, it is reinforced by constructive interference and remains stable (Fig. 10-12a). On the other hand, if the wave meets itself and interferes destructively, the wave rapidly cancels itself and disappears (Fig. 10-12b).

The condition that an electron wave fit exactly into an electron orbit is that the circumference of the orbit be equal to an integer number of wavelengths of the wave. Only in this way will the wave join smoothly onto itself. The orbit circumference and the electron velocity in the orbit can be calculated from Bohr's theory. Using the orbit velocity, the de Broglie wavelength of the electron wave can be computed. When

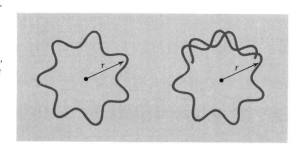

Figure 10-12 *Schematic representation of electron waves in atomic electron orbits. (a) If an integer number of wavelengths fit exactly into the circumference of the orbit, constructive interference preserves the orbit and it is stable. (b) If the wave does not join smoothly onto itself, destructive interference causes the wave to disappear. These orbits are not allowed.*

these two calculations are compared, it is found that the allowed Bohr orbits are of exactly the correct size to contain the corresponding electron waves. One wavelength fits into the first orbit ($n = 1$), two wavelengths fit into the second orbit ($n = 2$), and so forth. Thus, Bohr's idea of quantized states agrees perfectly with de Broglie's hypothesis of the wave properties of matter.

Stimulated Emission

We have already discussed how an atom that has an energy greater than that of the ground state (we say that such an atom is in an *excited* energy state) will spontaneously emit a photon and return to the ground state. We call this an *emission* or *de-excitation* process. How does an atom come to be in an excited state? There are several ways. The atom might absorb a photon of exactly the correct energy to raise the energy of the atom from that of the ground state to that corresponding to one of the excited states. Or, this energy might be supplied by a moving electron or another atom in a collision process. All of these possibilities are *excitation* processes. In every case, the atom absorbs only an amount of energy exactly equal to the energy difference between the ground state and the excited state. When the atom returns to the ground state, exactly this same energy is emitted in the form of a photon. These two types of transitions between atomic states are illustrated in Fig. 10-13.

In the description of atomic transitions according to quantum theory, no distinction is made between the excitation and de-excitation processes that connect a particular pair of states. The two types of transitions are described in equivalent terms. That is, the transition $A \rightarrow B$ (Fig. 10-13a) is equivalent to the transition $B \rightarrow A$ (Fig. 10-13b). The only difference between the two transitions is that energy is *absorbed* in one case and is *emitted* in the other case.

Suppose that we arrange to have an atom in state B (an excited state) when a photon with energy $E = hf = E_B - E_A$ is incident on the atom. The incident photon cannot excite the atom because it is already in the excited state. So the photon causes the equivalent de-excitation process, namely, $B \rightarrow A$. That is, the incident photon stimulates the atom to radiate and we call this process *stimulated emission* (Fig. 10-14).

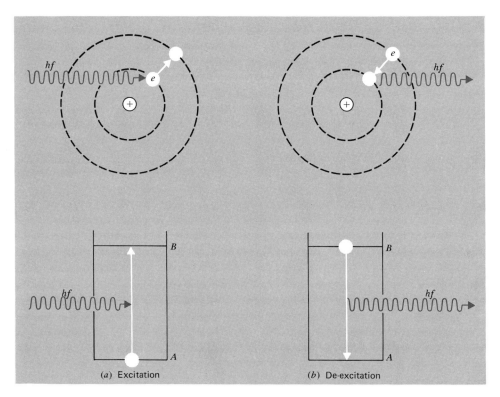

Figure 10-13 (a) *An incident photon of energy hf excites an atom by raising an electron to a higher energy state.* (b) *De-excitation occurs when the electron returns to its ground-state configuration and a photon of energy hf is emitted. According to the quantum theory of atomic transitions, these two processes are actually equivalent. The only difference between the two situations is that an energy* hf *is absorbed in excitation and an energy* hf *is emitted in de-excitation.*

Coherent and Incoherent Radiation

There is an important difference between ordinary *spontaneous* radiation and *stimulated* radiation. When a collection of atoms emits photons spontaneously, the photons leave the atoms in random directions. Moreover, there is no connection between the timing of the oscillations of one photon and another. Light from such a source is said to be *incoherent* (Fig. 10-15a). On the other hand, when an atom is stimulated to radiate a photon, the stimulated photon proceeds away from the atom in the same direction as the incident photon. Fur-

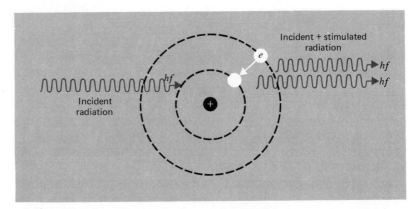

Figure 10-14 *Stimulated emission. The incident photon with energy* $E = hf$ *causes the atom to make a transition from a higher energy state to a lower energy state (the energy difference being exactly* E) *and to emit a photon with energy* $E = hf$. *The two photons proceed away from the atom with their oscillations in phase.*

thermore, the incident photon forces the stimulated photon to oscillate in step with its own oscillations. Thus, the two photons reinforce one another by constructive interference. We say that the photons are *in phase* and that the light is *coherent* (Fig. 10-15b). (The photons in incoherent light have random phases and there is no reinforcement.) Stimulated emission therefore increases (or *amplifies*) the intensity of light in the direction of the incident photon.

Suppose that we have a sample of atoms, some fraction of which are in the same excited state. Then, a single incident photon of the correct energy will begin triggering the de-excitation of these atoms by stimulated emission. Each stimulated photon will, in turn, cause another atom to emit a photon in a cascading increase of radiation (Fig. 10-16). (This process is similar to a nuclear chain reaction, as we will see in Chapter 13.) Thus, the light that would ordinarily be emitted spontaneously in all directions is concentrated by stimulated emission into an intense coherent beam. A device that accomplishes this is called a *laser*. (The term *laser* is an acronym for *l*ight *a*mplification by *s*timulated *e*mission of *r*adiation.)

In order to construct a practical laser, a number of problems must be solved. First, we must select an appropriate material and we must choose the pair of states between which the laser transition is to occur. Then, we must devise a

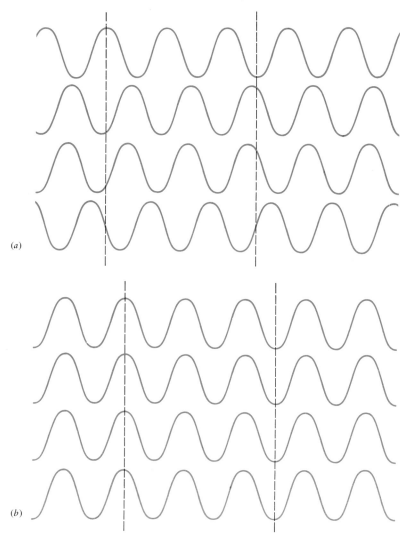

Figure 10-15 (a) *Incoherent radiation. The various waves have random phases. The dotted lines show that the peaks and the valleys of the first wave do not correspond to those of the other waves. There is no mutual reinforcement.* (b) *Coherent radiation. The various waves all have the same phase. The dotted lines show that the peaks and valleys of all of the waves occur together. The waves mutually reinforce one another.*

317

318   *Atoms and Atomic Radiations*

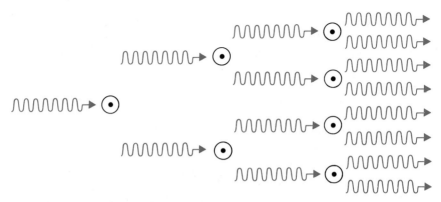

Figure 10-16 *By stimulated emission, a single initial photon can release a cascade of photons, all of which are in phase and travel in the same direction.*

method of exciting the atoms to the radiating state. Finally, we must arrange for the stimulated photons to be confined to a particular direction.

Population Inversion

Why is it that an ordinary light source does not emit stimulated radiation in a coherent beam? The reason is that in all ordinary light sources there are, at any instant, far more ground-state (unexcited) atoms than there are excited atoms. Consequently, as a photon proceeds through the material, it is much more likely to encounter a ground-state atom and be absorbed by that atom than it is to encounter an excited atom and stimulate radiation by that atom. Because of this absorption effect, there is no opportunity for the cascading of the stimulated emission to start.

In order to overcome absorption effects, we need a sample in which the number of excited atoms exceeds the number of ground-state atoms. Then, the stimulated emission process will be more likely and will dominate absorption. A sample in which this situation occurs is said to have a *population inversion*. That is, the normal population of states (many atoms in the ground state, few in the excited state) is inverted (many atoms in the excited state, few in the ground state). How can we do this? If the excited state has a lifetime typical of most excited states (a millionth of a second or so), then any atoms

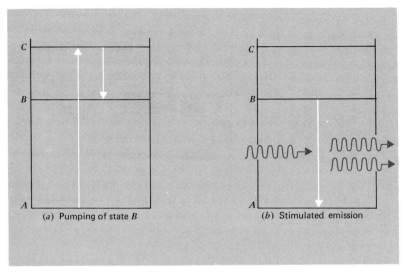

Figure 10-17 *A three-state (or three-level) laser. The long-lived state that exhibits laser action (state* B*) is pumped by first exciting state* C*. The lifetime of state* B *is sufficiently long that a population inversion can be established.*

in this state will quickly radiate away their excess energy and return to the ground state. This will happen before stimulated emission is effective. To promote laser action, we must use an excited state that has a lifetime much longer than normal—a few tenths of a second or so. In this way, the atoms will remain in the excited condition sufficiently long that a beam of radiation can build up.

One way to accomplish a population inversion is to *pump* the atoms through a third excited state. Figure 10-17 shows three energy states of an atom: state A is the ground state and states B and C are excited states. The first step is the excitation of state C by photon absorption, electrical discharge, or collision. State C is a typical excited state and has a short lifetime. But state C is chosen so that when it radiates, it leaves the atom in state B, which has a long lifetime. State B is called a *metastable state*. The transition $B \rightarrow A$ is the laser transition. When a photon with energy $E_B - E_A$ encounters the atom in state B, it stimulates emission and the two photons proceed away from the atom in phase.

(Why do we not simply pump state B directly? Remember, the two transitions, $A \rightarrow B$ and $B \rightarrow A$, are equivalent. If state B has a long lifetime, this means that the transition $B \rightarrow A$ is not a likely process and the atom remains in the state as long as possible. Consequently, the direct excitation transition $A \rightarrow B$ is also unlikely and is difficult to produce. It is much more efficient to excite state B by the two-step process, $A \rightarrow C \rightarrow B$.)

The Ruby Laser

How do we efficiently pump the atoms of the laser material into the radiating state? The point is this. If the high-energy state has a sharply defined energy (as does state C in Fig. 10-17), the pumping radiation must consist of photons with well-defined energy. In Fig. 10-17, the excitation photon must have an energy exactly equal to $E_C - E_A$. A source of ordinary *white* light would be very ineffective in raising the atoms to state C. The reason is that a source of white light emits photons with a wide range of photon energies. Therefore, only a few of the photons in white light have the proper energy to pump the atoms to state C. We could, of course, pump the atoms with the radiation of the correct energy from another sample of the same material. But then we have the problem of exciting the atoms in the second sample. This is the same problem!

In 1960, Charles Townes and Arthur Shawlow of Columbia University realized that an interesting property of ruby crystals appeared to offer a solution to this problem. The mineral ruby consists of aluminum oxide, a colorless substance, that contains a small amount of chromium as an impurity. This chromium impurity gives to ruby its characteristic red color. Figure 10-18 shows some of the energy states of the chromium atoms in ruby. The distinctive feature of this diagram is the fact that the energy states C and D are actually *bands;* that is, the individual atomic states in this region are so numerous and so closely spaced that they overlap. Thus, the atom is not limited to a single well-defined energy but can exist with any energy within a range centered about E_C or E_D. (We should note here that individual *atoms* can exist only in discrete energy states. Energy *bands* occur only when we consider the energy of a large collection of atoms, as in a crystal.)

Because the bands C and D are broad, the white light from a

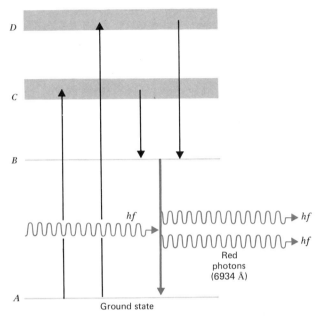

Figure 10-18 *Some of the energy states of chromium atoms in a ruby crystal. The pumping radiation (upward arrows) excite the two energy bands, C and D, which subsequently radiate to form the state B which exhibits laser action. The laser radiation (broad arrow) consists of red photons ($\lambda = 6934$ Å).*

pumping source will include a large number of photons whose energies are within the range that permits the excitation of these bands. When they de-excite spontaneously, each of the two bands radiates primarily to the metastable state B. This state can then be stimulated to radiate by a photon with the proper energy. Thus, the laser transition is $B \rightarrow A$, and for a ruby laser the corresponding radiation is in the red part of the spectrum at 6934 Å.

The problem of concentrating the laser radiation into a beam in a particular direction has been solved in the following way. A crystal of ruby is formed into a cylinder with the end surfaces accurately parallel (see Fig. 10-19). One of the ends is silvered to produce a mirror that will reflect all of the light incident on that end. The other end is given only a partial coating of silver so that about half of the radiation incident on this end will be reflected and about half will escape. Excitation of the chromium atoms is accomplished by a high-

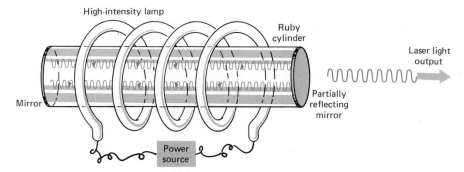

Figure 10-19 *Schematic of a ruby laser system. The pumping radiation is furnished by a high-intensity source of white light. The stimulated photons are reflected back and forth between the parallel mirrors and build up the intensity of the radiation. The beam is formed by photons escaping through the partially reflecting surface.*

intensity discharge lamp that spirals around the cylindrical crystal. The lamp is flashed to produce the pumping action. Then, the atoms begin to make the spontaneous transition, $B \to A$, and this triggers the laser action. Those photons that travel parallel to the cylinder axis are reflected from the ends and again traverse the crystal, stimulating the emission of additional photons. A fraction of this radiation escapes through the partially reflecting surface and constitutes the laser beam.

Most of the spontaneously emitted photons do not travel parallel to the crystal axis; these photons (and any photons they produce by stimulation) eventually escape through the sides of the crystal and do not contribute to the laser beam. Thus, only a small fraction of the energy that is pumped into the crystal actually emerges in the laser beam. But *all* of this output energy appears in a tiny coherent beam of small cross-sectional area that diverges hardly at all. (That is, the small cross-sectional area is maintained over large distances.) Because the conversion of input energy to laser beam energy is so small in the case of ruby crystals (and in most other solid laser materials), the lamp that supplies the pumping radiation is flashed instead of operated continuously. Otherwise, the crystal would become overheated and would crack apart.

The Helium-Neon Laser

The most common and least expensive type of laser in use today is the helium-neon laser. The method by which the laser

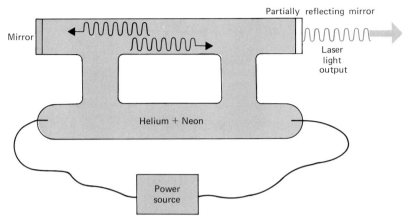

Figure 10-20 *Construction of a helium-neon laser. The mirror and partially reflecting mirror are arranged the same way as in a ruby laser (Fig. 10-19). The pumping action is produced by the electrical current that flows through the tube when the power source is switched on.*

radiation is produced in a helium-neon system is quite different from that in a ruby laser. Figure 10-20 shows in a schematic way the construction of a helium-neon laser. Basically, the laser consists of two glass tubes joined by a pair of short connecting tubes. The entire tube system is filled with a mixture of helium and neon gases at low pressure. One of the main tubes (the *laser tube*) has a mirror at one end and a partially reflecting mirror at the other end, just as in the case of a ruby laser. The other main tube (the *discharge tube*) has an electrode sealed in each end. These electrodes are connected to a power source.

When the power source is switched on, an electric current flows through the gas from one electrode to the other. The gas atoms are excited by impacts with the moving electrons, and the excited atoms radiate spontaneously. The gas in the tube is seen to glow with a red color. This is, in fact, exactly the way a neon sign tube is made to glow. The region of excitation of the gas atoms is not confined to the discharge tube and spreads easily into the laser tube. The entire tube system glows with the same red color.

The transition that exhibits the laser action is a transition between two states in neon. But the neon is not excited directly. As shown in Fig. 10-21, the impacts of the electrons on the helium atoms cause the excitation of state B. Because the gas atoms are continually in motion, there are frequent

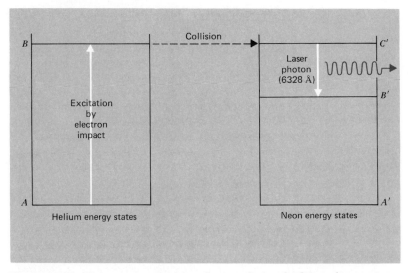

Figure 10-21 *The atomic transition that produces the laser photons in a helium-neon laser involves a pair of states in neon. The radiating state C' is excited by a collision with a helium atom which is in the state B. In the collision process, the excitation energy is transferred from the helium atom to the neon atom.*

collisions between atoms. If a helium atom in the excited state B collides with a neon atom in its ground state, the excitation energy of the helium atom can be transferred to the neon atom. Then, the neon atom is in the state C', which is a metastable state. This is the state that produces the laser photons when stimulated into emission. The laser transition $C' \rightarrow B'$ produces red photons with a wavelength of 6328 Å.

Unlike a ruby laser, a helium-neon laser can be operated continuously. (Excitation by electrical current is more efficient than by the absorption of light from a flash lamp, so the heating problem is not severe.) This is a definite advantage in many situations.

Laser Applications

Light from an ordinary source consists of spontaneous photons that are emitted in all directions. Various optical systems, consisting of mirrors and lenses, can be used to direct the light from such a source in a particular direction.

But this light beam is always broad and tends to spread out (or diverge) from the source position. Laser light, on the other hand, is a very special kind of light. The photons in a laser beam all have the same frequency and they are all in phase. As a result, the photons mutually reinforce one another by constructive interference. This produces a beam of a pure color that remains in a compact bundle even though the beam may be many miles in length.

Because a laser beam has a very small divergence, it is useful in a variety of surveying situations. The concentration of all of the light into a narrow beam means that the light from a relatively low-intensity source will be visible over a great distance. Lasers have even been used to measure the distance to the Moon, surely the most spectacular surveying job in history! In this experiment short bursts of laser radiation are projected through a telescope that points toward the Moon. The light pulse travels to the Moon where it strikes a special reflector that was placed on the Moon by the Apollo astronauts in 1969. Even after traveling nearly a quarter of a million miles to the Moon, the arriving laser beam has a diameter of only a few miles! A small fraction of the light in each pulse is reflected toward the Earth and is viewed by a highly sensitive photoelectric device at the focal position of the telescope. This detector records only a photon or two from each pulse. Because the velocity of light is known with high precision, the Earth-Moon distance can be determined by measuring the time for the laser photons to make the round trip from the Earth to the Moon and back. The uncertainty in this result is now about 10 cm, and it is expected that the accuracy will be substantially improved in the near future. The precise measurement of the Earth-Moon distance is providing important clues regarding the structures of the Earth and the Moon.

Lasers have also been found to be useful in a variety of industrial and medical applications. High-power lasers are used to weld materials that resist other methods, and microholes can be drilled into even the hardest substances by laser beams. It has been found that a laser can deposit just the right amount of energy to "weld" a detached retina in an eye onto the choroid surface that lies beneath it. This technique was first used in the treatment of human patients in 1964 and since then thousands of cases have been treated successfully. Lasers are also used in many other types of minor surgery, such as the removal of warts and cysts.

Laser light has a single, pure frequency. Some types of lasers can be *tuned* to a desired frequency—for example, the exact frequency at which a particular atomic or molecular species will absorb radiation. If such a beam is directed through air that does not contain the absorbing species, a detector placed some distance away will register very little attenuation of the beam. However, if the air does contain absorbing molecules, there will be a sharp drop in the intensity of light reaching the detector. Therefore, tunable lasers can be used as extremely sensitive detectors of pollutants in the atmosphere. Undesirable gases in automobile exhaust emissions or in smokestack effluents are detectable even though the concentrations of the offending gases may be only a few parts per billion. Experiments have shown that it is possible to detect the *hydroxyl* radical OH (which plays a crucial role in the production of smog) in concentrations down to 1 part in 10^{13} using laser techniques.

Another application of the unique characteristics of laser light is in the field of *holography*. It has been found possible to record on a single piece of film sufficient information to allow the reconstruction of a *three-dimensional* image of an object instead of the usual two-dimensional (or *flat*) image. By directing a laser beam through the special film, an image can be produced that stands lifelike in space. This technique, still in its infancy, is certain to be widely used in the future, not only in scientific fields but also in the entertainment industry.

A large number of different materials—solids, liquids, and gases—are now known to exhibit laser action. Some can produce continuous beams, whereas others (for example, ruby) must be flashed or pulsed because continuous operation would cause excessive heating. The highest power lasers now in operation can produce short bursts of light with a power rating in excess of 10 million million watts (10^{13} W)!

Obviously, the concentrated power in a laser beam can be dangerous. A laser of moderate power can cause skin burns, and even a low-power laser can produce eye damage. Direct exposure of the eye to a 2-mW (0.002 W) laser beam for 1 s is likely to cause a retinal burn. *Never look directly down a laser beam toward the laser*. Areas in which lasers are in operation are identified by the type of placard shown in Fig. 10-22.

In addition to excitation by photon absorption and electrical currents, it is also possible to pump laser states by chemical means. If two extremely reactive gases (such as hydrogen and fluorine) are suddenly combined, the product is a highly ex-

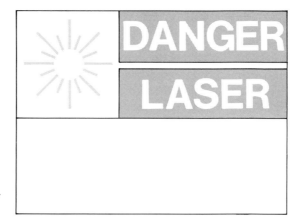

Figure 10-22 *Placard that is posted near operating lasers to warn of possible biological damage, particularly to eyes, due to the concentrated power in a laser beam.*

cited chemical compound (HF). By combining the gases in a chamber fitted with parallel mirrors in the usual way, the excited gas will exhibit laser action without any additional input of power. The advantage of this system is that a large amount of potential energy can be stored in the form of the reactive gases and then released in a burst of powerful laser radiation. The concentrated energy in the laser beam is sufficient to burn a hole through a steel plate. Such devices have obvious military applications as "death rays," but they are expected to produce considerable nonmilitary benefits as well.

Questions and Exercises

1. What facts regarding the photoelectric effect cannot be explained according to the nineteenth-century theories?
2. The threshold frequency for the emission of photoelectrons from atoms of element A corresponds to yellow light and that for element B corresponds to green light. In which atoms are the electrons more tightly bound?
3. Explain carefully why the occurrence of bright-line spectra guarantees that atoms cannot exist in arbitrary energy states.
4. The hydrogen atom contains only a single electron and yet the hydrogen spectrum contains many lines. Explain why this is so.
5. The Lyman, Balmer, and Paschen series of spectral lines occur in the spectrum of hydrogen. Do you expect that there

will be other such series? If so, in what part of the spectrum would you look for them?

6. Construct a diagram similar to Fig. 10-10 which shows the first four transitions in the Balmer series. (The wavelengths are given in Fig. 10-6.)
7. Sketch the de Broglie wave pattern similar to that in Fig. 10-12a for the case $n = 3$.
8. A laser is pumped by a certain light source. The laser beam is capable of burning a hole through a piece of metal, whereas the light from the pumping source could not do this. Does this mean that the laser "generates energy"? Explain.
9. Explain carefully why the state that exhibits laser action must have a long lifetime.
10. Explain why it is not possible to convert all of the energy used to excite a laser into the light of the laser beam.
11. When you play your phonograph records many times, they show an increasing amount of wear. Why? How might it be possible to use a laser beam to play records with *zero* wear?

Additional Details for Further Study

The de Broglie Wavelength

Consider a photon that has an energy $E = hf$. According to the Einstein mass-energy relation (see Chapter 9), a mass m has a mass-energy $E = mc^2$. Conversely, an entity that possesses an energy E has associated with it an *equivalent mass* $m = E/c^2$. Thus, we can say that a photon with an energy E has an equivalent mass equal to that energy divided by the square of the velocity of light.

Momentum, as we learned in Chapter 2, is equal to the product of mass and velocity:

momentum = (mass) × (velocity)

or

$p = m \times v$

Now, all electromagnetic radiation (including light photons) travels with a velocity c. Therefore, the momentum of a photon can be expressed as

photon momentum = (photon equivalent mass)
$\times$ (velocity of light)

or

$$p = \left(\frac{E}{c^2}\right) \times c = \frac{E}{c}$$

Next, we use the facts that $E = hf$ and $c = \lambda f$. Then

$$p = \frac{E}{c} = \frac{hf}{c} = \frac{hf}{\lambda f} = \frac{h}{\lambda}$$

Solving for λ, we can write

$$\lambda = \frac{h}{p}$$

This is the expression for the wavelength of a light photon in terms of its momentum. It was de Broglie's clever idea that the wavelength associated with a particle should be given by exactly the same expression.

Probability in Quantum Theory

According to the traditional view, if we know the position and the velocity of a particle at a particular instant and if we know the forces that act on the particle, then we can predict where the particle will be at all future times. However, if we extend this line of reasoning into the microscopic domain to make predictions about the behavior of photons and electrons, we find that it breaks down. Suppose that we direct a beam of photons toward a narrow slit, as in Fig. 10-11. Suppose also that we decrease the intensity of the beam until the photons pass through the slit one at a time. Where will the individual photons strike the screen or photographic plate?

Because we are directing the photons through a narrow slit, we expect to find an interference pattern such as that in Fig. 8-31. But Fig. 8-31 shows the intensity pattern for a *beam* of photons, and now we are considering *single* photons. Every individual photon will strike the photographic plate and will render developable a grain of the emulsion material. That is, when a photon interacts with the plate, it does so at a *particu-*

lar point. How can photons that strike the plate at particular points produce an intensity distribution such as that shown in Fig. 8-31?

When we consider individual photons proceeding through a slit, we have absolutely no way to predict where particular photons will strike the screen or photographic plate. We *do* know that if a very large number of photons pass through the apparatus, the distribution of impacts will duplicate the intensity pattern of Fig. 8-31. For an individual photon, we can only state the *probability* or the likelihood that it will strike the screen at a particular point. For each and every photon (they are all identical and cannot be distinguished), the probability is greatest that it will strike the screen at the maximum of the wave-theory intensity curve. Of course, not every photon will strike at this position, but more photons will be detected here than at any other point across the screen. The probability of striking any particular point is proportional to the magnitude of the intensity distribution at that point. When a large number of photons have been collected, the resulting pattern of impacts is the same as the wave-theory intensity curve. That is, the patterns of Fig. 10-11 and Fig. 8-31 are identical. (In Fig. 10-11 we considered *electrons,* but the pattern is the same for photons with the same wavelength.)

All of the remarks we have just made for photons apply equally for electrons. When we consider the behavior of individual quantum entities such as photons and electrons, we do not deal with *certainties* but only with *probabilities.* If we wish to make a precise prediction, we must be content with predicting the *average* of a large number of individual identical events. Thus, we *can* predict with high precision the location along a screen of the various intensity maxima and minima, but we cannot predict where any individual photon or electron will strike the screen. In the same way, if we prepare a sample of atoms in the same excited state, we cannot predict the instant at which any particular atom will spontaneously radiate. But we can predict the average time that the atoms will remain in that state; this we call the *average lifetime* of the state.

The Uncertainty Principle

Suppose that we wish to determine both the position and the momentum of an electron at a particular instant. In order to make the necessary measurements, we must arrange for some interaction with the electron to occur. That is, we must probe for the electron with a photon or another electron or some other entity. Only in this way can we detect the presence of the electron. But when the measuring interaction takes place, the very act of measurement disturbs the electron. There is no conceivable way of detecting the electron without altering its condition.

A thorough analysis of the situation shows that if we attempt to locate the electron with higher and higher precision, then we lose information about its motion. That is, if we make *smaller* the uncertainty in the *position* of the electron, we simultaneously make *larger* the uncertainty in its *momentum*. The converse is also true. By making the momentum uncertainty small, the position uncertainty increases. These statements are embodied in the *uncertainty principle,* first stated in 1927 by the German theorist Werner Heisenberg (1901–   ). The Heisenberg principle can be summarized in the following way:

(uncertainty in position) × (uncertainty in momentum)
    is approximately equal to (Planck's constant)

In symbols, this statement is usually written as

$(\Delta x) \times (\Delta p) \cong h$

The uncertainty principle should not be thought of as some mysterious device conceived by Nature to prevent us from probing too deeply into her method of making atoms and other microscopic entities behave properly. Rather, the uncertainty principle is just one result of the wave-particle duality of radiation and matter. Indeed, the recognition that the uncertainty principle is a fundamental rule of Nature has permitted the development of a highly successful theory of quantum behavior based on probabilities instead of certainties.

Atomic Electrons in Quantum Theory

We now see why the original idea of Bohr concerning electron orbits in atoms has not survived. When we speak of an "orbit," we refer to a definite path along which an electron moves with a definite velocity. But the uncertainty principle tells us that motion along a definite path with a definite velocity is not a meaningful concept. Moreover, we know that an electron has a wave-like character within an atom and we know that waves cannot be localized in space.

How, then, can we state our knowledge about the location of an electron within an atom? All that is possible is to state the *probability* that an electron will be found at a particular position if we make a measurement. It therefore makes no sense at all to speak of electron "orbits." Atomic electrons must be described in terms of *probability distributions*. We can imagine an electron as represented by a cloud of varying density surrounding an atomic nucleus. The electron is most likely to be found where this cloud has maximum density. In quantum theory, the idea of an electron as a tiny rigid ball loses all significance. An electron is a quantum thing, and the best we can do is to represent it as a distribution of probability density.

Additional Exercises

1. Suppose that you are presented with a beam of electrons and a beam of photons, and suppose that the electrons and the photons have the same wavelength. What experiments could you perform to distinguish the electrons from the photons? What experiments would *not* distinguish between the electrons and the photons?
2. Suppose that an electron is "at rest." What is the uncertainty in momentum? (What does "at rest" mean?) What is the uncertainty in position? What would be the wavelength of such an electron? What do these answers imply about the possibility of determining the location of the electron?
3. In this chapter there are three different kinds of pictorial representations of electrons in atoms (Figs. 10-4, 10-7, and 10-12a). Which is the most realistic? Why?

How is matter put together?

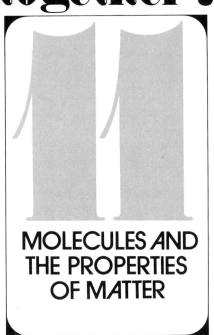

11 MOLECULES AND THE PROPERTIES OF MATTER

The *matter* that we see in the world around us occurs in a seemingly endless variety of forms. We see matter as solids and liquids, and although we cannot *see* the air, we know that it too contains matter in gaseous form. We also know that all matter consists of atoms. How are these atoms assembled into the forms of matter that make up our world? What are the rules that Nature follows in putting together atoms to make molecules and bulk matter?

In this chapter we will first discuss some additional features of atomic structure. Then we will look at the ways atoms combine to form molecules and bulk matter. We will see how some unusual forms of bulk matter—polymers and semiconductors—have been developed into products of extraordinary usefulness.

Atomic Structure

In the preceding chapter we learned that all atoms consist of tiny nuclear cores surrounded by electrons. We know also that atoms of different types have different properties. For example, copper is a good conductor of electricity, but sulfur is not; iron is a dense solid, but oxygen is a gas; sodium is very active chemically, but neon is inert. Why do atoms behave in such different ways? First, we can rule out the effects of the atomic nuclei. Buried beneath the atomic electrons, nuclei do not influence or participate in any of the ordinary chemical or electrical processes. Therefore, the differences among the various species of atoms must be due entirely to the atomic electrons. That is, whether an atom is a good electrical conductor or whether it is chemically active must depend on the arrangement of the electrons that circulate around the nucleus.

The atoms of each chemical element have a characteristic number of electrons. We call this number the *atomic number* of the element and denote it by the letter Z. A hydrogen atom has one electron, so $Z = 1$ for hydrogen. Helium has two atomic electrons and $Z = 2$. The next elements in line are lithium ($Z = 3$), beryllium ($Z = 4$), boron ($Z = 5$), carbon ($Z = 6$), and nitrogen ($Z = 7$). Every normal atom is electrically neutral, so that negative charge represented by the atomic electrons must be exactly balanced by an equal amount of positive charge carried by the nucleus. The nucleus of a carbon atom, for example, contains six positive charges (protons).

The chemical properties of an atom cannot be due simply to the *number* of electrons it contains. If this were the case, we would expect that elements with almost equal atomic numbers (for example, $Z = 18$ and $Z = 19$) would behave in almost the same way. But argon ($Z = 18$) is an inert gas, whereas potassium ($Z = 19$) is a chemically active metal. Therefore, to explain the properties of the various elements we must look to some additional feature of the atomic electrons.

In the preceding chapter, we began the discussion of atomic structure by considering the atomic electrons to move around the nuclear cores in well-defined orbits. This was the early picture developed by Niels Bohr. In the modern theory, we speak of "electron states" instead of "electron orbits," and we describe electrons in terms of waves and probabilities instead of precise positions and velocities. Although the *wave* description is correct and the *orbit* description is wrong, it is often easier to form a mental picture of an atom in terms of electron orbits. In fact, we can characterize each quantum state of an electron by the size, shape, and orientation of the corresponding electron orbit. (The "shape" of an orbit refers to whether it is circular or elliptical. The "orientation" of an orbit means the direction of the plane of the orbit with respect to some specified direction, such as that of an external magnetic or electric field.)

At the time that quantum theory was emerging as the proper description of atoms (the 1920s), it had been known for many years that various groups of chemical elements exhibit similar properties. For example, lithium ($Z = 3$) is chemically similar to sodium ($Z = 11$) but not to helium ($Z = 2$) or beryllium ($Z = 4$). By recognizing this feature of the properties of elements, a *periodic table of the elements* had been constructed (as early as 1871) that organized the elements into groups that exhibit similar chemical behavior. But no one knew *why* this should be possible. Clearly, some fundamental physical principle was at work, a principle that had not yet been identified.

The problem was solved in 1925 when the German theorist Wolfgang Pauli (1900–1958) found that the elusive new principle could be stated in a very simple way. Pauli's statement was:

> No more than two electrons in an atom can exist in the same orbit.

That is, atomic electrons are excluded from orbits that are already occupied by a pair of electrons. This principle is called the *exclusion principle* (or sometimes, the *Pauli principle*).

Why does the exclusion principle specify *two* electrons per orbit? The reason has to do with a new property of the electron that was discovered shortly before Pauli announced the exclusion principle. Certain features of the atomic spectra of elements had been observed that can be explained only if it is assumed that electrons not only revolve around the atomic nuclei but also rotate around their own axes. That is, electrons were found to have a *spin* motion in addition to their *orbital* motion. (We are continuing to use the classical description of electrons in terms of discrete particles that follow well-defined orbits.)

The spinning electrons in atoms can rotate in two possible directions with respect to their atomic orbits, as shown in Fig. 11-1. Each orbit can accommodate one electron with spin "up" and one electron with spin "down." When two electrons have been placed in an orbit with opposite spins, the orbit is "filled," and the addition of further electrons is for-

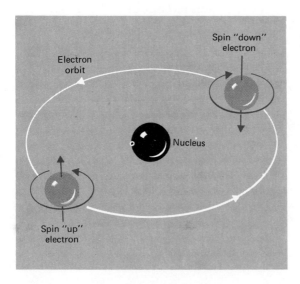

Figure 11-1 *Each atomic orbit can accommodate two electrons, one with spin "up" and one with spin "down."*

bidden by the exclusion principle. Thus, every electron in an atom is specified in a unique way: Each electron is in a particular orbit and has either spin "up" or spin "down."

Building the Elements

How does the exclusion principle, together with the idea of electron spin, explain the similarity of chemical properties of groups of elements? Let us imagine that we "build" the elements by adding electrons one at a time. (Of course, we also add positive charge to the nucleus so that each atom is electrically neutral.) The first element ($Z = 1$) is, of course, hydrogen. The next electron to be added can be placed in the same orbit as the first electron if the two spins are opposite; this element is helium ($Z = 2$). The two electrons that fill the first orbit form a particularly stable system. These electrons are so tightly bound to the nucleus that they do not participate in chemical reactions. Helium is therefore chemically inert. The single electron in a hydrogen atom, however, does not have a partner with which it can form a stable pair. The hydrogen electron can interact with the electrons of other elements; hydrogen is chemically active.

The next *four* orbits into which electrons can be placed are all very similar. A total of eight electrons can be placed in these orbits, representing the elements from $Z = 3$ (lithium) to $Z = 10$ (neon). When the last electron has been placed in its position, the entire set of orbits is filled and again a particularly stable system results: Neon is chemically inert just as is helium. Because these four orbits produce such a stable system when filled, we group the set of orbits together and call the collection an electron *shell*. The first orbit, which contains the electrons for hydrogen and helium, constitutes the first electron shell, and the next four orbits constitute the second electron shell. Similarly, orbits 5 through 8 ($Z = 11$ through $Z = 18$) constitute the third shell, which terminates with the inert gas argon.

Figure 11-2 shows the way that electrons fill the first three shells. Notice how the elements are arranged in the diagram. At the left, each element has a single electron in its outermost shell. At the right, each element has a completely filled outermost shell. The various vertical columns, labeled I through VIII, are called *groups*. The Group I elements all have a single

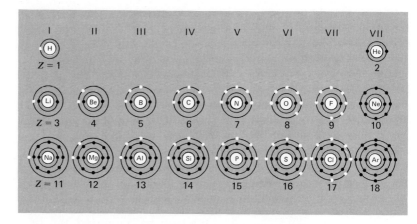

Figure 11-2 *Electron placement in the first three shells. Electrons in the shell that is being filled as Z increases are shown as open circles. Electrons in closed or filled shells are shown as solid circles. The first shell can accommodate two electrons; the next two shells can each accommodate eight electrons. (This diagram is schematic only; electrons do not actually exist in definite orbits.) When the elements are arranged in this way, the chart is called the* periodic table of the elements.

electron in the outermost shell. The Group II elements have two electrons in the outermost shell. And so forth.

Electrons that are in filled (or *closed*) shells are tightly bound and do not participate in chemical processes. Only the electrons in the outermost shell are available to take part in chemical reactions. The *number* of electrons that an element has in its outermost shell (chemists call these the *valence* electrons) determines the chemical activity of the element. All of the elements in a particular group have the same number of valence electrons and therefore have similar chemical properties. Thus, lithium ($Z = 3$) and sodium ($Z = 11$) are chemically similar, as are fluorine ($Z = 9$) and chlorine ($Z = 17$).

As we proceed to build the elements, we find that the next shells contain *eighteen* electrons each. Therefore, the fourth shell terminates at $Z = 36$ (the inert gas, krypton), and the fifth shell terminates at $Z = 54$ (the inert gas xenon). The sixth shell ends with the inert gas radon ($Z = 86$). The chemical similarity in the various groups extends throughout the entire periodic table of the elements. A total of 106 different elements is now known; 92 of these are found in Nature and the remainder can be produced artificially in the laboratory.

Molecules and Chemical Compounds

Most of the matter in our world does not exist in the form of *atoms*. Instead, we usually find atoms joined together to form *molecules*. Sometimes, the atoms in a molecule are of the same species. For example, the oxygen in the atmosphere consists of two oxygen atoms bound together into an oxygen molecule. We represent this form of oxygen as O_2, where the subscript indicates the number of atoms in a molecule. Similarly, the gases hydrogen, nitrogen, fluorine, chlorine, and others exist in Nature as two-atom molecules (H_2, N_2, F_2, Cl_2). The inert gases (He, Ne, Ar, Kr, Xe) do not form molecules, and these elements exist in Nature as atoms.

We also find molecules that are composed of various numbers of atoms of different kinds. These are called *chemical compounds*. For example, the molecule of carbon dioxide consists of a carbon atom joined with two oxygen atoms: The chemical formula for carbon dioxide is CO_2. Also, common table salt (sodium chloride) is composed of sodium atoms and chlorine atoms that are bound together to form NaCl. The gas propane consists of molecules formed from four carbon atoms and ten hydrogen atoms, C_4H_{10}. Complicated protein molecules contain millions of atoms, primarily carbon, hydrogen, oxygen, and nitrogen.

The properties of a chemical compound are generally unrelated to the properties of the constituent atoms. For example, sodium chloride is a compound that is necessary for our bodies to function properly. But sodium metal is such a reactive substance that it could not be ingested without serious, probably fatal consequences. Moreover, chlorine is an extremely poisonous gas. Thus, two substances that are highly injurious join together to produce a compound that is necessary for life.

How do atoms bind together to form stable molecules? Actually, there are several different types of molecular binding. Depending on the atomic species that are involved, one binding mechanism or another will be primarily responsible for holding the atoms together in the molecule. We will examine the various types of molecular binding in turn and we will indicate the characteristics of the bulk matter that results for each binding mechanism.

Ionic Binding

Let us consider first the binding together of sodium and chlorine atoms to form sodium chloride. According to Fig. 11-2, the first two electron shells of a sodium atom are completely filled and there is a single electron in the third shell. Now, we know that closed electron shells are particularly stable. The electron in the third shell is the only electron outside the stable configuration of closed shells, and, consequently, this electron is loosely bound to the atom. This means that it is relatively easy to remove the outermost electron from sodium and convert a neutral atom into a positively charged ion (Fig. 11-3). We represent this process as

$$Na \rightarrow Na^+ + e^-$$

The chlorine atom, as shown in Fig. 11-2 and in Fig. 11-4a, lacks one electron to fill its outermost shell. If the atom acquires this electron it will become a negatively charged ion (Fig. 11-4b). This process is

$$Cl + e^- \rightarrow Cl^-$$

When a sodium atom gives up an electron or a chlorine atom acquires an electron, a particularly stable structure results because, in each case, the outermost electron shell is then completely filled. We can bring about this doubly stable condition by transferring an electron from a sodium atom to a chlorine atom, thereby forming Na^+ and Cl^- ions. That is,

$$Na + Cl \rightarrow Na^+ + Cl^-$$

The oppositely charged Na^+ and Cl^- ions that result from the electron transfer are attracted to one another and are held

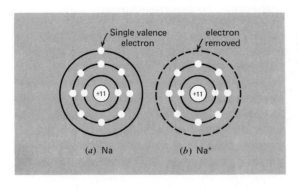

Figure 11-3 *The outermost electron of a neutral sodium atom* (a) *is easily removed to produce a sodium ion, Na^+* (b).

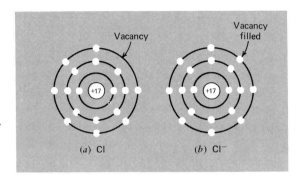

Figure 11-4 *If a neutral chlorine atom (a) acquires an electron, the outermost shell will be filled and the atom will become a chlorine ion, Cl^- (b).*

together by the mutual electric force. We call this *ionic binding*. Although we have discussed this process in terms of single atoms of sodium and chlorine binding together, we do not find isolated molecules of sodium chloride in Nature. Instead, we find sodium chloride only in the form of a bulk solid (grains of salt) or in the form of an ionic solution (salt dissolved in water). In the latter case, the Na^+ and Cl^- ions move independently in the solution and are not bound together at all. In the bulk solid, the ions possess a high degree of order. Every grain of salt consists of a regular array of Na^+ and Cl^- ions bound by electrical forces into a cubic lattice structure, as shown in Fig. 11-5. An orderly solid such as this is called a *crystal*. Notice that in the sodium chloride crystal, each sodium ion is surrounded by six chlorine ions and that each chlorine ion is surrounded by six sodium ions. This is a particularly stable arrangement of the ions, and consequently, the crystal is a tightly bound and rigid structure.

Many other chemical compounds occur as ionic crystals. Every Group I element has a single electron in its outermost shell and every Group VII element needs one additional electron to fill its outermost shell. Therefore, each Group I-Group VII pair of elements will form an ionic compound similar to NaCl. Some of these compounds are (refer to Fig. 11-2) LiF, LiCl, and NaF.

Notice also that the Group II elements, such as beryllium and magnesium, have *two* electrons that they can contribute to a Group VI element in order to fill the two vacancies in the outermost shell. Then, we have compounds such as BeO, MgO, and MgS.

If two sodium atoms each contribute an electron to fill the

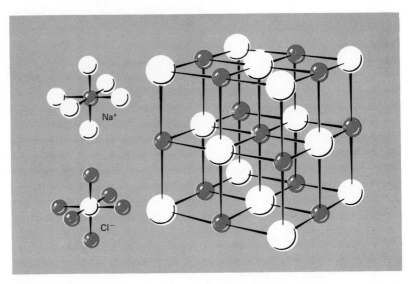

Figure 11-5 *Schematic arrangement of Na^+ and Cl^- ions in a crystal of sodium chloride. Each ion is surrounded by 6 ions of the other type. (The sizes of the ions relative to their separations have been reduced in order to show clearly the lattice structure of the crystal.)*

two vacancies in the outermost shell of a sulfur atom, then the three atoms bind together as Na_2S. Similarly, the two outer electrons in a magnesium atom can fill the shells of two atoms of chlorine; the resulting compound is $MgCl_2$.

It is apparent that a large number of chemical compounds can occur as ionic crystals. Indeed, most of the minerals that we find in Nature are crystalline solids that are held together by ionic bonds. Some minerals have very simple structures (for example, the mineral *halite* is the natural form of sodium chloride), whereas others can be extremely complex [for example, *hornblende* has the formula $Ca(Mg,Fe)_3(SiO_2)_4Al_2(Mg,Fe)_2(AlO_3)_2(SiO_3)_2Fe(Mg,Fe)_2(FeO_3)_2(SiO_3)_2$].

Covalent Binding

Chlorine gas occurs naturally as a two-atom molecule, Cl_2. How are the atoms bound together in this molecule? The mechanism must be quite different from that which binds sodium to chlorine because, in this case, the atoms are identical and each requires an additional electron to complete its

Covalent Binding   343

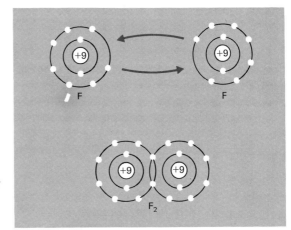

Figure 11-6 *Two fluorine atoms share electrons in order to complete their outermost shells and form an F_2 molecule.*

outermost shell. The chlorine atoms solve the problem in the following way: Each atom *shares* one of its outer electrons with the other atom of the pair. By *sharing* an electron instead of *donating* it to the other atom, each atom manages to complete its outermost shell. This process is illustrated in Fig. 11-6 for the case of fluorine. The element fluorine is, like chlorine, a member of Group VII and therefore has an outer-shell structure that is the same as that of chlorine. The sharing of electrons to form molecular bonds is called *covalent binding*.

The hydrogen molecule, H_2, is also held together by covalent bonds. In Fig. 11-7 we show the orbit representation of the H_2 molecule. However, we should not lose sight of the fact that atomic electrons do not really move in well-defined orbits. Figure 11-8 shows a more realistic representation of

Figure 11-7 *Two hydrogen atoms share their electrons, thereby filling the first shell for both atoms and forming a strong covalent molecular bond.*

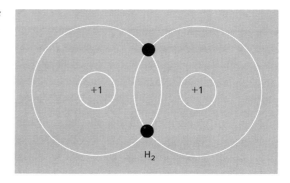

344 Molecules and the Properties of Matter

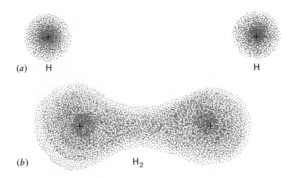

Figure 11-8 *Electron clouds* (probability *clouds*) *for* (a) *a separated pair of hydrogen atoms and* (b) *a hydrogen molecule,* H_2.

the situation. In Fig. 11-8a we see the spherical electron cloud (the *probability* cloud) that surrounds the nuclear proton in each of the separated hydrogen atoms. In Fig. 11-8b we see the merging of the electron clouds when the H_2 molecule is formed. Each nuclear proton is attracted toward the concentration of negative charge between the atoms, thereby producing a strong molecular bond.

The element carbon enters into an enormous number of compounds via covalent bonds. In fact, the entire class of substances that we call *organic compounds* contain covalently bonded carbon atoms. As we can see in Fig. 11-2, a carbon atom has four electrons and four vacancies in its outermost electron shell. Therefore, a carbon atom can supply four separate covalent bonds in forming a molecule.

One of the simplest compounds that contains carbon is *methane* or natural gas. In the methane molecule, each of the carbon bonds is used to bind a hydrogen atom to the molecule. The orbit representation of a methane molecule is shown in Fig. 11-9. Usually, we simplify this picture to

$$CH_4: \quad H-\underset{\underset{H}{|}}{\overset{\overset{H}{|}}{C}}-H$$

where each of the short lines stands for a covalent bond, namely, a pair of electrons that are being shared.

In the molecule of carbon dioxide, CO_2, the carbon bonds are used in pairs instead of singly as in methane. As indicated schematically in Fig. 11-10, each of the oxygen atoms is joined to the carbon atom by *two* covalent bonds. Each of the curved bonding "sticks" in the diagram represents a pair of shared

Covalent Binding   345

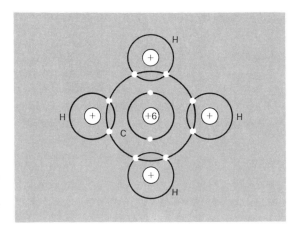

Figure 11-9 *Orbit picture of the methane molecule, CH_4. The carbon atom participates in four covalent bonds.*

electrons. We represent these *double bonds* by a pair of short lines:

CO_2:   O = C = O

Carbon forms two different types of crystal structures by utilizing its covalent bonds in different ways. In the form of carbon called *graphite*, the atoms are arranged in planes of interconnecting hexagons, as shown in Fig. 11-11. The binding between adjacent planes is very weak. Consequently, the planes slip easily over one another and graphite feels "greasy" to the touch. Because of this slippery property, graphite is often used as a lubricant, particularly in the form of a powder that can be blown into inaccessible places, such as door locks.

In the *diamond* form of carbon, the atoms do not lie in planes. Instead, the atoms are joined together in a three-dimensional structure that has great strength (Fig. 11-12). Indeed, diamond is the hardest substance known (and graphite is one of the softest!).

Figure 11-10 *In the carbon dioxide molecule, CO_2, each oxygen atom is attached to the carbon atom by two covalent bonds.*

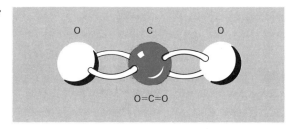

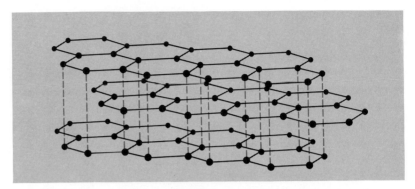

Figure 11-11 *In the graphite form of carbon, the atoms are joined together to form planes. The binding between planes (dashed lines) is weak, so that the planes slip easily over one another.*

Metallic Binding

The forces that hold atoms together within metals constitute a third type of binding and produce a third type of crystal structure. It is a characteristic feature of metals that, in bulk form, the electric fields in which the atoms find themselves are such that the outer electrons are no longer bound to the atoms.

Figure 11-12 *In the diamond form of carbon, the atoms are bound together in a rigid three-dimensional structure.*

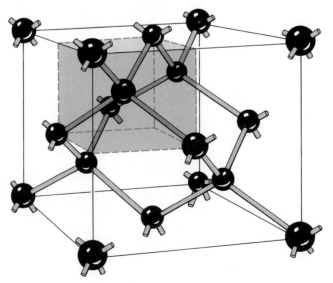

Figure 11-13 *A typical metal crystal. The electrons (small solid circles) are detached from their parent atoms and are free to move throughout the crystal. The positively charged metallic ions are bound in the "sea" of negative electricity.*

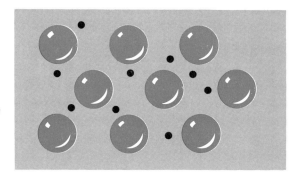

These electrons are free to move throughout the material. The free electrons constitute a kind of "sea" of negative electricity in which the positively charged metallic ions are bound, as illustrated schematically in Fig. 11-13. In ionic crystals and in covalent crystals, each electron is associated with a particular atom or pair of atoms; there are no electrons that are unattached to atoms. Therefore, such crystals as sodium chloride or graphite have no free electrons and are not good conductors of electricity. Metals, with free electrons in the interatomic spaces, are good conductors.

Polar Binding

Hydrogen is a unique substance because when the single electron in a hydrogen atom is pulled away from the nucleus to participate in a molecular bond, the nuclear proton is left almost completely exposed. In the case of the water molecule (Fig. 11-14), the hydrogen electrons are displaced to the extent that the hydrogen portions of the molecule carry a positive charge and, in compensation, the oxygen portion carries a

Figure 11-14 *In the water molecule (H_2O), the hydrogen electrons are displaced toward the oxygen atom. This results in a separation of the charge in the molecule. The molecule is said to be a* polar *molecule.*

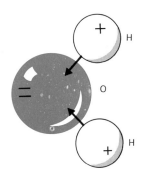

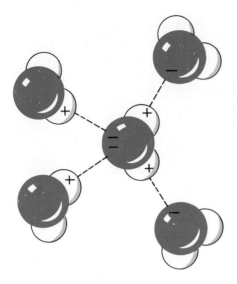

Figure 11-15 *Because of its polar nature, a water molecule attracts to itself the oppositely charged portions of other water molecules.*

negative charge. Thus, there is a separation of charge in the water molecule, even though the molecule as a whole remains electrically neutral. Such molecules are called *polar molecules*.

When a water molecule is in the presence of other water molecules, it attracts to itself the oppositely charged portions of these other molecules (Fig. 11-15). Even in the liquid state, a certain degree of order exists among the water molecules. When water freezes and becomes ice, the orderliness brought about by the electrical attraction among the molecules becomes particularly evident. As shown in Fig. 11-16, the molecules in ice are arranged in a hexagonal structure and are held in place in this type of crystalline lattice by the electrical forces connecting the oppositely charged portions of the polar molecules (solid lines). These intermolecular forces are called *polar bonds* (or *hydrogen bonds*). Polar bonds are also responsible for holding together the strands of the double-helix molecule, DNA (deoxyribonucleic acid), which carries the genetic code.

In Fig. 11-16, notice the large void space in the center of the crystal's characteristic hexagonal ring. This space is larger than the corresponding volume in liquid water (see Fig. 11-15). Consequently, when water freezes, it *expands*. Ice therefore has a lower density than water; ice will *float*.

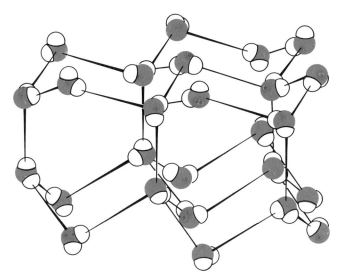

Figure 11-16 *Polar bonds (solid lines) hold water molecules in the crystal lattice of ice.*

Crystals and Noncrystals

Thus far in our discussion we have been concerned only with substances that form crystals in the solid state. Although crystalline matter is an important part of our world, many of the materials that we see and use every day do not occur as crystals. How do we distinguish crystals from noncrystals? A crystal is characterized by a regular arrangement of the constituent atoms. But the crystalline nature of a substance may or may not be evident to the eye. A diamond is clearly a crystal. And with a magnifying glass, you can easily see the tiny crystals of table salt. But can you tell whether baking soda is crystalline? (Actually, it is.) Or can you tell whether glass is crystalline? (Actually, it is not, even though some high-quality glass is called "crystal.") A piece of iron does not appear to be crystalline; but if you clean a portion of the surface and etch it with acid, you will see that the iron is actually composed of an enormous number of tiny crystals welded together.

There must be some satisfactory tests for crystals other than appearance. A crystal consists of a highly ordered lattice, and because of this fact, the structure of a crystal is exactly the same everywhere throughout its volume. This has important consequences for the mechanical, electrical, optical, and thermal properties of the substance. For example, one characteristic property of a crystal is that it can be *cleaved*. That is, when a crystal is split apart, it tends to break smoothly along one of the planes defined by the arrangement of the atoms. Noncrystals do not break in this way; instead, they tend to shatter.

Also, consider the change of a crystalline material from the solid to the liquid state. As heat is added to the solid, the temperature increases until a point is reached at which the material melts. For a crystalline substance, the melting process is sharp and the melting temperature is well defined. Thus, ice melts at a temperature of 0°C, not sometimes at +1°C and sometimes at −1°C. The reason that there is a well-defined melting point for a crystal is due to the fact that the crystal has the same structural features throughout its volume. The change from solid to liquid occurs when sufficient energy is supplied to the substance to break the interatomic or intermolecular bonds that hold the atoms or molecules in the crystal lattice. These bonds are the same at every corresponding point within the crystal. Therefore, when the temperature of the crystal has been raised to the point that the bonds break at one position in the crystal, the bonds will break at all positions.

Noncrystals do not behave in this way. In a piece of glass or a piece of tar, the bonds that connect the atoms and molecules are not all the same. Some of the bonds will break at one temperature, whereas others will break at different temperatures. Therefore, when a piece of glass or a piece of tar is heated, the substance gradually softens to the liquid state and does not exhibit a well-defined melting point. Noncrystalline substances are said to be *amorphous*. Also in this category we find such materials as wood, fibers, and plastics.

Plastics

Some of the most useful materials in our modern world are *plastics*. The variety of plastics available today is so wide that we find these materials used for everything from food wrappings to floor coverings, from furniture to Formica. Thousands upon thousands of items are now manufactured entirely or partially from plastics.

A plastic is a relatively tough, noncrystalline material that consists of a chain of identical units, each of which is a collection of atoms bound together in a particular way. The basic unit of a plastic might contain three atoms or thirty, or even more. Most plastics can be heated and molded into various shapes. Some plastics are soft and pliable; these are used to make toys, furniture coverings, and wrapping materials. Other plastics are hard and brittle; these are used to make tool handles, telephone cases, and nonbreakable substitutes for glass. Still other plastics can be formed into long, elastic fibers; these are used to make various kinds of clothing and carpeting materials.

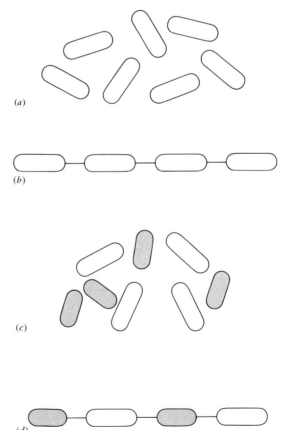

Figure 11-17 *Identical monomer units* (a) *are joined together to form the polymer* (b). *In* (c) *two different types of monomers are linked to produce the alternating polymer* (d).

352   *Molecules and the Properties of Matter*

All plastics are manufactured in the same basic way by joining together small molecules to form long chains of identical molecular units. The small molecular building blocks are called *monomers*. When many monomers are combined into a long-chain molecule, the resulting structure is called a *polymer*. The process that converts monomers into polymers is called *polymerization*.

Figure 11-17a shows a collection of identical monomers which can be polymerized into the long-chain structure shown in Fig. 11-17b. To form certain types of plastics, two different kinds of monomers are used (Fig. 11-17c), and these are polymerized into a chain of alternating units, as shown in Fig. 11-17d.

The process of polymerization *by addition* involves the use of a monomer that has a double carbon bond. This double bond can be broken and converted into two active bonds by the application of heat or by irradiation with ultraviolet light or electrons. This process is shown schematically in Fig.

Figure 11-18 *A photon is absorbed by a molecule of ethylene (a) and the double bond is broken, producing two active bonds (b). Many similar activated ethylene units combine to form polyethylene (c).*

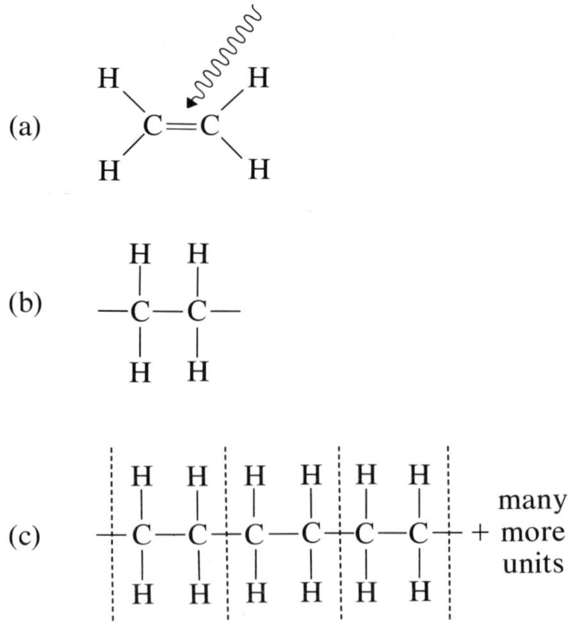

Plastics 353

11-18a, where a photon is incident on the double bond between the two carbon atoms in a molecule of *ethylene* (C_2H_4). The absorption of the photon causes the double bond to break, thereby creating two active bonding sites in the molecule (Fig. 11-18b). When two activated molecules come close to one another, the newly available bonds link together. In this way a long chain of C_2H_4 units is formed (Fig. 11-18c). The resulting polymer is called *polyethylene,* the formula for which is usually written as $(C_2H_4)_n$, where the subscript n simply means "many more identical units."

A substance that is structurally identical to polyethylene is *Teflon.* The monomer unit of Teflon is *tetrafluorethylene,* C_2F_4, which is simply ethylene with a fluorine atom substituted for each hydrogen atom (Fig. 11-19a). The polymerization is carried out in the same way as for polyethylene and produces the long-chain fluorocarbon called Teflon (Fig. 11-19b).

Teflon has a number of unusual properties that make it an extremely useful material. Teflon is a good electrical insulator. It is completely chemically inert; because Teflon will not combine with oxygen, it cannot support life and is therefore immune to attack by molds, fungi, or pests. Teflon will not dissolve in any liquid; it cannot be corroded; dirt and grease will not stick to it; and it does not begin to melt until a temperature of 330°C is reached. Because of these properties, cooking utensils coated with Teflon are easy to clean, and snow will not adhere to shovels that are Teflon-coated Industrial applications of Teflon include the preparation of elec-

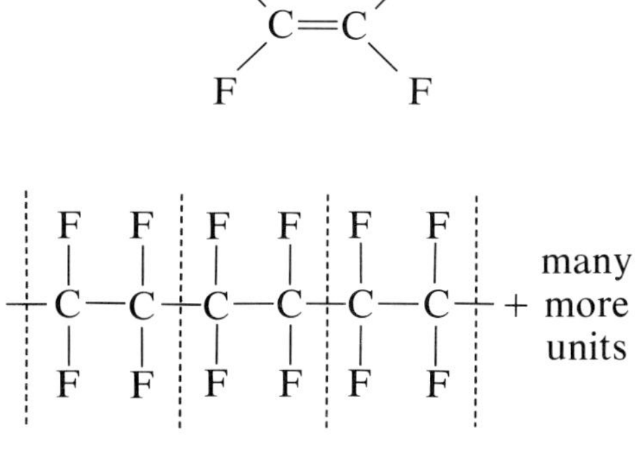

Figure 11-19 *The polymerization of the* tetrafluoroethane *molecule* (a) *produces* teflon (b).

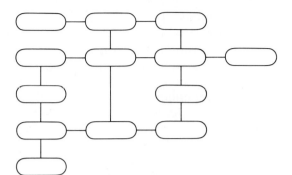

Figure 11-20 *Cross-linked polymers have much greater strength than long-chain polymers (Fig. 11-17b,d).*

trical insulation and grease-resistant gaskets for fuel and lubricating systems.

Polymers such as polyethylene and Teflon become much more useful materials if they are given increased rigidity through *cross-linking*. This process involves the production of bonding sites on the *sides* of the monomer units as well as on the *ends*. When these various bonding sites are linked together, as in Fig. 11-20, a much stronger material results. Long-chain polymers are useful as synthetic fibers, but cross-linked polymers are necessary when rigid structures are required.

All of the various cross-linked polymers have different properties, and they are used in different ways to take advantage of specific qualities. When cross-linked, polyethylene remains relatively soft and pliable, and it is widely used as a packaging and insulating material and in various kinds of molded products. Polystyrene, on the other hand, is much harder and more rigid when cross-linked; polystyrene is therefore used in situations that require greater structural strength.

Polyethylene, Teflon, and similar materials are formed by *addition polymerization*. Another class of polymers is formed by *condensation polymerization*, in which a small molecule (usually water) is eliminated at the bonding site. Figure 11-21a shows two different monomers: One monomer has OH units at both ends of its molecule and the other has hydrogen atoms at both ends. If these monomers are brought together in such a way that the OH and H units join to form a water molecule, two new bonding sites are formed (Fig. 11-21b). These, in turn, can join together to produce a new and larger molecule (Fig. 11-21c), which then acts as a monomer for further po-

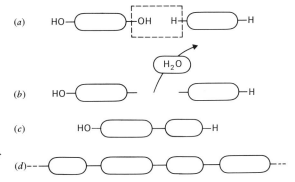

Figure 11-21 *A water molecule is eliminated between two different molecules (a, b) to produce a new and larger molecule (c) that acts as a monomer for the formation of a long-chain polymer (d).*

lymerization. The result is a long-chain polymer in which the original monomer units alternate (Fig. 11-21d). This is the way that such synthetic fibers as Dacron, Nylon, and Acrilan are formed. Figure 11-22 shows the condensation polymerization process that forms the fiber known as Nylon 66.

When fibers are made from these long-chain polymers, the molecules lie side by side, with their long dimensions approximately parallel. If the fiber is stretched, the molecules slide past one another and the material does not rupture. Synthetic fibers generally have greater elasticity than cotton (but less than rubber). All have high strength, light weight, low mois-

Figure 11-22 *Condensation polymerization of two complex organic compounds to form Nylon 66. (The designation 66 refers to the fact that each monomer unit contains 6 carbon atoms.)*

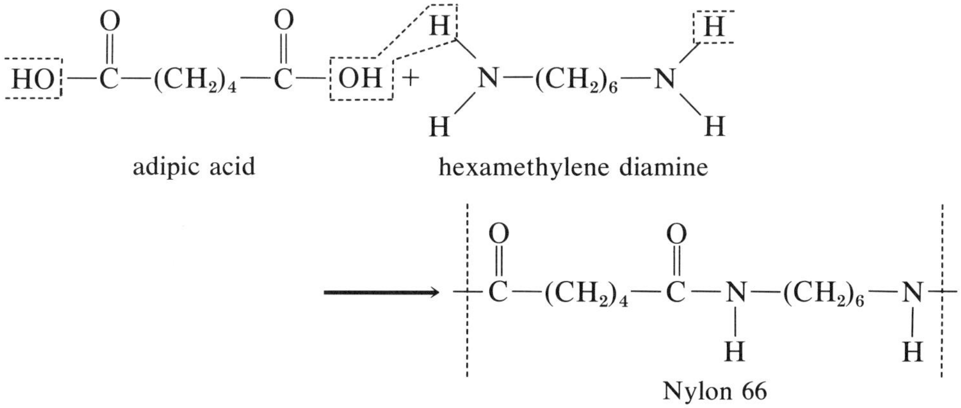

$$HO-\overset{O}{\underset{\|}{C}}-(CH_2)_4-\overset{O}{\underset{\|}{C}}-OH + \underset{H}{\overset{H}{N}}-(CH_2)_6-\underset{H}{\overset{H}{N}}$$

adipic acid   hexamethylene diamine

$$\longrightarrow \quad -\overset{O}{\underset{\|}{C}}-(CH_2)_4-\overset{O}{\underset{\|}{C}}-\underset{H}{N}-(CH_2)_6-\underset{H}{N}-$$

Nylon 66

ture absorption, and will not rot or mildew as will natural fibers.

Semiconductors

A good electrical *conductor* is a material that has many free electrons that will flow through the material when a potential difference is applied. A good electrical *insulator* is a material that has no (or, at least, very few) free electrons, so that no current will flow when a potential difference is applied. Some materials, such as silicon and germanium, can be classified neither as good conductors nor as good insulators. In these materials, the normal thermal motion of the atoms will break loose some of the electrons, thereby permitting the material to conduct electricity to a small extent. Such materials are called *semiconductors*.

By adding a small and carefully controlled amount of an impurity element to a semiconductor, its ability to conduct electricity can be enormously increased. To see how this comes about, refer to Fig. 11-23. Silicon is a Group IV element and therefore has an electron shell structure very similar to that of carbon (see Fig. 11-2). (All of the remarks here concerning silicon apply equally to germanium, which is the Group IV element in the next shell higher than silicon.) In particular, silicon has four electrons that can participate in bonding, so silicon forms a crystal structure similar to the diamond form of carbon (Fig. 11-23a). A crystal of pure silicon

Figure 11-23 (a) *The diamond-like crystal structure of silicon.* (b) *The substitution of a phosphorous atom for a silicon atom in the crystal releases an electron, leaving a P^+ ion in the lattice. This material is an n-type semiconductor.* (c) *A boron atom that is substituted in the crystal will "steal" an electron from a neighboring silicon atom, thereby producing an electron "hole." This material is a p-type semiconductor.*

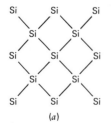
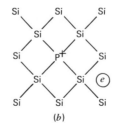
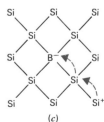

will be only a very weak conductor because it is dependent on thermal agitation to produce conduction electrons.

Now, suppose that we replace one of the silicon atoms in a crystal with an atom of phosphorous. Phosphorous is a Group V element and therefore has five electrons in its outermost shell. Only four of these electrons are needed to duplicate the bonds of the silicon atom, so there is one "extra" electron. When the phosphorous atom is bound in the lattice, this extra electron is no longer attached to the atom and it becomes a free electron that can participate in the conduction of electricity through the crystal (Fig. 11-23b). Therefore, by *doping* silicon with a small amount of phosphorous, the conductivity of the material is greatly increased. Because of the conduction of electricity in silicon that is doped with a Group V element takes place by *negative* charge carriers (namely, *electrons*), these materials are called *n-type* semiconductors.

Next, look at Fig. 11-23c, which shows what happens when a silicon atom is replaced by an atom of boron. Boron is a Group III element and therefore has three electrons in its outermost shell. When boron is substituted for silicon in a crystal, each boron atom finds itself lacking one electron to complete the four bonds that are available to it from neighboring silicon atoms. To remedy this situation, a boron atom "steals" an electron from a nearby silicon atom, thereby completing its bonds and becoming a B⁻ ion in the process. The silicon atom now finds itself one electron short of its normal complement, so it steals an electron from the next atom. Each electron vacancy is called an electron *hole,* and the successive stealing of electrons by the atoms amounts to the propagation of the hole through the crystal. We can see this in the block diagram in Fig. 11-24. Here, 20 blocks are arranged in a 3 × 7 rectangle that leaves one hole in the group. If the block imme-

Figure 11-24 *An array of blocks illustrates the movement of an electron hole in a crystal. As each block (or electron) moves to the right to fill the hole, the hole itself moves to the left.*

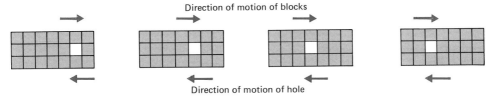

diately to the left of the hole is pushed into the hole, the hole moves one unit to the left. The motion of each block (in a crystal, these would be the electrons) in filling the hole is to the *right,* and this is entirely equivalent to the motion of the hole to the *left.*

An electron hole in a crystal is the *absence* of an electron, so the hole is the equivalent of a *positive* charge carrier. For this reason, the crystal is called a *p-type* semiconductor when a Group III dopant is used. Notice that the direction of the motion of positive charge (that is, the holes) in a *p*-type material is always opposite to that of the electrons.

The great usefulness of semiconductor materials is the result of the interesting effects that occur when *n*-type and *p*-type materials are placed in contact. Figure 11-25a shows the two types of material with their charge carriers when separated. (Remember, the materials are electrically neutral as a whole; the *plus* and *minus* signs indicate only the charge carri-

Figure 11-25 (a) *The charge carriers of* p-*type and* n-*type materials when separated.* (b) *When the two materials are in contact, some of the charge carriers diffuse across the boundary into the other material.*

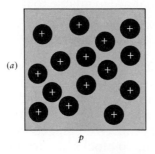
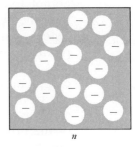

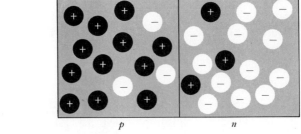

ers.) When the two types of material are placed in contact, thereby forming a *p-n junction* (Fig. 11-25b), some of the carriers of each sign diffuse across the boundary. This migration of electrons into the *p*-type material and holes into the *n*-type material builds up a positive charge in the *n* region and a negative charge in the *p* region. The amount of charge that can be accumulated in each region is limited by the fact that the free electrons tend to fill the holes, thereby eliminating both charge carriers. This recombination process does not really diminish the number of electrons and holes because the thermal agitation in the material continually produces new electrons and holes. At equilibrium, the rate of thermal excitation is equal to the rate of recombination. The net result is that there exists a constant potential difference across the *p-n* junction.

Next, suppose that we connect a *p-n* junction to the terminals of a battery. How will the charge accumulation due to diffusion affect the flow of current through the crystal? Figure 11-26a shows what happens when the positive terminal of the battery is connected to the *p*-type material and the negative terminal is connected to the *n*-type material. The electrons are driven to the left (toward the positive terminal) and the holes are driven to the right (toward the negative terminal). The increased rate at which the charge carriers now move across the boundary means that the rate of recombination is much greater than in the no-voltage equilibrium condition. To offset these losses, new electrons and holes must be produced in both parts of the junction. The disappearance of electrons due to recombination in the *n* region is compensated by the inflow of electrons from the battery. Similarly, the holes that are lost in the *p* region are replaced by the electrons leaving the region on their way to the positive terminal of the battery.

If we now reverse the battery terminals and apply the opposite polarity to the junction, we find that the effect is quite different (Fig. 11-26b). In this case, the holes are drawn toward the negative terminal and the electrons are drawn toward the positive terminal. The region near the boundary is depleted of charge carriers and any new carriers that are created will not diffuse across the boundary because the applied voltage pulls the carriers in the opposite direction. Consequently, there is no current flow when the negative terminal of the battery is connected to the *p*-type material and the positive terminal is

360   *Molecules and the Properties of Matter*

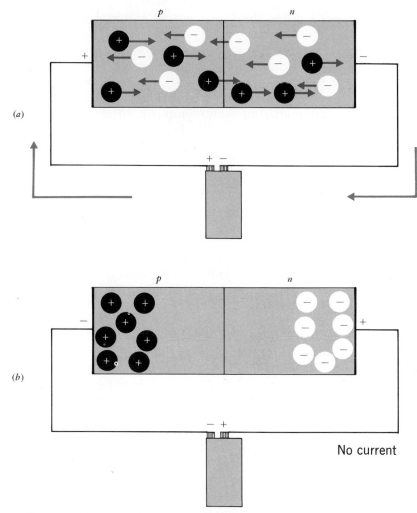

Figure 11-26 (a) *With a battery connected to a* p-n *junction in this way, a current will flow through the crystal.* (b) *With the battery reversed, no current flows. A* p-n *junction will allow current to flow only in one direction.*

connected to the *n*-type material. We conclude that a *p-n* junction will pass current only in one direction. Such a device is called a *diode*.

In Chapter 5, we learned how electric generators produce alternating current (see Fig. 5-27). It is AC that is delivered through the power distribution network to our homes and

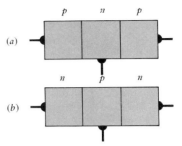

Figure 11-27 (a) *A* p-n-p *transistor.* (b) *An* n-p-n *transistor.*

businesses. For many applications, however, direct current (DC) is required, not AC. Diodes made from semiconductor *p-n* junctions can be used to convert (*rectify*) alternating current to direct current by permitting the current to flow only in one direction. Diodes are used in a long list of electrical and electronic devices, from complex computers to blending machines.

A *transistor* is a device that consists of three semiconductor elements arranged in the order *p-n-p* or *n-p-n* (Fig. 11-27). These devices are used in a wide variety of electronic circuits to amplify and control electrical signals. For example, the voltage applied to the *p* region of an *n-p-n* transistor (Fig. 11-27b) will alter the amount of accumulated charge in this region; therefore, this applied voltage can control the flow of current through the transistor from one n region to the other.

Transistors are extremely reliable because they are solid devices with no flimsy parts and because they generate very little heat. Vacuum tubes have neither of these advantages and they have now almost completely given way to transistors. About the only applications in which vacuum tubes continue to be used are in high-voltage circuits, where transistors tend to break down.

The development of semiconductor devices, begun in 1949, has completely revolutionized the field of electronics. The communications and computer industries now rely almost completely on semiconductor components. A tremendous additional advance has been made in the miniaturization of electronic devices by the development of *integrated circuits*. These tiny modules (called *ICs*) are actually complete electronic subassemblies that consist of a large number of circuit elements (diodes, transistors, resistors, and capacitors) placed together on a single wafer of semiconductor material. Thousands of circuit elements can be placed on a single semiconductor chip that is only 1 or 2 mm square! One type of digital wrist watch has a 3 mm × 3 mm IC that contains more than 2000 transistors. High-speed digital computers once required thousands of vacuum tubes and thousands of square feet of floor space. But now, by using ICs, these computers can be made no bigger than a breadbox. Sophisticated mathematical computations can now be carried out on IC-equipped calculators that will fit in your pocket. (These calculators could easily be made much smaller, but then you would not be able to manipulate the keys with a finger!)

Questions and Exercises

1. The amount of energy required to remove an outer electron from an atom and thereby to convert the atom into an ion is called the *ionization energy* for the particular element. The elements in which group of the periodic table should have the lowest ionization energies? Which elements should have the highest ionization energies?
2. Why are neutral atoms of lithium more chemically active than Li^+ ions?
3. Sodium and chlorine combine to produce NaCl, and chlorine combines with itself to produce Cl_2. Why does sodium not exist in the molecular form, Na_2?
4. Find the values of n and m for the following ionic compounds (refer to Fig. 11-2): Al_nO_m, Mg_nF_m, Li_nO_m.
5. Construct an orbital picture to show the simplest way that chlorine can combine with carbon. Compare the structure of this molecule (which is carbon tetrachloride) with that of the methane molecule.
6. Can the molecule He_2 exist? Explain.
7. The molecule H_3 does not exist in Nature, but the ion H_3^+ *does* exist. Explain why.
8. How are the carbon bonds utilized in the *acetylene* molecule (C_2H_2)? (Compare the molecule of ethylene, C_2H_4, shown in Fig. 11-18a.)
9. The water molecule (Fig. 11-14) is a polar molecule. Do you expect the carbon dioxide molecule (Fig. 11-10) to be a polar molecule? Explain.

10. List some of the consequences of the fact that ice crystals have a large void space (Fig. 11-16).
11. Show the way in which the synthetic fiber Orlon is formed from the monomer $H_2C = (CH) - C \equiv N$. What type of polymerization process is this?
12. The Group III, IV, and V elements in the fourth electron shell are, respectively, gallium (Ga), germanium (Ge), and arsenic (As). Which element is a semiconductor? What dopant would be used to produce an *n*-type material? A *p*-type material?
13. Do you think that the Apollo space program could have been carried out successfully before the development of transistor technology? Explain.

Additional Details for Further Study

Superconductors

When a battery is applied to a conducting material, the free electrons are set into motion and a current flows. We know that all ordinary materials offer a resistance to the flow of electricity. In the case of a metallic crystal, this resistance is caused in part by the electrons colliding with impurities or with points of imperfection in the crystal lattice. However, even in an ideal crystal that is absolutely pure and has perfect lattice structure, the motion of the electrons will be impeded by the thermal vibrations of the atoms in the crystal. As the temperature is decreased, the thermal agitation of the atoms is lessened and the motion of the electrons is affected to a smaller degree. Thus, we expect the electrical resistance of a metal to decrease with temperature. Indeed, all materials demonstrate this effect. The resistance of silver, a good conductor, is shown in Fig. 11-28. Although the resistance of silver decreases with temperature, it does not decrease below a certain value, even near absolute zero.

Figure 11-28 also shows the extraordinary behavior of the electrical resistance of lead. The resistance decreases slowly with temperature until a temperature of 7.2°K is reached. At this point, the resistance of lead suddenly drops to zero! At temperatures below 7.2°K lead conducts electricity without any resistance whatsoever; lead is a *superconductor*. Several elements and many alloys (over 1000 are known) have been

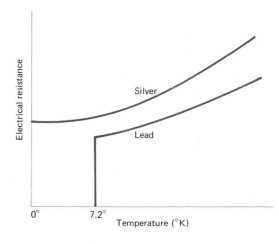

Figure 11-28 *The electrical resistance as a function of temperature for silver (a normal conductor) and for lead (a superconductor).*

found to exhibit superconductivity. Each material has a characteristic temperature at which the transition between the normal and the superconducting state occurs.

How can the electrical resistance of a material suddenly become *zero*? Earlier in this chapter we learned that the atomic structure of the elements is a result of the fact that electrons obey the exclusion principle. It is also true that particles without spin do *not* obey the exclusion principle. That is, because individual electrons have spin, a maximum of two electrons can be placed into an atomic orbit or into *any* quantum state. On the other hand, if a certain type of particle does not possess spin, any number of these particles can be placed into a given quantum state. When the temperature of a superconducting material is lowered, the interaction of the free electrons with the vibrations of the atoms in the crystal lattice overcomes the mutual repulsive force that exists between electrons and causes the electrons to group together into pairs. Each of these pairs consists of a spin-up electron and a spin-down electron (Fig. 11-29). The electron pair therefore has zero spin and the pair, acting as a unit, does not obey the exclusion principle.

The net attraction between the electrons in a pair is quite small. Consequently, a pair can easily be broken by the effects of thermal agitation. Therefore, it is only at very low temperatures that pairs can exist. Because they do not obey the exclusion principle, the electron pairs all tend to collect in the lowest possible energy state. When the temperature is

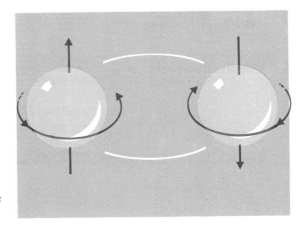

Figure 11-29 *A spin-up, spin-down electron pair has zero net spin; consequently, the pair does not obey the exclusion principle.*

lowered to the transition temperature (7.2°K for lead), *all* of the pairs are in the lowest state. No pair can then give any energy to another pair; there can be no energy losses because every electron pair is already in the lowest possible energy state. Consequently, there can be no energy dissipation and no electrical resistance.

One of the applications of superconductors is in the production of strong magnetic fields. In a conventional electromagnet, the field is produced by causing a current to flow in a wire that is wrapped around an iron core. If a high field strength is desired, the current must be large and the resistive losses can be quite severe. Moreover, huge amounts of electric power must be continually supplied to the magnet merely to maintain the field. Now, consider an electromagnet that is made with a coil of superconducting wire. Once a current is induced to flow in the coil and a magnetic field is established, the current and the field will persist forever because there are no resistive losses. In a practical situation, however, some energy will be lost from the system, and this amount of energy must be supplied to the magnet by "pumping up" the current in the coil. But is is necessary to provide the magnet only with an amount of energy equal to the energy used or lost—no additional energy input is required simply to maintain the field as is the case with conventional electromagnets.

Superconducting magnets are now being used in many research situations that require high magnetic field strengths. Also, new electric generating systems are being designed to incorporate superconducting magnets. These generators are expected to be only one-tenth the size of conventional generators for the same power output, and the losses due to resistive effects are expected to be only about one-third as great.

The transportation of electrical energy from the generating plant to the consumer involves the use of ordinary wires carried on high-voltage towers. Substantial losses occur in these wires because of resistive effects. If the resistance losses in the transport of electrical energy could be eliminated or substantially reduced, enormous savings in the cost of power would be realized. Therefore, the possibility of using superconducting materials for the construction of electrical transmission lines is of great economic importance. Perhaps someday we will begin to replace the huge and unsightly towers that carry out electrical power with underground superconducting electrical lines.

Additional Exercises

1. List some of the items that will involve substantial costs in constructing an underground superconductor power distribution system. How much land now occupied by transmission towers in the United States can be recovered by converting to a superconducting system? (There are approximately 300,000 miles of transmission lines in use and the average right-of-way is 100 ft in width.) [Ans. About 4 million acres or 6000 square miles, an area slightly greater than that of Connecticut and Rhode Island combined.]

2. According to the theory of superconductivity, it may be possible to synthesize some material that will become a superconductor at a temperature near room temperature. Would this be an important development? Suggest some ways in which such a material could be used.

What will radiation do FOR you and TO you?

12
RADIATION AND ITS EFFECTS

We live in a sea of radiation. All around us there is a continual flow of radiation of various sorts. Most of these radiations are quite familiar: we are accustomed to heat radiation, light, radio and television radiation, as well as ultraviolet and X radiation. Although we call these radiations by different names, they are, in fact, only different forms of the same basic phenomenon, namely, electromagnetic waves. Heat radiation and the radiation from a television or radar transmitter are all electromagnetic waves—they differ only in wavelength. Similarly, an X ray is distinguished from a photon of visible light only by its wavelength.

Even though the various kinds of electromagnetic radiation are basically the same, they do differ in the ways in which they interact with matter (because the nature of this interaction generally depends on the wavelength of the radiation). For example, a radar signal will be reflected from a metallic surface, whereas an X ray (which has a much shorter wavelength) can pass completely through a sheet of metal. Our eyes will respond only to radiation that falls in the narrow band of wavelengths we call visible light. We can feel heat radiation through the thermal sensors in our skin, but our bodies cannot sense long-wavelength radio waves in any way.

The differences in the way that radiations of various kinds interact with matter has important consequences in our handling and use of radiation. In this chapter we will concentrate on those types of radiations that are capable of producing significant effects in chemical and biological systems. We will not be concerned, for example, with the effects of solar radiation, although an overexposure can have painful effects in terms of a sunburn. Instead, we will emphasize the more potent forms of radiation such as the X radiation from X-ray machines and the gamma radiation from natural and artifically produced radioactivity. Again, we point out that X rays and gamma rays (or γ rays) are the same as other forms of electromagnetic radiation except that they have shorter wavelengths and higher energies than light or radio waves. The fact that X rays and γ rays carry high energies makes these radiations particularly effective in producing chemical and biological changes.

When we turn to the topic of radioactivity, we will learn that radioactive processes involve the emission of *particles* as well as the emission of radiation (that is, γ rays). In discussions of radioactivity, it has become customary to use the term "radiation" to mean both the emitted particles and the emitted γ rays. We will see that the particles emitted in radioactive decay events can also produce important changes in chemical and biological systems.

Are the effects caused by radiation *good* or *bad?* In recent years there seems to have been a greater emphasis on the harmful aspects of radiation (of which there are many) than on the positive benefits (of which there are also many). In our expanding technological society, every scientific or technical advance is accompanied by hazards as well as benefits. The automobile, for example, has become an indispensable part of modern life, but the automobile has also brought us air pollution and an enormous loss of life due to accidents. Still, on balance, we view the automobile as a beneficial development. Can we say the same for radiation? In this chapter we will study both the positive and the negative aspects of radiation so that you will be able to approach this question in a sensible way.

A New Unit of Energy

As we deal with radiation processes in this chapter and with nuclear transmutations in the following chapter, it will sometimes be necessary to mention the *energy* involved in a particular situation. Because we are now dealing with the behavior of individual atoms and nuclei, the energies associated with these processes are extremely small. For this reason, a new energy unit of convenient size is usually chosen to describe atomic and nuclear events.

The new energy unit is defined in the following way. Suppose that an electron is released from rest and allowed to accelerate through a potential difference of 1 V, as in Fig. 12-1. The electric force acting on the electron will cause it to be accelerated to a certain velocity (actually, the velocity will be 600 km/s when it reaches the positive plate). The electron therefore possesses a certain kinetic energy at the end of the acceleration process. We call this amount of energy 1 *electronvolt*, which we abbreviate as 1 eV.

If the electron is accelerated through 10 V, its energy will be 10 eV. If the potential difference is 1000 V, the energy will

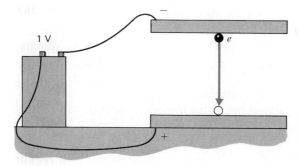

Figure 12-1 *If an electron e is released from rest at the upper plate, it will have a kinetic energy of 1 eV when it reaches the lower plate.*

be 1000 eV or 10^3 eV, which we can also write as 1 keV (1 kiloelectronvolt). Similarly, 1 000 000 eV = 10^6 eV or 1 MeV (1 megaelectronvolt). Typical energies involved in atomic processes are in the range from an electronvolt to a few tens of electronvolts. For example, the energy necessary to remove the electron from a hydrogen atom is 13.6 eV. In the violent processes that produce X radiation, the energies may be a few or many kiloelectronvolts. And in nuclear changes, the energies are typically 0.1 MeV to 10 MeV. The *fission* of a nucleus, which we will study in the next chapter, releases an exceptionally large amount of energy—about 200 MeV.

The *electronvolt* is simply a unit of energy and is not necessarily associated with charged particles being accelerated through potential differences. We could, for example, measure the energy of an electrically neutral particle in electronvolts even though the particle will experience no force at all in an electric field. Also, we measure the energies of photons (visible light, X rays, γ rays) in terms of electronvolts. Some typical photon wavelengths and energies are listed in Table 12-1.

We must keep in mind that 1 eV, or even 1 MeV, is a very small amount of energy by all ordinary standards. For example, suppose that you cut a 1 cm × 1 cm square from a sheet of household aluminum foil. Now, you allow this piece of aluminum to fall through a height of 1 mm (that is, a distance about equal to the width of the letter *s*). The kinetic energy acquired by the aluminum square in this fall will be approximately 100 billion eV! To equal an energy of 1 kWh would require 10^{18} (a billion billion) γ rays, each with an energy of 20 million eV (20 MeV).

Table 12-1 Wavelengths and Energies of Various Types of Radiations

Radiation	Wavelength Å	m	Energy
Infrared	1 000 000	10^{-4}	0.012 eV
	100 000	10^{-5}	0.12 eV
	10 000	10^{-6}	1.2 eV
Visible light	4000–7500		1.6–3 eV
Ultraviolet	1000	10^{-7}	12 eV
	100	10^{-8}	120 eV = 0.12 keV
X rays	10	10^{-9}	1.2 keV
	1	10^{-10}	12 keV
	0.1	10^{-11}	120 keV = 0.12 MeV
γ rays	0.01	10^{-12}	1.2 MeV
	0.001	10^{-13}	12 MeV

X Radiation

An energy of 13.6 eV is required to remove the electron from an atom of hydrogen and create a hydrogen ion. As the atomic number of the substance increases, the energy that is necessary to remove an electron from the innermost shell of an atom also increases. To remove such an electron from a lead atom requires 88 000 eV or 88 keV. After an electron has been removed from an atom of hydrogen or lead or any other substance, the atom will not remain long in the ionized condition. The atom will quickly capture an electron from its surroundings and return to its normal electrically neutral condition. In this process, energy is emitted in the form of radiation.

When an H^+ ion captures an electron, ultraviolet or visible light is emitted. But in the process that fills an electron vacancy in the first orbit of a lead atom, the emitted radiation is much more energetic. Look at Fig. 12-2. Here we see represented, in Fig. 12-2a, the removal of an inner electron from an atom in a collision with a high-speed electron. In Fig. 12-2b we see that the electron vacancy in the innermost orbit is filled by an electron making a transition from the next higher shell; an energetic X ray is emitted in this process. This new vacancy is filled, in turn, by an electron from a higher shell,

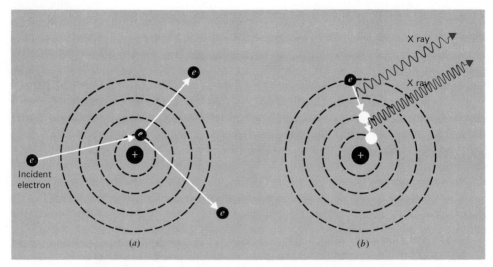

Figure 12-2 (a) *A high-speed electron is incident on an atom and removes one of the inner electrons in a collision.* (b) *The vacancy in the innermost orbit produced by the collision is filled by an electron from the next shell and this new vacancy is filled by an electron from a still higher shell. Each electron transition is accompanied by the emission of an X ray.*

and this transition is also accompanied by the emission of an X ray. Eventually, the vacancy is transferred to the outermost shell and an electron is captured from the surroundings thereby returning the atom to the neutral condition.

We conclude, therefore, that X rays can be produced through the bombardment of a material that has a high atomic number by energetic electrons. Figure 12-3 shows in a schematic way how this can be accomplished. Two metal electrodes are brought through the ends of a glass tube which is then evacuated and sealed. The electrodes are connected to the terminals of a high-voltage supply. Electrons released at the negative electrode are accelerated through the potential difference that exists between the two electrodes and then strike the positive electrode with high speeds. When the electrons crash into the positive electrode, X rays are produced in the way shown in Fig. 12-2. Because the power dissipated in the positive electrode can be quite high in some X-ray tubes, this electrode is usually a block of metal that has a high

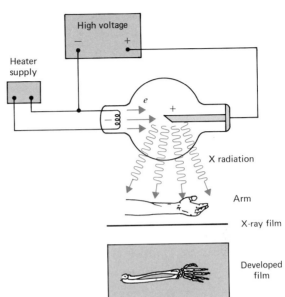

Figure 12-3 *Schematic diagram of an X-ray system. Electrons are accelerated through the high voltage that exists between the two electrodes at the ends of the evacuated tube. (The current that passes through the negative electrode from the heater supply raises the temperature of the electrode and facilitates the release of electrons.) The high-speed electrons that strike the positive electrode produce X rays (see Fig. 12-2). The X rays pass through the object being studied (here, a broken arm) and leave a developable image in a special film.*

melting point (tungsten is often used). In high-power X-ray systems, the positive electrode is water-cooled.

How do we use X rays? Probably the most familiar application of X radiation is in the medical and dental professions. If a person suffers a bone injury, such as a broken arm, the physician needs to know the exact position and nature of the fracture in order to give the proper treatment. Figure 12-3 shows how an X-ray photograph of a broken arm is made. We must remember that X rays have higher energies than does visible light (see Table 12-1). Therefore, even though light is unable to penetrate flesh or bone, X rays can pass through a limited thickness of such material. (High-energy X rays can penetrate a greater thickness of a substance than can lower-energy X rays.) The X rays used for medical and dental diagnostic purposes can penetrate 10 cm or so of fleshy material, but they are absorbed by most bony structures. Consequently, when X rays are directed toward an arm, much of the radiation that does not encounter bony material will pass completely through the arm and will then strike a special X-ray film (Fig. 12-3). On the other hand, those X rays that are incident on the bones will be absorbed to a much greater extent and a relatively small number will reach the film. As a result, the film will be exposed in those areas corresponding to the fleshy material but not in those areas that are shadowed by the bones. When the film is developed, the contrasting dark (fleshy) and light (bony) arcas will reveal the nature of the fracture.

Although X radiation is quite important to the physician and dentist, there are hazards involved in using X rays and caution must always be exercised. The same property of X rays that makes them useful—namely, their ability to penetrate into the body—also makes them potentially dangerous. Excessive doses of X radiation can have serious detrimental effects on the body. For this reason, X-ray photographs should never be taken needlessly. We'will return to the discussion of radiation hazards later in this chapter.

In addition to the examination of humans, X rays are used to probe hard-to-reach places in a wide variety of situations. Using X rays, metal castings can be inspected for flaws, aircraft parts can be examined for fractures that might lead to failure in flight, and suitcases can be quickly searched by airlines or customs officials. In many applications of X rays, a television-type viewer is used instead of film to speed up the examination process.

Radioactivity

A new era in the study of the way Nature behaves was ushered in by the discovery in 1896 of the phenomenon of *radioactivity*. As was mentioned in Chapter 10, Henri Becquerel's accidental discovery was quickly followed up by a number of scientists investigating the new radiations. The substance that Becquerel had identified as the source of the radiations in his experiments was *uranium*. Within a short time, the element *thorium* was also found to be radioactive. And by 1898, the investigation of the sources of radioactivity had led Marie and Pierre Curie to the discovery of two new elements, *radium* and *polonium*.

Other scientists continued and expanded these investigations by studying the way the new radiations behave in electric and magnetic fields. An experiment of this type is shown schematically in Fig. 12-4. Here, a sample of radium is placed at the bottom of a hole in a lead block. The radiations from

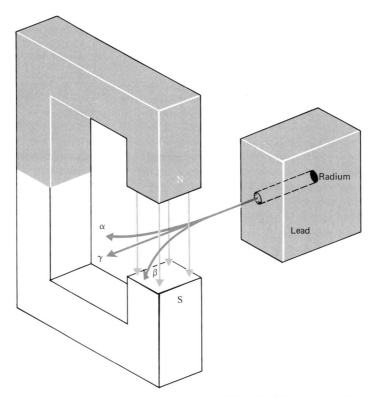

Figure 12-4 *The radiations from a sample of radiations are directed into a magnetic field. One part of the beam is bent to the right because these rays (the α rays) carry a positive electrical charge. Another part is bent to the left because these rays (the β rays) carry a negative electrical charge. The third component (the γ rays) carry no electrical charge and pass through the field undeviated.*

radium emerge from the hole and are directed into a magnetic field. The effect of the field is to separate the beam of radiation into three parts. One part consists of rays that are bent in a direction that indicates they carry a *positive* electrical charge; these rays are called *alpha* (α) *rays*. Another part consists of rays that are bent in the direction opposite to that of the α rays; these rays carry a *negative* charge and are called *beta* (β) *rays*. The third component of the beam passes directly through the field without deviation; these rays carry no electrical charge and are called *gamma* (γ) *rays*.

Further experimentation revealed the true identities of these new radiations:

α rays (or α *particles*) are identical to the nuclei of helium atoms and carry a positive charge of two units ($+2e$).

β rays (or β *particles*) are identical to the atomic electrons found in the shells of all atoms and carry a negative charge of one unit ($-e$).

γ rays are identical to photons of light or X rays except that they have much higher energies (and correspondingly shorter wavelengths); γ rays carry no electrical charge.

It must be emphasized that all of these radiations are the result of *nuclear* changes: α particles, β particles, and γ rays all emerge directly from the nucleus of the atom. When an α particle is emitted by a nucleus, two units of electrical charge are lost by the nucleus and it becomes the nucleus of a different element. For example, when a nucleus of radium ($Z = 88$) emits an α particle, a nucleus of radon ($Z = 86$) remains. When a negative charge is removed from a nucleus by the emission of a β particle, the atomic number increases by one. For example, when radioactive carbon ($Z = 6$) undergoes β decay, a nitrogen nucleus ($Z = 7$) is formed. Because γ rays are electrically neutral, no change in atomic number is involved in the emission of γ radiation.

Isotopes—Stable and Radioactive

The atomic number Z of an element is equal to the number of electrons possessed by a normal, electrically neutral atom of the substance. The atomic number is also equal to the number of positive charges (or *protons*) in each nucleus of the substance. Indeed, the nucleus of the element hydrogen consists of *one proton*. The nucleus of an atom of helium ($Z = 2$) contains *two protons*. And so forth.

If nuclei consisted entirely of protons, we would expect that the mass of a nucleus with atomic number Z would be Z times the mass of a hydrogen nucleus. But experiments show that this is not the case. In fact, the mass of a helium nucleus is *four* times the mass of a hydrogen nucleus. What else, in addition to protons, is contained within nuclei? The answer to this question was provided in 1932 when James Chadwick (1891–1974) discovered the *neutron,* a nuclear particle that has approximately the same mass as a proton but carries no electrical charge. The composition of nuclei could then be explained. The nucleus of hydrogen consists of a single proton, and the nucleus of helium consists of two protons and two neutrons. In fact, the nuclei of *all* elements, with the exception of hydrogen, contain neutrons.

By adding the number of protons Z and the number of neutrons N in a nucleus, we arrive at the *mass number A* of the nucleus. We usually write the mass number of a nucleus as a superscript to the element symbol. Thus, hydrogen is abbreviated by ^{1}H and helium by ^{4}He. The nucleus of oxygen contains eight protons and eight neutrons, so we use the symbol ^{16}O for oxygen.

If we look very closely at the nuclei of hydrogen, helium, and oxygen, we discover that we must modify some of the statements just made. For example, *most* of the hydrogen nuclei that we find in Nature do consist of a single proton and nothing else. But a small fraction of hydrogen nuclei (0.015 percent) contain one proton plus one neutron. These nuclei have mass number $A = 2$, and we abbreviate them as ^{2}H. This "heavy hydrogen" is often called *deuterium*. Also, we find that some oxygen nuclei contain nine or ten neutrons instead of the eight neutrons found in the common variety of oxygen. Thus, oxygen occurs in three forms in Nature: ^{16}O, ^{17}O, and ^{18}O. Helium occurs in two forms: ^{3}He and ^{4}He. Nuclei of the same element that differ only in the number of neutrons are called *isotopes* of the element. That is, natural hydrogen consists of two isotopes, ^{1}H and ^{2}H; natural oxygen consists of three isotopes, ^{16}O, ^{17}O, and ^{18}O; and similarly for other elements.

The electron shell structure of an atom depends only on the number of protons in the nucleus (that is, on Z) and not on the number of neutrons. Therefore, the chemical properties of the isotopes of a particular element are identical (but the nuclear properties are not). Thus, the isotopes ^{16}O, ^{17}O, and ^{18}O cannot be distinguished by chemical means.

Most elements have several isotopes that occur in Nature, but a few elements have only a single natural isotope. Tin has ten isotopes, but fluorine, aluminum, cobalt, gold, and a few other elements occur naturally in only one isotopic form.

We can separate all isotopes into two categories. Those isotopes that do not exhibit radioactivity are called *stable* isotopes. All stable isotopes are found in Nature. Isotopes that undergo radioactive decay are called *radioisotopes*. Radioisotopes exist for *all* of the elements. For example, there is an isotope of hydrogen that consists of one proton and two neutrons, ^{3}H. This isotope, called *tritium*, is radioactive and undergoes β decay to form ^{3}He. As we will see in the next section, tritium exists only a few years before it decays away. Therefore, tritium does not occur naturally in the Earth and it must be prepared artificially. This is the case with most radioisotopes: The lifetimes are generally too short for the isotopes to be found in Nature, but they can be produced in the laboratory. Some of the stable and radioactive isotopes of the light elements are listed in Table 12-2.

Some radioisotopes do occur naturally in the Earth. The isotopes ^{235}U and ^{238}U (uranium) and ^{232}Th (thorium) have lifetimes of a billion years or more, so quantities of these isotopes remain from the time the Earth was formed, about $4\frac{1}{2}$ billion years ago. Not only do we find uranium and thorium radioactivity in the Earth's crust, but we also find their radioactive decay products. That is, when ^{238}U undergoes α-particle decay, radioactive ^{234}Th is formed; ^{234}Th undergoes β-particle decay, producing ^{234}Pa (protactinium), which is also radioactive. This chain of radioactive decays continues until a stable product is formed; the end product of ^{238}U radioactivity is ^{206}Pb (lead). The sequence of decays that begins with ^{238}U and ends with ^{206}Pb involves eight α decays and six β decays. (The complete decay scheme is given in the section on *Additional Details* at the end of this chapter.)

Table 12-2 *Properties of Some Isotopes of Light Elements*

Element	Z	A	Symbol	Radioactivity type or relative natural abundance
Hydrogen	1	1	^{1}H	99.985%
(Deuterium)	1	2	^{2}H	0.015%
(Tritium)	1	3	^{3}H	β radioactive
Helium	2	3	^{3}He	0.00015%
	2	4	^{4}He	99.99985%
	2	6	^{6}He	β radioactive
Lithium	3	6	^{6}Li	7.52%
	3	7	^{7}Li	92.48%
	3	8	^{8}Li	β radioactive
Beryllium	4	8	^{8}Be	α radioactive
	4	9	^{9}Be	100%
	4	10	^{10}Be	β radioactive

α, β, and γ Decays

We can now understand in more detail how α- and β-decay events take place and why these processes are so different. The nucleus ^{226}Ra consists of 88 protons and 138 neutrons. When ^{226}Ra undergoes α decay, two of these protons and two of these neutrons join together to form an α particle which is then ejected from the nucleus. The remaining nucleus contains 86 protons and 136 neutrons, and this nucleus is an isotope of radon (^{222}Rn). This decay process is shown schematically in Fig. 12-5a.

Figure 12-5 *Schematic representation of the two different types of radioactivity processes. (a) The α decay of ^{226}Ra to form ^{222}Rn. (b) The β decay of ^{14}C to form ^{14}N. Notice that one of the nuclear neutrons is converted into a proton in this decay event.*

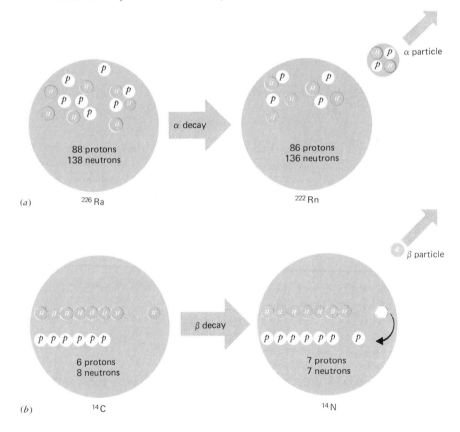

The change that takes place in a β-decay event is quite different from the change involved in the emission of an α particle. Figure 12-5b shows the β decay of the radioactive carbon isotope, ^{14}C. The original nucleus consists of six protons and eight neutrons. In the β-decay process, one of the nuclear neutrons is converted into a proton and, simultaneously, an electron (β particle) is emitted. The final nucleus therefore consists of seven protons and seven neutrons; that is, the final nucleus is an isotope of nitrogen ($Z = 7$). In α decay, the protons and neutrons that are emitted in the form of an α particle pre-exist in the original nucleus. In β decay, however, the emitted electron does *not* pre-exist in the original nucleus; instead, the electron is created at the instant of decay and does not remain in the nucleus.

In α decay, both the atomic number and the mass number of the nucleus change. In β decay, the atomic number of the nucleus changes but the mass number does not. In γ decay, radiation is emitted but no particles are emitted; in these processes neither A nor Z changes. The emission of γ rays usually *follows* α or β decay as the protons and neutrons in the final nucleus rearrange themselves into the lowest possible energy state. For example, when ^{60}Co (cobalt) undergoes β decay, a ^{60}Ni (nickel) nucleus is formed. Immediately following the decay process, the protons and neutrons in the ^{60}Ni nucleus readjust themselves in two steps and two γ rays are emitted. It is these γ rays (and not the decay β particles) that are the desired radiations in most medical and industrial applications of ^{60}Co.

The Half-Life

The radioisotope ^{131}I (iodine) is often used in diagnosing and treating abnormal thyroid conditions. One of the reasons for using this particular isotope of iodone is that it retains its radioactivity for a convenient period of time—not too short a time to limit its usefulness and not too long a time to represent a hazardous burden to the body. Radioactive iodine, or *any* radioisotope, does not suddenly lose its activity, nor does the activity decrease steadily to zero at a certain time. Instead, radioactive substances obey a different kind of decay law. Let us examine this decay law for the case of ^{131}I.

The radioisotope ^{131}I undergoes β decay to form ^{131}Xe

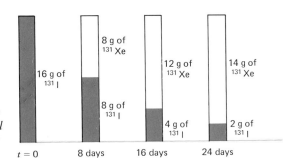

Figure 12-6 *The radioactive decay of a sample of ^{131}I. In each interval of 8 days, the amount of ^{131}I in the sample decreases by one-half.*

(xenon). Suppose that we begin with a 16-gram (g) sample of ^{131}I. Over a period of time we measure the amounts of iodine and xenon in the sample. After 8 days, we find that the sample contains 8 g of ^{131}I and 8 g of ^{131}Xe. After 16 days, we find that the amount of iodine has decreased to 4 g and that the xenon has increased to 12 g. After 24 days, we find only 2 g of ^{131}I and 14 g of ^{131}Xe. At later times we continue to find a buildup of ^{131}Xe and a decrease in the amount of ^{131}I in the sample. The relative amounts of iodine and xenon in the sample at various times are shown in Fig. 12-6. The law obeyed by the

Figure 12-7 *The radioactive decay curve for ^{131}I. The half-life of ^{131}I is 8 days.*

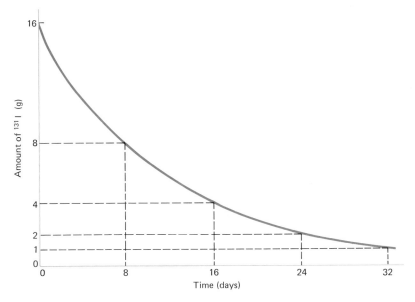

decaying ^{131}I is this: At the end of *any* 8-day interval, the amount of ^{131}I in the sample is exactly one-half the amount at the beginning of the interval. It does not matter when we choose to begin an 8-day interval, there is always a decrease by one-half during this period. For example, if we were to measure the amount of ^{131}I in the sample after 5 days, we would find 10.4 g; another measurement 8 days later (at $t = 13$ days) would show that 5.2 g of ^{131}I remained. A graph of the amount of ^{131}I remaining at any time is shown in Fig. 12-7. The time period of 8 days is characteristic of ^{131}I and is called the *half-life* of the radioisotope.

All radioisotopes follow the same type of decay curve that describes the behavior of ^{131}I. Each radioisotope has its own particular half-life, and these vary from many billions of years to tiny fractions of a second. Any sample of radioactivity will eventually decay away, but some substances will decay much more rapidly than others. We must remember that matter is not lost in the decay process; the decay of one isotope means the formation of another. A few interesting radioisotopes and their half-lives are listed in Table 12-3.

The half-life of a radioisotope is an important factor in determining how effective the radiations will be in particular situations. For example, if the half-life of ^{131}I were 8 minutes instead of 8 days, any radioactive iodine injected into a patient would decay away before it could concentrate in the desired region, namely, the thyroid gland. On the other hand, if the half-life were 8 months, then a much larger amount of the isotope must be injected into the patient to provide the desired level of activity; this increased long-term radioactive burden may produce undesirable effects.

Another situation in which the half-life is important is in dealing with radioactive waste materials from nuclear power reactors. As we will learn in the next chapter, the fission events that take place in a reactor produce a large number of radioisotopes. When the rods that carry the nuclear fuel have become depleted in uranium, they are removed from the reactor, and the radioactive residue must be disposed of. Many of the radioisotopes contained within the fuel rods have short half-lives (up to a few weeks). Therefore, the safest procedure for dealing with the radioactive wastes from nuclear reactors is simply to place the used fuel rods in a remote or shielded location and to allow them to "cool off" for several weeks. After the short-lived activities have effectively decayed away,

Table 12-3 Properties of Some Radioisotopes

Isotope	Type of Radioactivity	Half-Life	Remarks
^{235}U (uranium)	α	0.7×10^9 y	Used in nuclear power reactors and in some weapons
^{14}C (carbon)	β	5730 y	Used in archeological dating and in biological tracer studies
^{226}Ra (radium)	α	1620 y	α particles from this source first identified as helium nuclei (Rutherford, 1909)
^{90}Sr (strontium)	β	28.8 y	Fission product prominent in fallout from nuclear weapons tests
^{3}H (hydrogen)	β	12.3 y	Used in nuclear fusion devices (H bombs)
^{60}Co (cobalt)	β	5.3 y	Widely used in cancer treatments and for industrial irradiations
^{131}I (iodine)	β	8.0 days	Used in medical diagnostics and therapeutics, particularly in thyroid cases
^{24}Na (sodium)	β	15.0 h	Used in medical diagnostics to follow the flow of blood in the body

the remaining waste materials can be handled and treated much more easily.

Ionization by Radiation

What happens when radiation passes through matter? There are three types of radiation that we need to consider: α particles, β particles (electrons), and electromagnetic radiation (X rays and γ rays). The common feature of these various radiations is that they all produce *ionization* in passing through matter. But the different radiations have distinctive ionization characteristics.

An α particle has a mass that is about 7000 times greater than the mass of an electron (or β particle). Therefore, if an α particle and an electron have the same kinetic energy, the speed of the electron will be much greater than that of the heavier α particle. In fact, any electron that has an energy of more than a few kiloelectronvolts will have a speed greater than the most energetic α particle emitted from a radioisotope (about 5 MeV).

When an α particle passes through matter, it moves relatively slowly and therefore has sufficient time to interact with and ionize the individual atoms along its path. A rapidly moving electron, on the other hand, has less time to interact with the atomic electrons and to remove them from the atoms. A fast electron therefore produces ions much less frequently than does an α particle. Thus, there are many more ions per centimeter in the track of an α particle moving through matter than there are in the wake of an electron.

An α particle is much more massive than the atomic electrons with which it collides in passing through matter. Consequently, an α particle is deflected no more in an ionization collision than a bowling ball would be in a collision with a golf ball. Thus, an α particle plows through matter in an almost straight path, leaving a high density of ions behind (see Fig. 12-8a). An electron produces a far smaller density of ions. Moreover, because the electron has the same mass as the atomic electrons with which it collides, the electron's direction of motion is frequently changed in these collisions. Therefore, when an electron passes through matter, it leaves a diffuse collection of ions along an erratic path (see Fig. 12-8b).

From this discussion of how ions are produced, we can see one other important difference between the ways in which α particles and electrons interact with matter. Each ion that is

Figure 12-8 (a) *An α particle leaves a dense track of ions.* (b) *An electron leaves a diffuse collection of ions along an erratic path.*

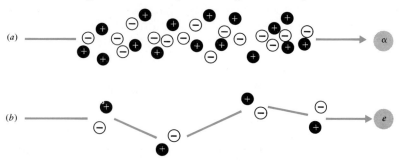

produced in a particular substance requires the expenditure of energy by the ionizing particle. This energy is the same whether the ionizing particle is an α particle or an electron. A particle will lose speed in each ionization collision until it finally comes to rest. Therefore, an electron with a certain energy will move a considerably greater distance in matter before coming to rest than will an α particle with the same energy (or larger energy). We say that electrons are more *penetrating* than are α particles. In fact, electrons from many radioactive substances can easily pass through clothing and skin, and can then penetrate into the body. On the other hand, α particles are stopped by even a thin sheet of paper. Generally, α particles can enter the body only if the radioactive material is ingested or breathed.

Finally, we must consider how X rays and γ rays interact with matter. These radiations do not carry any electrical charge, so they cannot produce ions in the same way that α particles and electrons do. In Chapter 10 we learned that photons can eject electrons from atoms by the photoelectric effect. This same process is one of the important ways that X rays and γ rays produce ionization in matter. In the photoelectric effect, the photon is absorbed and disappears; the energy of the incident photon (less the electron binding energy) is carried away by the electron. It is also possible for a photon to be deflected by an atomic electron in much the same way that one billiard ball is deflected by another billiard ball. In this process, some of the energy of the incident ball (the photon) is transferred to the other ball (the atomic electron). The incident photon does not disappear, but a sufficient amount of energy is transferred to the electron to separate it from the atom and to give it a high speed. The incident photon, now degraded somewhat in energy, moves away to interact again at another site. This type of interaction is called the *Compton effect*, after the American physicist Arthur H. Compton (1892–1962), who studied the process in the early 1920s.

The high-speed electrons that are produced in the photoelectric and Compton processes by X rays and γ rays move through the material and cause further ionization. This secondary ionization is in no way different from that produced by β particles from radioisotopes.

Because X rays and γ rays are not charged, they can penetrate deeply into matter before they interact to release electrons. High-energy γ radiation is very difficult to stop com-

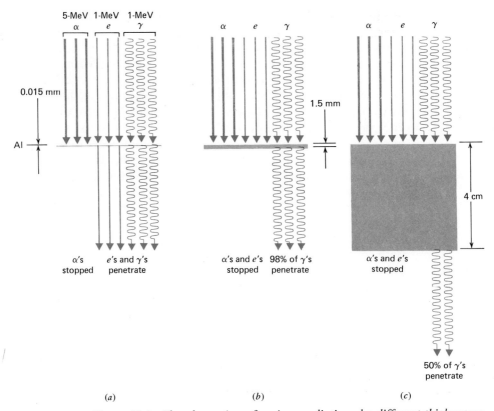

Figure 12-9 *The absorption of various radiations by different thicknesses of aluminum.*

pletely. Thick shielding walls are necessary to prevent energetic γ rays from entering areas that must be maintained radiation-free.

The relative penetrating powers of the three types of radiation are illustrated in Fig. 12-9. Here, we imagine that 5-MeV α particles, 1-MeV electrons, and 1-MeV γ rays (typical radioactive emissions) are incident on various thicknesses of aluminum. An aluminum sheet only 0.015 mm thick is sufficient to stop the α particles completely, but this amount of material has essentially no effect on the electrons and γ rays. If the thickness is increased to 1.5 mm, the electrons will be stopped, but only about 2 percent of the γ rays will be absorbed. Finally, a 4-cm-thick piece of aluminum is seen to reduce the γ-ray intensity only to 50 percent. Lead absorbs γ radiation much more efficiently than does aluminum: A 4-cm-thick piece of lead will reduce the intensity of 1-MeV γ rays to about 2.5 percent.

Chemical Effects of Radiation

The basic effect that all types of radiation have on matter is to produce ionization within the matter. It is this internal ionization that makes radiation both extremely useful and potentially hazardous. Let us consider what will happen if we irradiate a sample of a very simple chemical compound such as methane, CH_4. We could use any type of radiation, but because of the limited penetrating power of α particles, either electrons or electromagnetic radiation is more convenient. When energetic electrons or γ rays pass through methane, we expect that CH_4^+ ions will be produced, and indeed they are. But even more interesting is the fact that many other types of ions are also produced. The reason is that molecules are rather fragile structures whose bonds can be broken by the absorption of only a few electronvolts of energy. Therefore, in the irradiated sample we find a variety of ionic forms that are produced from methane: CH_4^+, CH_3^+, CH_2^+, CH^+, C^+, and H^+. These ions are chemically active and can combine among themselves or with methane molecules to produce a number of new molecules. Figure 12-10 shows one such

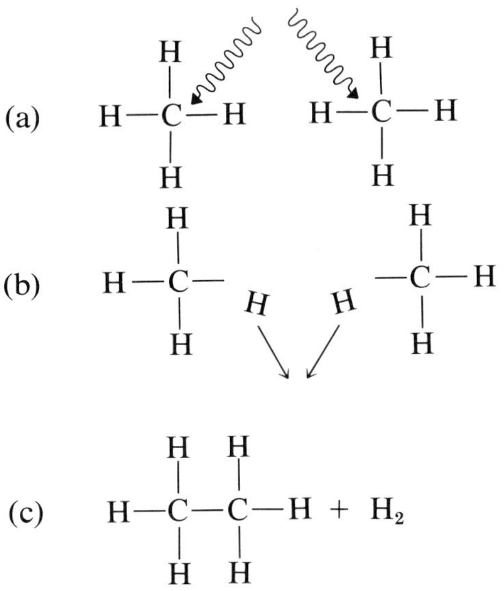

Figure 12-10 *The formation of ethane, C_2H_6, from irradiated methane. (a) γ radiation is incident on two methane molecules. (b) Absorption of the radiation produces two CH_3^+ ions. (c) The two CH_3 units combine to form C_2H_6. The two hydrogen atoms that were broken off the CH_4 molecules can combine to form hydrogen gas, H_2.*

process, namely, the formation of ethane, C_2H_6, from irradiated methane.

When methane is actually irradiated, dozens of new molecular species are found in the sample. These range from H_2, C_2H_6, C_2H_4 (ethylene), and C_3H_8 (propane) to long-chain hydrocarbons with as many as 20 carbon atoms per molecule. When more complex compounds (or mixtures of compounds) are irradiated, a wide variety of new molecules can be formed. Several different industrial chemical processes depend on radiation effects to produce the desired material. For example, the Dow Chemical Company each year produces about a million pounds of ethyl bromide (a gasoline additive) by ^{60}Co irradiations.

In the preceding chapter we learned that the cross-linking of polymers produces increased rigidity in plastic materials. Cross-linking can be accomplished by irradiation with electrons or some form of electromagnetic radiation (ultraviolet, X-ray, or γ radiation). Figure 12-11 shows in a schematic way how two long-chain polymers become cross-linked. Irradiation of the polymer sample tends to break hydrogen atoms off the chains. Frequently, a detached hydrogen atom will wander to a site that lacks a hydrogen atom and recombination will take place. But if two active sites on a pair of polymers find themselves in close proximity, as in Fig. 12-11, they will join together, producing a cross-link between the molecules. Cross-linking can also be accomplished by heat treatment (this was the earliest method used), but for commercial production of plastic materials, radiation techniques are faster, cheaper, easier to control, and do not suffer from the undesirable effects caused by elevated temperatures.

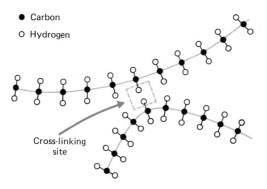

Figure 12-11 *The cross-linking of a pair of polyethylene molecules by irradiation. The radiation removes a hydrogen atom from each molecule and the two active sites bond together to produce the cross-linking.*

The cross-linking of polymers is important in the curing of certain types of paints in commercial production facilities. Any one of a number of special paints can be applied to a metal or plastic surface which is then conveyed past a radiation source. An irradiation of only a few seconds is all that is required to cross-link the polymers and thereby to cure the paint. This method of curing is particularly useful when paint must be bonded to a plastic surface because ordinary heat treatment would tend to damage the material.

Radiation Applications

In addition to promoting chemical activity, radiation techniques have found numerous applications in other fields. In this section we will mention only a few of these applications; in the following section we will see how radiation methods are used in medical diagnostics and therapeutics.

Radiation techniques can be used for sterilization processes because microbial activity is easily destroyed by an intense dose of ionizing radiation. Medical sutures (which are used for closing wounds and incisions) are packaged individually because they are generally used one at a time. The manufacturer must sterilize each suture so that it will be absolutely free of microbes when it is removed from its package. The most convenient and efficient method of sterilizing packaged sutures is to pass them through a beam of electrons or γ radiation. This procedure is much faster than conventional heat treatment and is completely effective in destroying microbial activity. Most of the sutures now used in the United States are sterilized in this way.

Canned foodstuffs are always submitted to a heat treatment to kill the bacteria and other microorganisms that would otherwise attack the food and cause it to "spoil." Fresh produce and meats cannot be treated in this way, and they usually deteriorate within a few days or weeks. However, by irradiating fresh foods, the undesirable microorganisms can be destroyed without the necessity of heating the foods. Unfortunately, the large doses of radiation that are necessary for food preservation often affect the color, flavor, odor, and texture of the food. In some cases, the irradiation may produce undesirable and potentially dangerous chemical compounds. For these reasons, the use of radiation methods for food preservation has been proceeding cautiously. The U.S. Food and Drug Administration (FDA) has so far approved the use of radiation only to prevent sprouting of potatoes that are stored for long periods of time and to kill insects in various grains.

One of the problems associated with the manufacture of goods on a mass-production basis is to ensure that each item leaving the production line meets specifications. For example, a pharmaceutical manufacturer must determine whether each vial or capsule has been filled with the proper amount of material. Visual observations are often not possible (nor sufficiently reliable). One method of monitoring a production line with radiation techniques is illustrated in Fig. 12-12. Here, we wish to check the level to which a vial has been filled with a liquid. On one side of the production line we position a radiation source (which could be a radioisotope that emits an appropriate radiation); on the opposite side we place a radiation detector. When a vial containing the proper filling passes between the source and the detector (Fig. 12-12a), the meter indicates a certain amount of radiation received by the detector. The equipment is adjusted so that whenever the meter

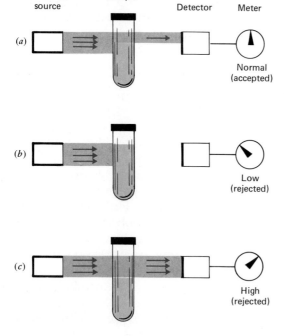

Figure 12-12 *Control of vials on a production line by radiation methods. An improperly filled vial (b, c) results in a higher-than-normal or lower-than normal amount of radiation reaching the detector, thereby triggering a rejection mechanism.*

indicates this value within narrow limits, the vial then being checked is allowed to pass through the system and be packed for shipment. However, if a vial has received too much liquid, the amount of radiation reaching the detector is less than normal (Fig. 12-12b) and the low meter reading triggers a mechanism to reject this particular vial. Similarly, if the vial has too little liquid, the meter reading will be high (Fig. 12-12c) and this vial will also be rejected. By inspecting every unit on the production line, an extremely high level of quality control can be achieved. (The traditional method of inspection calls for removing one item in a hundred or a thousand for checking. This leads to a low degree of quality control.)

Suppose that a certain object must be tested to determine whether it contains a certain element. If a chemical method is used, a small sample must be cut from the object in order to carry out the tests. In many situations (for example, in examining works of art), the removal of a test chip from the object is undesirable or impossible. Various radiation techniques can be used for the nondestructive examination of objects. One of these methods involves the use of X-ray photography (explained earlier) to examine underlying layers of paint. Another technique is known as *neutron activation analysis*. For example, if ^{23}Na (the only stable isotope of sodium) is irradiated with neutrons, some of the nuclei will absorb a neutron and become the radioactive isotope ^{24}Na. Now, ^{24}Na undergoes β decay with a half-life of 15 h and in the process emits γ rays with energies of 2.75 and 1.37 MeV. This half-life and these γ-ray energies are unique to ^{24}Na. In fact, every radioisotope has its own characteristic decay signature in terms of its half-life and radiation energy. Therefore, if a certain object is irradiated with neutrons and if the object then emits 2.75- and 1.37-MeV γ rays with a half-life of 15 h, we can be certain that the object contains sodium. Only a tiny fraction of the nuclei in a sample are transmuted in this type of testing procedure. Consequently, neutron activation analysis is not only *nondestructive* but it gives a highly specific indication of the elements in the sample. The authenticity of works of art have been tested by performing activation analyses of the paints used. If the composition of a paint shows it to have been manufactured only recently, long after the death of the artist, this is sufficient reason to pronounce the work a fake.

The process of neutron activation actually causes radioactivity to be produced within the irradiated sample. This is *not* true, however, for irradiations by X rays, γ rays, and electrons (at least, not in the energy range we are considering). Thus, there is no danger that a pharmaceutical or any other product that emerges from a testing procedure involving radiation can itself be radioactive. Foods that have been given radiation doses for preservation cannot become radioactive because of the irradiation.

Biological and Medical Uses of Radiation

All living matter functions by the continual production of an enormous number of complex organic compounds. Our understanding of how these compounds are formed and how they participate in life processes is largely derived from studies performed with radioactive *tracer* isotopes. Organic compounds consist primarily of the elements carbon, hydrogen, and oxygen, with important contributions from nitrogen, potassium, sulfur, iron, and several other elements. It is possible to prepare radioactive isotopes of all of these elements. These isotopes, when introduced into a living system, will participate in chemical reactions in exactly the same way as will their stable counterparts. If various chemical compounds are extracted from the system and tested for radioactivity, it is possible to determine, in many instances, how the radioisotope became incorporated into the compounds. From this kind of information we can deduce the nature of the chemical reactions involved in the life processes in plants and animals.

Another important application of radioisotopes in living systems is in diagnostic and therapeutic uses with human patients. We have already mentioned the medical use of radioactive iodine (^{131}I). The element iodine is selectively absorbed by the thyroid gland. Therefore, when a small amount of ^{131}I is ingested by a patient, the isotope quickly migrates toward and is deposited in the thyroid gland. A radiation detector is used to monitor the rate at which the radioactive iodine accumulates in the thyroid. The rate of uptake of iodine is greater for cancerous tissue than for normal tissue. Consequently, the rate measurement serves to identify unhealthy tissue. Once a carcinoma has been diagnosed by using ^{131}I, the same radioisotope can be used therapeutically to attack the diseased area. When ^{131}I undergoes β decay, the emitted radiation can destroy tissue in its vicinity by virtue of its ionization and disruption of chemical bonds in the tissue molecules. The radiation destroys both healthy and unhealthy tissue. But because the iodine concentrates in the unhealthy tissue, more of the diseased material is attacked than is the normal material. The use of radioactive iodine is now a routine procedure for diagnosing and treating thyroid conditions.

More than a dozen radioisotopes are currently being used in standard medical procedures. The isotopes used are chosen for their specificity, that is, their tendency to concentrate selectively in certain organs or areas of the body, and for their radiation characteristics (half-life, type and energy of radiation). For example, small tumors in the eyeball can be located by using ^{32}P (phosphorus); ovarian cancer can be treated with ^{198}Au (gold); and ^{90}Y (yttrium) is used when tissue in the pituitary gland must be destroyed.

Certain types of cancers do not lend themselves to treatment with radioisotopes administered internally. These cases are usually treated with external sources of radiation. The γ rays that result from the decay of ^{60}Co are used in the treatment of a variety of cancers. The radioactive cobalt is contained in a large lead housing that has a hole through which the γ rays emerge in a narrow beam. The beam is directed toward the cancerous region. To prevent an excess of radiation exposure to healthy tissue, the patient is usually rotated or rocked about the point of treatment. In this way the cancerous region receives a maximum dosage, whereas the surrounding parts of the body receive a minimal exposure. In some cases, a small, needle-shaped sample of cobalt is inserted surgically into the region to be irradiated. After the cancerous growth has been arrested or destroyed, the cobalt needle is removed.

Radiation Doses

Radiation is useful in the treatment of cancers because the intense ionization produced by the radiation leads to the destruction of the diseased tissue. But it is also true that radiation is capable of producing severe damage in healthy tissue. Although radioactivity has many important uses in medicine, science, and industry, it must be recognized that these beneficial materials are also potentially hazardous.

The radiations to which the general public is exposed are almost exclusively in the form of X rays or γ rays. Individuals who work with radioactivity sometimes come into contact with materials that emit α and β particles. All types of ionizing radiation can produce biological damage in living tissue. When some part of a biological system is damaged (by radiation or in some other way), the organism tends to heal itself by producing new cells. If the irradiation causes only weak and scattered ionization (as would be the case for a low-level exposure to electrons or γ rays), the organism will be able to repair itself and there will be a relatively small amount of permanent damage, or perhaps there will be no permanent effect at all. However, when an α particle passes through matter, it leaves an exceptionally dense track of ions in its wake. A biological system is much less able to cope with concentrated ionization than it is with diffuse ionization. Consequently, the biological damage produced by an α particle is about 10 times as severe as that produced by an electron or γ ray of the same energy. (We say that an α particle has a relative biological effectiveness or *quality factor* that is 10 times that of an electron or γ ray.)

In order to specify the radiation dosage received by an object or individual, we use a unit that specifies the amount of radiation energy absorbed per unit mass. This unit is the *rad*, which stands for *radiation absorbed dose*. The amount of energy that is deposited in a 1-kg object when the dose is 1 rad is actually quite small by ordinary standards:

$$1 \text{ rad} = \frac{1}{420\ 000} \text{ Cal/kg}$$

This means that a dose of 420 000 rad (a huge dose!) is required to deposit 1 Cal in 1 kg of matter. (Remember from our discussion in Chapter 6 that 1 Cal is the amount of energy required to raise the temperature of 1 kg of water by 1 Celsius degree.) Because the human body consists primarily of water, its thermal properties are very similar to those of water. Thus, if an individual received a dose of 1000 rad uniformly over his body, the body temperature would be increased by only 0.0024 Celsius degree. But this dose would be fatal! The effectiveness of radiation in producing biological damage is enormously greater than would be expected simply on the basis of its ability to produce heating within the body. Ionization causes the molecules to break apart, whereas heating merely causes them to jiggle about more rapidly. In the next section we will discuss some of the damaging effects of radiation.

Radiation doses are now usually specified in terms of a unit called the *rem*. This unit measures the *dose equivalent* of different types of radiation and takes into account the greater effectiveness of α particles compared to electrons and γ rays for producing biological damage. For electrons and γ rays a dose of 1 rad is the same as a dose of 1 rem, but for α particles a dose of 1 rad is equivalent to a dose of 10 rem.

When considering the biological effects of radiation, we must remember that the unit of absorbed dose—the *rem*—refers to the amount of energy absorbed *per kilogram*. Thus, the amount of radiation energy absorbed by an individual who receives a 1-rem dose to his arm is much less than the amount that would be absorbed if he received a 1-rem dose to his entire body.

Radiation Damage in Humans

The effects of radiation on humans can be classified as *somatic* (effect on the body cells of the individual exposed) or *genetic* (effect on the offspring of the individual exposed). We must also distinguish between *acute* and *chronic* exposures to radiation. If a dose is delivered to an individual within a relatively brief interval of time, the exposure is said to be acute. If the dose is accumulated over a relatively long period of time by continuous exposure to low levels of radiation, the exposure is said to be chronic. The distinction is important because the repair mechanism of the body can cope with a long-term, low-level exposure, but it will be swamped by a large dose delivered in a short time. Thus, an acute exposure is more effective in producing biological damage than is a chronic exposure that delivers the same dose. A 1000-rem whole-body dose to an individual will be fatal if it is acute, but the same dose delivered over a 40-year period might produce no observable pathological effects.

The somatic effects on an individual exposed to radiation depend on many factors—the age and general health of the individual, the particular part of the body exposed, and the size of the dose. An acute, whole-body dose of 25 rem delivered to a healthy young adult usually will produce no observable effects. A dose of 100 rem will cause some changes in the blood, but no illness. A dose of 300 rem will produce radiation sickness, but most individuals will recover completely. A very large exposure of 800 rem almost certainly will be fatal. The somatic effects of acute exposures at various dose levels are listed in Table 12-4. In addition to these immediate somatic effects, there are also long-term effects such as increased susceptibility to leukemia, bone cancer, and eye cataracts, general shortening of lifespan, and abnormal growth patterns. Thus, a person who receives an acute, whole-body dose of 40 rem would suffer some mild effects of radiation sickness and he would recover completely. But his life expectancy would be reduced by about a year due to the exposure, and he might contract some type of radiation-induced cancer during the next 20 to 30 years.

The genetic effects of human exposure to radiation are much more subtle and less well understood than the somatic effects. We still know relatively little about the way in which the hereditary information carried by DNA molecules is affected by radiation. It is clear, however, that genetic damage does occur, and it cannot be assumed that there is a radiation dose level below which these effects are unimportant.

What is the level of radiation to which we are exposed? The

Table 12-4 Somatic Effects of Acute Radiation Exposure

γ-Ray Whole-Body Dose (rem)	Effects	Remarks
0–25	None detectable	
25–100	Some changes in blood but no great discomfort, mild nausea.	Some damage to bone marrow, lymph nodes, and spleen.
100–300	Blood changes, vomiting, fatigue, generally poor feeling.	Complete recovery expected; antibiotic treatment.
300–600	Above effects plus infection, hemorrhaging, temporary sterility.	Treatment involves blood transfusions and antibiotics; severe cases may require bone marrow transplants. Expected recovery about 50 percent at 500 rem.
Greater than 600	Above effects plus damage to central nervous system.	Death inevitable if does greater than 800 rem.

largest part of the radiation exposure received by an individual who is not a radiation worker is from natural sources—cosmic rays and the radioactivity that occurs naturally in the Earth. The amount of this natural radiation received annually by an individual varies between about 0.090 and 0.150 rem in this country; the average dose is approximately 0.100 rem/y. Medical and dental X rays account for 0.05 to 0.100 rem/y; the average is 0.075 rem/y. Fallout from weapons tests contributes about 0.005 rem/y (and this figure will gradually decrease with time because the major nuclear nations no longer carry out tests in the atmosphere). The total average dose from all these sources is approximately 0.180 rem/y.

To this figure we must add an average exposure of about 0.000 003 rem/y for the various radiation effects attributable to the nuclear power industry. This exposure appears so small by comparison with the other sources that we might be tempted to dismiss it altogether. However, the proliferation of nuclear power plants will cause this exposure to increase to about 0.003 rem/y by the year 2000. Moreover, this is an *average* exposure; individuals who live close to a nuclear facility will be likely to receive several times more radiation than those who live a considerable distance away. Even so, the maximum probable exposure that any individual might suffer due to nuclear power plant operations is small compared to the exposure from natural sources.

The estimates of current and future exposures due to radiations released in nuclear power plant operations are based on the assumption of *normal* operations. There is considerable controversy regarding the likelihood of nuclear accidents and the severity of radiation exposures that would result from such accidents. However, even the most pessimistic estimates indicate that there is only a very tiny probability that a nuclear accident will injure any substantial number of persons. We shall discuss this important topic further in the next chapter.

One very striking way to present the biological effects of radiation is to give the number of radiation-induced cancers that would be produced in the United States due to the particular radiation source. This information is given in Table 12-5. If we divide the number in the last column by the U.S. population (220 million), we obtain the likelihood per year that any individual will suffer a radiation-induced cancer. Thus, there

Table 12-5 Radiation Doses and Effects in the United States

Source	Average per Capita Dose (rem/y)	Radiation Cancers per Year if Entire U.S. Population Received Average Dose
Natural background at sea level	0.100	5000
Medical and dental X rays	0.075	3750
Fallout	0.005	250
Nuclear power operations:		
1976	0.000 003	0.15
2000	0.003	150

is a chance of about one in a million that you will contract a cancer this year due to the radiation from weapons test fallout. Notice the extremely small cancer rate due to the present nuclear power program.

Because there are long-term somatic effects of radiation exposure, and because we do not completely understand the genetic effects, the recommendations for maximum permissible radiation exposures have been set very much lower than the level at which somatic effects become evident. The maximum allowable whole-body exposure for an individual in the

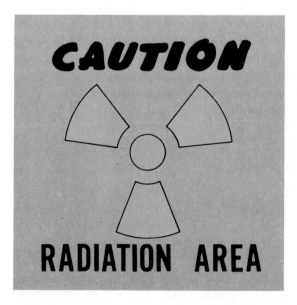

Figure 12-13 *This symbol is internationally recognized as a warning that radioactive materials are being handled or that artificial radiations are being produced in the vicinity.*

general public (not a radiation worker) has been set at 0.5 rem of electron or γ radiation. Very few individuals are actually monitored to determine the extent of exposure. Therefore, a limit of 0.170 rem/y (over and above the dose due to medical radiations) has been set for a *typical* individual in the population. It is generally believed that this level of radiation exposure represents an acceptable risk that is balanced by the benefits of our utilization of radiation.

Not everyone subscribes to this view, however. Some persons argue that steps should be taken to reduce the amount of radiation from all non-natural sources. This could be accomplished by stopping all weapons testing, by curtailing the use of X rays for nonessential diagnostic purposes, by reducing the radiation from television sets and microwave ovens, and by preventing the further development of the nuclear power industry. Some of these measures seem to be desirable, whereas the general public may be unwilling to accept the additional inconvenience and expense associated with others.

Questions and Exercises

1. The Earth was formed about $4\frac{1}{2}$ billion years ago. After a period of cooling, water condensed and covered some fraction of the Earth's surface. These waters contained a rich broth of chemicals. The Earth's atmosphere was not as thick as it is today and did not shield the surface from the intense ultraviolet and higher-energy radiations from the Sun. Speculate on how the irradiation of the chemical-laden waters of the Earth might have produced complex chemical compounds that eventually became able to reproduce themselves.
2. What isotope is formed by the β decay of ^{12}B (boron)? By the α decay of ^{8}Be (beryllium)?
3. A sample of a certain radioisotope is placed near a Geiger counter (a type of radiation detector). The counter registers 400 decay events per second. Twenty-four hours later, the detector counts at a rate of 50 per second. What is the half-life of the radioisotope? [Ans. 8 h.]
4. The manufacturer of a certain detergent wishes to compare the effectiveness of his product for removing dirt from clothes

to that of a competing product. Suggest how a radioisotope might be used in such an investigation.

5. In a steel mill, huge sheets of steel are formed by passing the hot metal through a pair of rollers. Explain how radioactivity methods might be used to determine whether the rollers maintain their proper spacing during the course of a production run.

6. Two of the radioisotopes that are found in the fallout from nuclear weapons tests and in the waste products of nuclear power plants are ^{90}Sr (strontium) and ^{137}Cs (cesium). Cesium is chemically similar to potassium and strontium is chemically similar to calcium. Why are these similarities important?

7. Bricks and concrete usually contain small amounts of radioactivity, particularly ^{40}K (potassium) and ^{226}Ra (radium); wood, on the other hand, contains very little radioactivity. The fire hazard in a brick and concrete house is much smaller than in a frame house. Do you believe that the added radiation dose that would be received in a brick and concrete house (perhaps 0.050 rem/y) is worth the increased benefit in terms of fire protection?

Additional Details for Further Study

Radioactive Decay Chains

The heaviest elements found in Nature are uranium (U, $Z = 92$), protactinium (Pa, $Z = 91$), and thorium (Th, $Z = 90$). All of the isotopes of these elements are radioactive but each element has at least one isotope with a sufficiently long half-life that the element still exists in Nature. For example, ^{238}U has a half-life of 1.4 billion years. When these nuclei decay, they form new daughter elements that are also radioactive. Some of these nuclei are β radioactive and others emit α particles. A few can even decay by either α or β emission. A series of successive radioactive decays takes place that continues until a stable isotope of either lead (Pb, $Z = 82$) or bismuth (Bi, $Z = 83$) is formed. The stable isotopes of lead are ^{206}Pb, ^{207}Pb, and ^{208}Pb; only ^{209}Bi is stable. These four nuclei are the termination points for all of the radioactive decay chains that originate with the long-lived heavy elements. One

Radioactive Decay Chains   401

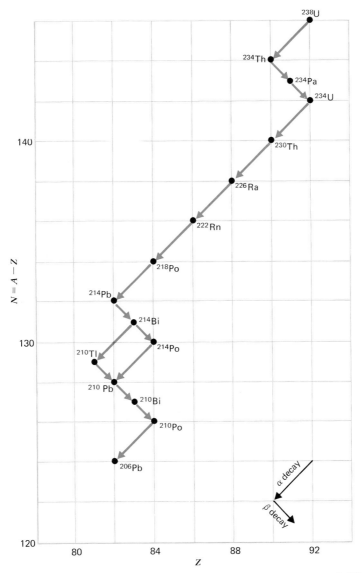

Figure 12-14 *The radioactive decay chain that originates with ^{238}U and ends with ^{206}Pb. Notice that ^{214}Bi can undergo either α or β decay and that after one additional decay, both branches lead to ^{210}Pb. Each ^{214}Bi nucleus has a certain probability for decay by β-emission and a certain probability for decay by α-particle emission.*

such decay chain begins with ^{238}U and ends with ^{206}Pb; this series of α and β decays is shown in Fig. 12-14.

Additional Exercise

1. ^{232}Th (thorium) is radioactive and decays by α-particle emission to form ^{228}Ra (radium). The decay chain continues until ^{208}Pb is formed. The sequence of decays that carries ^{232}Th into ^{208}Pb is: $\alpha, \beta, \beta, \alpha, \alpha, \alpha, \alpha, \beta, \beta, \alpha$. Prepare a diagram similar to Fig. 12-14 that illustrates this decay chain.

How do nuclear reactors work?

13
NUCLEI AND NUCLEAR POWER

In 1945 the Nuclear Age burst upon the world with an awesome display of the destructive power contained within nuclei. The nuclear weapons that were detonated over Japan caused 68 000 fatalities in the city of Hiroshima and 38 000 in Nagasaki. The terrible effects of these weapons precipitated the collapse of the Japanese war effort and brought World War II to an end.

Nuclear weapons are still with us; they form a critical part of the armament systems of the world's major powers. Indeed, there are nearly a hundred thousand nuclear weapons of various sorts now in existence throughout the world! However, not all of the effects of the release of nuclear energy are destructive. The thirty-odd years that have passed since the obliteration of Hiroshima and Nagasaki by nuclear weapons have seen the continuing development of the controlled release of nuclear energy for peaceful purposes. We now have a large and growing sector of the electrical power industry that generates electricity from the same physical process that takes place in the explosion of nuclear weapons. There are now more than 50 nuclear reactors in the United States that contribute their output to the nationwide electrical power grid. In future years we can expect increases in the number of nuclear power stations and an increased reliance on nuclear energy as the production of electricity from other sources lags behind our demands.

There is at present a great deal of controversy regarding the role that nuclear reactors should play in generating the nation's electricity. Various questions are being asked: How likely is an accident in a nuclear power station? What would be the result of such an accident? Are nuclear reactors safe *enough*? How does the *benefit* compare with the *risk*? Do we really *need* nuclear reactors to meet our future energy needs?

In this chapter we will learn *why* nuclear reactors work and *how* they work. We will discuss the importance of nuclear power in the overall energy situation, and we will see what risks are involved in utilizing this new source of energy.

Nuclear Masses

You have probably heard that the energy released in a nuclear reactor or in a nuclear explosion is due to the *fission* of uranium or plutonium nuclei. When a nucleus undergoes fission, it breaks apart into two more-or-less equal fragments. Why is energy released in such a process? In order to answer this question, we need to examine the way nuclear masses vary as we proceed from light to heavy nuclei. And we need to remember that Einstein's mass-energy equation tells us how *mass changes* are related to *energy changes*. This important relationship, as we learned in Chapter 9, is expressed as $E = mc^2$.

Let us look first at the simplest nucleus that consists of more than a single particle. This is deuterium, ^{2}H or ^{2}D—one proton and one neutron bound together. Now, the masses of the free proton, the free neutron, and the deuterium nucleus have all been measured accurately. If we compare the mass of the deuterium nucleus, m_d, with the combined mass of a free proton and a free neutron, $m_p + m_n$, we find an interesting result:

$$\begin{array}{r} m_p = 1.6729 \times 10^{-27} \text{ kg} \\ m_n = 1.6749 \times 10^{-27} \text{ kg} \\ \hline m_p + m_n = 3.3475 \times 10^{-27} \text{ kg} \end{array} \qquad m_d = 3.3435 \times 10^{-27} \text{ kg}$$

We see that m_d is *less* than $m_p + m_n$ (Fig. 13-1). In fact,

$$(m_p + m_n) - m_d = 0.0040 \times 10^{-27} \text{ kg}.$$

How can we interpret this result? Let us imagine that a free

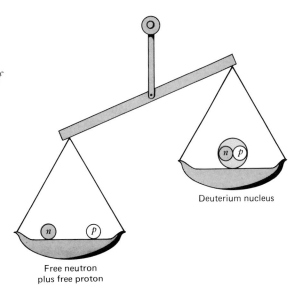

Figure 13-1 *The mass of a deuterium nucleus is less than the combined masses of the particles that make up the nucleus. All nuclei have this property: the mass of any nucleus is less than the total mass of the constituent neutrons and protons in the free condition.*

Deuterium nucleus

Free neutron plus free proton

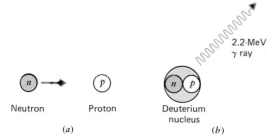

Figure 13-2 When a neutron and a proton come together to form a nucleus of deuterium, there is a decrease in the mass of the system and the release of an equivalent amount of energy in the form of a γ ray.

proton and a free neutron come together to form a nucleus of deuterium, as shown schematically in Fig. 13-2. The mass *before* the capture event takes place is $m_p + m_n$; the mass *afterwards* is m_d. According to the result obtained above, the formation of the deuterium nucleus causes some mass (in fact, 0.0040×10^{-27} kg) to *disappear*. From Einstein's relativity theory we know that the combined *mass plus energy* of any system must be conserved. Thus, if some amount of mass in a system *disappears,* an equivalent amount of energy must *appear*. In the case of a proton and a neutron coming together to form a deuterium nucleus, the mass decrease is accompanied by the release of energy in the form of a γ ray (Fig. 13-2b). The energy equivalent to the mass decrease of 0.0040×10^{-27} kg is 2.2 MeV, and a γ ray with this energy is emitted in the process $n + p \rightarrow d$.

The deuterium nucleus is a stable system—it will never spontaneously separate into a proton and a neutron. Such a breakup cannot occur because it would involve an *increase* in the total mass-energy of the system. However, the breakup, $d \rightarrow n + p$, *can* take place if we supply to the nucleus an amount of energy equivalent to the mass increase that occurs in the process. This amount of energy is 2.2 MeV, exactly the amount that is released in the inverse process, $n + p \rightarrow d$. That is, the particles in the deuterium nucleus are bound together by an energy of 2.2 MeV, and the nucleus can be broken apart if we supply to it an energy of 2.2 MeV. We say that 2.2 MeV is the *binding energy* of the deuterium nucleus.

Does the binding effect we have found for the deuterium nucleus also occur for other nuclei? Indeed it does; indeed it *must*. We can recognize this from a very simple fact: Nuclei do not spontaneously separate into protons and neutrons. The stability of nuclei means that every nucleus is bound together with a certain binding energy. The very massive nuclei, such as lead and uranium, contain more than 200 protons and neutrons. Consequently, a large amount of energy is required to separate such a nucleus into free protons and free neutrons. For example, the binding energy of ^{235}U is 1784 MeV, compared to only 2.2 MeV for ^{2}H and 92 MeV for ^{12}C.

Because of the large variation in binding energy between the lightest and the heaviest nuclei, it proves convenient to give binding energies in units of *megaelectronvolts per nuclear particle*. That is, for ^{12}C we have 92 MeV ÷ 12 = 7.7 MeV per particle, and for ^{235}U we have 1784 MeV ÷ 235 = 7.6 MeV per particle. Thus, the binding energy per particle is essentially the same for ^{12}C and ^{235}U. In fact, except for the very light nuclei, all nuclear binding energies per particle fall within the narrow range from 7.5 MeV to 8.8 MeV. As shown in Fig. 13-3, the binding energy per particle

Figure 13-3 *The binding energy of nuclei in units of MeV per particle. Some representative measured binding energies are shown as dots in the diagram. The point for ^{4}He is far above those for neighboring nuclei because the ^{4}He nucleus (the α particle) is an exceptionally tightly bound group of particles.*

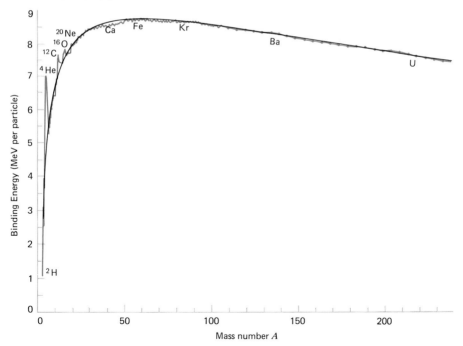

increases with mass number A to a peak of 8.8 MeV at iron (Fe) and then decreases smoothly to a value of 7.5 MeV for the heaviest nuclei. This behavior of the binding energy curve leads to the release of energy in the *fission* process (as we will see in the next section) and in the *fusion* process (as we will see later in this chapter).

Nuclear Fission

The name of Ernest Rutherford is associated with most of the significant early advances in the field of nuclear physics. He proved that the α particles emitted in radioactive decay are identical to helium atoms with two electrons removed (1909); he discovered the atomic nucleus (1911); he produced the first artificial nuclear transmutation (1919); and he was awarded the 1908 Nobel Prize (in chemistry) for his pioneering work in unraveling the mysteries of radioactivity. Rutherford had remarkable insight regarding physical phenomena, and his intuition was almost always correct. However, shortly before his death in 1937, Rutherford stated that "the outlook for gaining useful energy from the atoms by artificial processes of transformation does not look very promising." Within a few years, a series of scientific and technological advances had shown Rutherford's view to be incorrect—incorrect, in fact, to an astonishing degree.

In 1939, the German radiochemist, Otto Hahn (1879–1968), in collaboration with Fritz Strassman (1902–   ), exposed uranium to a beam of neutrons and performed very careful chemical tests on the radioactive substances that were formed as a result of the exposure. They found that among the products of neutron absorption by uranium there was radioactive barium ($Z = 56$), an element much less massive than the original uranium. How could such a light element be formed from uranium? The puzzle was soon solved by Lise Meitner and Otto Frisch, German scientists working then in Sweden as refugees from Nazi Germany, who suggested that neutron absorption by uranium produces a breakup (or *fission*) of the nucleus into two more-or-less equal fragments:

$$U + n \longrightarrow Ba + Kr$$

uranium   neutron   barium   krypton
$Z = 92$ $Z = 56$ $Z = 36$

As soon as it was realized that uranium responds to neutron bombardment by breaking up in this way, it became apparent that here is a source of energy with enormous potential. To see this, one has only to look at the binding energy curve, Fig. 13-3. The binding energy of uranium is 7.6 MeV per particle. The binding energies of barium and krypton are both approximately 8.4 MeV per particle. Therefore, when uranium breaks apart to form barium and krypton, the binding energy in the system *increases* by approximately 0.8 MeV per particle. An increase in the binding energy means that the particles in the system are more tightly bound *after* the fission event than *before*. In other words, the combined mass of the barium and krypton nuclei is *less* than the mass of the original uranium nucleus. This decrease in mass must be balanced by an increase in energy. When the uranium nucleus breaks apart, this energy appears in the form of the kinetic energy of the barium and krypton nuclei as they speed away from the fission site (Fig. 13-4).

How much energy is released in a fission event? The change in binding energy for the system is 0.8 MeV per particle, and

Figure 13-4 *The fission of uranium into barium and krypton. (a) A neutron is incident on and absorbed by a uranium nucleus. (b) The uranium nucleus breaks apart into nuclei of barium and krypton. The energy released in the process appears in the form of the kinetic energy of the fission fragments. Notice that two (sometimes three) neutrons are also released in the fission event.*

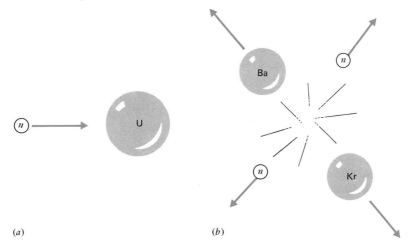

(a) (b)

there are 236 particles involved (^{235}U + n). Therefore, the total energy release is 236 × 0.8 MeV = 190 MeV. Even though the change in mass of the system is only about 0.1 percent, the amount of energy released is truly enormous. For comparison, when natural gas (methane) is burned, the energy released is only about 0.5 eV per particle. Thus, the energy released per nuclear particle in the system is more than a *million* (10^6) times greater for the fission of uranium than for the combustion of methane.

In order to appreciate just how much energy is released in the fission process, consider a 1-g sample of uranium. (This is the amount in a cube 4 mm or $\frac{1}{6}$ in. on a side.) If every nucleus in the sample undergoes fission, the energy released would be sufficient to heat 200 000 gallons of water (about 10 home-sized swimming pools) from room temperature to the boiling point. The energy released by the sample would be equivalent to that obtained by burning 2.7 tons of coal *or* 13.5 barrels (567 gallons) of oil *or* 78 000 ft^3 of natural gas.

Chain Reactions

When uranium undergoes fission, nuclei of barium and krypton can be produced. But other pairs of nuclei are also possible, the limitation being that the number of protons and the number of neutrons in the pair remain the same. Thus, the pair of fission fragments can be barium ($Z = 56$) and krypton ($Z = 36$), or lanthanum ($Z = 57$) and bromine ($Z = 35$), or cesium ($Z = 55$) and rubidium ($Z = 37$), or other similar pairs. It is also found that in each fission event, two or three neutrons are released (see Fig. 13-4). Therefore, some of the possible results of the fission of ^{235}U by a neutron are

$$^{235}\text{U} + n \rightarrow \begin{cases} ^{137}\text{Cs} + {}^{96}\text{Rb} + 3n \\ ^{139}\text{Ba} + {}^{95}\text{Kr} + 2n \\ ^{144}\text{La} + {}^{89}\text{Br} + 3n \end{cases}$$

As soon as experiments revealed that neutrons are released in the fission process, it became clear how to tap the tremendous supply of energy stored within nuclei. The reasoning is very simple. One neutron is required to cause a uranium nucleus to undergo fission. But this fission event itself releases two or three neutrons. If a neutron from each fission event is captured by a uranium nucleus and causes another fission

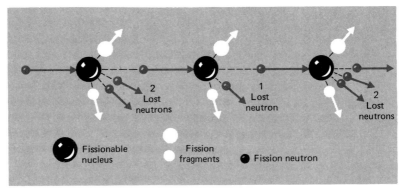

Figure 13-5 *A controlled fission chain reaction in which one neutron from each fission event triggers another event. One or two neutrons from each fission event escape the system and are "lost."*

event, the process will sustain itself. This is called a *chain reaction* and is illustrated schematically in Fig. 13-5.

If the uranium sample is arranged in such a way that exactly one neutron from each fission event causes another fission event, the chain reaction is said to be *controlled*. That is, the amount of energy released per second (the *power*) remains constant. This is exactly the situation we want for a nuclear power station. The device that accomplishes this is called a *nuclear reactor*.

In a controlled fission chain reaction, one or two of the neutrons produced in each fission event escape from the system (Fig. 13-5). Suppose that we alter the geometry of the uranium sample to decrease the number of "lost" neutrons. In particular, suppose that *two* neutrons from each fission event cause additional fission events. Figure 13-6 shows the result of this situation. The rapid and uncontrolled multiplication of the number of fissioning nuclei leads to the explosive release of fission energy. This is the principle of the atomic bomb (which is, of course, actually a *nuclear* bomb).

In order for the uncontrolled release of fission energy to take place, as many of the fission neutrons as is possible must be kept within the material. Unless the fissionable material has a mass greater than a certain value (called the *critical mass*), too many neutrons will escape the system and the rate of energy release will be too slow for an explosion to occur. The problem in constructing an atomic bomb, therefore, is to bring together into a small volume an amount of fissionable

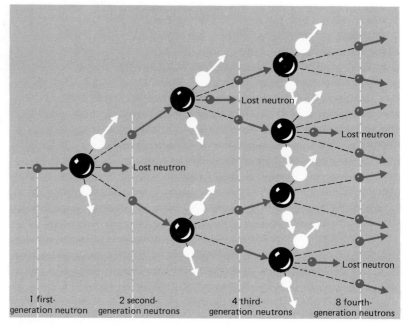

Figure 13-6 *An uncontrolled series of fission events. The rapid release of the fission energy in such a system leads to an explosion.*

material at least as great as the critical mass. This assembly of the critical mass must be accomplished in an extremely short period of time (less than about 0.001 s), because otherwise a slow, nonexplosive series of fission events will take place. One of the methods devised to overcome this problem is to drive together two or more subcritical masses by means of a conventional (chemical) explosion. The original atomic bombs of 1945 contained several kilograms of fissionable material and were detonated in this way.

Nuclear Reactors

What are the features that a nuclear reactor must have in order to exploit the release of energy in the fission process? First, there is the matter of the proper fuel. The uranium isotope ^{235}U readily undergoes fission when it absorbs a slowly moving neutron; ^{235}U is a good nuclear fuel. Unfortunately, ^{235}U constitutes only a small part (0.7 percent) of natural uranium. The abundant uranium isotope, ^{238}U, does not undergo fission when struck by a slow neutron and so this isotope is not useful in the types of reactors that employ ^{235}U. Because ^{235}U is such a small fraction of natural uranium (about 1 atom in 150), elaborate measures must be taken to enrich the ^{235}U content of any uranium sample before it becomes useful as a nuclear fuel. The design and construction of an enrichment facility was one of the major problems that had to be solved by the scientists and engineers of the Manhattan Project (the atomic bomb project) during World War II. Several techniques were devised to remove preferentially some of the ^{238}U from a sample, thereby increasing the fraction that consists of ^{235}U. The methods developed at that time are still used today to provide uranium enriched in ^{235}U for nuclear reactors. Atomic weapons and some special types of reactors require uranium that is enriched to 90 percent or more of ^{235}U. However, most commercial power reactors operate with fuel that contains no more than about 3 percent ^{235}U.

The fission of ^{235}U is most efficient when the neutrons that are absorbed have relatively low speeds and correspondingly small energies. Most fission events in a sample of ^{235}U are triggered by neutrons that have energies below 0.1 eV. On the other hand, the neutrons that are released in the fission process have energies of several MeV. Therefore, a nuclear reactor must contain some material that will slow down (or *moderate*) the rapidly moving fission neutrons to the point that they will efficiently trigger additional fission events. What material should be used as a neutron moderator? Think about a more familiar situation. Suppose that you wish to slow down a medium-size ball, for example, a billiard ball. Would you have this ball collide with a small object (a golf ball), a large object (a bowling ball), or with an object of the same size? If a moving billiard ball strikes a stationary golf ball, the golf ball will be set into motion, but the billiard ball will continue forward with only a small reduction in speed, as shown in Fig. 13-7a. (Application of the laws of energy conservation and momentum conservation gives this result and the following results.) Furthermore, if the billiard ball strikes a stationary bowling ball, the bowling ball will be given a small forward speed, but the billiard ball will rebound with almost its original speed (Fig. 13-7c). Thus, neither a collision with a smaller ball nor a collision with a larger ball is an efficient way to slow down a moving ball.

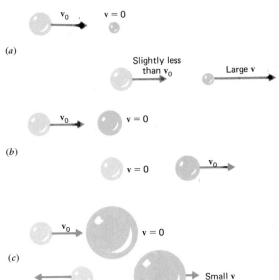

Figure 13-7 *Collision of a moving ball with stationary balls having different masses. A moving ball will come to rest if it makes a head-on collision with a ball of equal mass* (b).

Next, look at Fig. 13-7b and think about what happens when you shoot a cue ball directly toward an object ball. The cue ball comes to rest and the object ball moves off with a speed equal to the original speed of the cue ball. In a direct collision between a moving object and a stationary object with the same mass, all of the kinetic energy of the incident object is transferred to the struck object. Of course, if the collision is not head-on, only a fraction of the kinetic energy will be transferred; nevertheless, this fraction will always be larger for collisions between objects with the same mass than for collisions between objects with unequal masses. We must conclude that the most efficient way to slow down a moving object is through collisions with objects that have the same mass.

In order to moderate fission neutrons efficiently, we need in the reactor a material that contains particles with a mass equal to that of a neutron. Suitable particles are protons—hydrogen nuclei. Pure hydrogen would be satisfactory, except for the fact that we need a higher density for the moderator than that provided by a gas. Instead of hydrogen gas, we use the hydrogen contained in ordinary water.

Although the mass of the hydrogen nucleus is proper to act as an efficient neutron moderator, there is a difficulty in using hydrogen in a reactor. When a neutron collides with a hydrogen nucleus, it is usually deflected and slowed down. Sometimes, however, the neutron is captured by the hydrogen nuclear proton to form a deuterium nucleus (see Fig. 13-2). Such a capture process removes a neutron that would otherwise be available to induce a new fission event. Neutron capture therefore decreases the efficiency of energy production in the reactor. To compensate for the loss of neutrons, more fissionable nuclei must be presented to the surviving neutrons. In fact, it is because of the capture of neutrons by hydrogen nuclei in the moderator that fuel enriched in ^{235}U must be used.

One class of nuclear reactors uses deuterated water, D_2O, as the moderator instead of ordinary water, H_2O, because deuterium nuclei do not absorb neutrons as readily as do the nuclei of ordinary hydrogen. The neutron losses are minimized in such reactors and they can operate with natural, unenriched uranium. These reactors are called *heavy-water* reactors; those that use ordinary water are called *light-water* reactors. In the United States, all commercial nuclear power reactors are light-water reactors. At present, heavy-water reactors are being built only in Canada. One of the problems with the commercial use of these reactors is the high cost of the heavy-water moderator.

Water is used in reactors not only to moderate the fission neutrons but also to transfer the fission energy to the electrical generators. When a fission event takes place, the fragments move away from the fission site with high speeds (Fig. 13-4). These fragments collide with and set into motion the atoms along their paths. Thus, the binding energy of the uranium nucleus is first changed into the kinetic energy of the fragments and then into the thermal energy of the system. Water is pumped through the region of the reactor where the fission events are taking place (the reactor *core*) and absorbs some of the heat that is being produced. The steam that is produced by the heating of the water is used to drive turbine electrical generators. Thus, the end product of a reactor is *electricity*.

There is one final ingredient that is necessary for a reactor to operate productively and safely. If the power output of a reactor is to remain constant, then we must ensure that exactly one neutron from each fission event triggers a new fission event. If this number (which is called the neutron *multiplica-*

tion factor) is slightly less than one, the power output will gradually decrease; similarly, if the multiplication factor is slightly greater than one, the power output will grow. Neither of these situations is desirable, so some control mechanism must be provided to maintain the multiplication factor at exactly one. This is done by introducing into the reactor a number of movable *control rods*. The rods contain an element (usually boron) that readily absorbs neutrons. If the multiplication factor begins to exceed one (rising power level), the control rods are lowered into the reactor core by a small amount. The increased absorption of neutrons by the rods causes the multiplication factor to decrease. Conversely, if the multiplication factor falls below one, the rods are withdrawn slightly and fewer neutrons are absorbed. By inserting the control rods fully into the reactor core, the reactor can be shut down for routine maintenance or in an emergency situation.

Reactor Construction

As we have seen, nuclear reactors consist of four basic parts: the fuel, the moderator, the coolant, and the control rods. In many reactors, the moderator and the coolant are the same. Figure 13-8 shows in a schematic way where these various parts are located in a reactor. First, notice that the uranium fuel is in the form of long, thin rods (about 1 cm in diameter) and that these rods are spaced apart with moderating material between. There are two reasons for using this geometry. When a reactor is operating at any reasonable power level, an enormous amount of heat is being generated by the fission events within the fuel rods. The rods are made thin so that this heat can be transferred quickly to the coolant material (usually water). If the rods were too thick, the heat could not be removed rapidly enough, and the temperature would rise to the point at which the rods would melt. Also, the thinness of the rods permits the fission neutrons to escape into the moderator where they are slowed down. The moderated neutrons then drift back into the fuel rods and trigger new fission events. (Incidentally, spacing apart the fuel rods in this way means that too many neutrons escape from the fissionable material for an explosion to take place; a reactor can never explode as can an atomic bomb.) In a typical commercial power reactor, there will be about 40 000 fuel rods, each containing about 2 kg of uranium oxide, UO_2. Each fuel rod has a use-

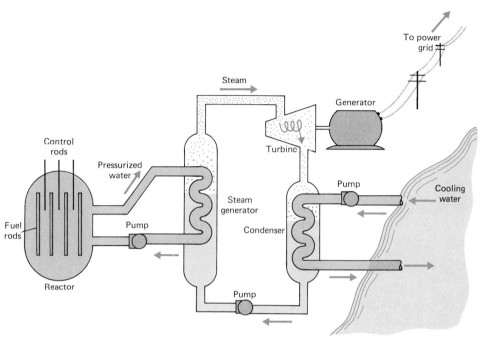

Figure 13-8 *Schematic diagram of a nuclear power plant using a pressurized water reactor (PWR).*

ful life of about three years in the reactor before it must be replaced.

Next, look at the cooling system. Water in a closed system is circulated through the reactor by means of a pump. In the type of reactor illustrated in Fig. 13-8, the cooling water is at high temperature (about 600°F) and at high pressure. This type of reactor is called a *pressurized water reactor* (PWR). When the high-temperature water passes through the steam generator, heat is transferred to the water in the second closed system and steam is produced. The steam is forced through a turbine which drives an electrical generator. Finally, the steam is condensed by cooling water drawn from an external reservoir.

The basic method of producing electrical power in a nuclear power plant is the same as that in a conventional oil- or coal-burning plant. In each case, fuel is consumed and heat is produced. This heat converts water into steam and the steam drives an electrical generator. A nuclear reactor is simply a device for generating heat!

Reactor Efficiency

One of the bothersome features of any type of electrical power plant is the relatively low efficiency with which the stored energy in the fuel is converted into useful electrical energy. A typical oil- or coal-burning plant has an efficiency of about 40 percent. In such a plant, heat may be generated at a rate of 1000 MW (megawatts), but the electrical output will be only 400 MW. Thus, about 600 MW of heat energy is wasted in the process. What becomes of this waste heat? It is dissipated into the surroundings, raising by a small amount the temperature of the air or of a nearby body of water. If the waste heat from a power plant is carelessly dumped into the environment, damage to the local ecology can result. For this reason, we say that power plants produce *thermal pollution*. Effective measures are now being taken to minimize the impact of the thermal emissions of electrical power plants. However, the long-term effects of increasing thermal pollution are not known; careful studies of the thermal problems associated with power plants (of all types) must be continued.

Although the operating efficiency of an oil- or coal-burning power plant is only 40 percent, the efficiency of the power reactors in use today is even lower, about 32 percent. This means that a nuclear plant will exhaust about 40 percent more heat to its surroundings than will a conventional plant with the same output (see Table 13-1). Therefore, dealing with the waste heat from a nuclear power plant is a significantly greater problem than that associated with a conventional plant. Consequently, special precautions are taken to ensure that the effects on the environment are minimized. New types of reactors, which should become part of the electrical power network in about 15 years, will have efficiencies of approxi-

Table 13-1 Comparison of Efficiencies of Conventional and Nuclear Power Plants

Type of Plant	Thermal Power Produced (MW)	Efficiency	Electrical Power Output (MW)	Waste Thermal Power Exhausted (MW)
Conventional (oil, coal)	2500	0.40	1000	1500
Nuclear	3125	0.32	1000	2125

mately 40 percent and will therefore represent thermal problems that are no more severe than those of conventional plants.

The Nuclear Power Industry

In 1957, twelve years after the first use of uncontrolled nuclear energy, this same source of energy was put to controlled use on a commercial basis. The first nuclear power reactor to be operated by private industry was put into service by the Duquesne Light Company at Shippingport, Pennsylvania, and produced 90 MW of power. In 1976, we have 60 nuclear power plants operating in the United States with a total generating capacity of nearly 42 000 MW. In addition, there are more than 60 plants in various stages of construction and over 100 under firm order in the United States. (In 41 countries outside the United States, there are more than 100 operating commercial power reactors, with an additional 450 units under construction or on order.) Many of the commercial units now being built or designed have power capacities in the range from 1000 to 1500 MW.

In the United States we now produce about 8 percent of our electrical power with nuclear reactors. In certain parts of the country, the percentage is considerably higher. For example, in the Chicago area, about 30 percent of the electrical power is nuclear. It is anticipated that the fraction of the total U.S. electrical generating capacity represented by nuclear plants will be 15 percent in 1980, 30 percent in 1985, and more than 50 percent by the year 2000. If the demand for electricity grows according to the present estimates, we may need as many as a thousand nuclear power plants early in the next century.

Electrical power plants are expensive. The cost today to build a coal-fired plant with an output capacity of 1000 MW is about $500 million; the figure for a nuclear plant with the same output is about $600 million. The plants that must be built in the future to satisfy our increasing demands for electrical power will, of course, cost even more. These increased costs will be reflected in our electric bills.

Although a nuclear plant costs about 20 percent more than a conventional plant, this difference is more than offset during the lifetime of the plant because of the smaller fuel and operating costs of a nuclear facility. One reason for this is the expense involved in dealing with the sulfur problem. All coal contains a certain amount of sulfur. When the coal is burned, sulfur dioxide is formed and this noxious gas is exhausted along with the other fumes from the plant. Sulfur dioxide is known to cause respiratory problems in humans and to affect vegetation adversely. Consequently, there are federal regulations now in effect limiting the amount of sulfur dioxide that can be exhausted into the atmosphere. The removal of sulfur dioxide from a plant's exhaust gases is an expensive operation.

Oil usually contains much less sulfur than does coal, and many of our generating plants operate on oil. But as the oil supply becomes more critical, essentially all of our oil-fired plants will be converted to coal. As we increase our reliance on coal, the costs involved in sulfur removal will become even greater. Then, the fuel-cost advantage that nuclear power plants hold over coal-burning plants will be accentuated.

How Safe Are Nuclear Power Plants?

Nuclear power plants have several apparent advantages compared to conventional plants. They require no bulky shipments of fuel; they emit no particulate matter or noxious fumes; and, as we will see, the fuel supply is assured, whereas our oil and coal resources are definitely limited. Why, then, are we not quickly converting our electrical power industry completely to nuclear reactors? This question is actually a very complicated one, but basically the answer centers about the fear of radioactivity.

In the preceding chapter we learned about the hazards of radiation exposures. How can we receive a radiation dose due to the radioactivity produced in a nuclear reactor? There are several possibilities. First, during routine operations a small amount of radioactivity will escape from a nuclear power plant. This amount is exceedingly small, however, and, as we learned in Chapter 12, will cause an average exposure to an individual of only about 0.003 rem per year in the year 2000 when we may have 500 or so operating reactors. Second, radioactivity can escape into the environment during the mining and processing of uranium ore and in the reprocessing of the partially used fuel removed from reactors. Again, the doses that the general public could receive from these operations are quite small and, except in special situations, cannot be considered to represent a major problem.

There are, however, two important ways in which the general public could be exposed to a significant amount of reactor-produced radiation. (a) Radioactivity could be released in a catastrophic reactor accident and distributed over the countryside. (b) The stored radioactive wastes from used fuel rods could be released into the atmosphere or into underground water that eventually becomes part of the public water supply. We will look at each of these possibilities.

First, we must again state that a nuclear reactor cannot explode in the manner of an atomic bomb. The geometrical arrangement and the enrichment of the fuel rods in a reactor are simply not correct for the explosive release of nuclear energy. Even though a reactor cannot *explode,* there is the possibility that it could *melt down.* If a reactor's cooling system were to fail, the control rods would automatically shut down the reactor. But there would still be so much heat generated by the radioactivity in the fuel rods that the hot radioactive material could melt its way through the reactor vessel. Even if the vessel were breached, there is a double shielding wall surrounding the reactor (Fig. 13-9). The first of these is a thick concrete shield immediately around the reactor vessel. The second is a very large reinforced concrete dome that is lined with steel plate that encloses the reactor and all of the heat exchange equipment. The primary purpose of this outer containment vessel is to hold any radioactive dust or gas that might be released in an accident. This vessel is strong enough to withstand the impact of a jetliner. In the unlikely event that both of these shielding walls are ruptured, then the radioactive material could escape to the outside world.

The key question to ask with regard to the possible melt down of a reactor is: How likely is the failure of the cooling system? Every reactor is constructed with a cooling system that has the highest possible reliability. In addition, this system is backed by an emergency core cooling system (ECCS) designed to cope with a series of potential but unlikely accidents that could result in the loss of coolant from the primary cooling system. The ECCS is a highly redundant system; no simple failure of pumps or valves would prevent its operation.

An exhaustive study has been made by a task force of ex-

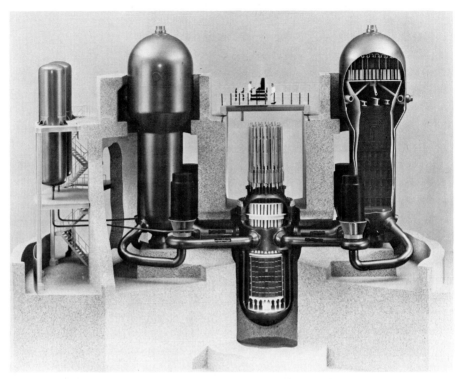

Figure 13-9 *Cutaway view of a nuclear steam supply system. The reactor vessel is in the center and is surrounded by a thick concrete shield. Large pipes connect the reactor vessel with two steam generators on either side. The steam generators are large heat exchangers in which hot, high-pressure water heats a second source of water which boils to steam and leaves through the top to drive a turbine-generator. Note the size comparison of the men on the refueling crane over the reactor vessel. (Photo courtesy of C-E Power Systems, Combustion Engineering, Inc.)*

perts concerning the various sequences of failures that could possibly lead to a loss-of-coolant accident and the melt down of a reactor. Estimates were also made of the amounts of radioactivity that might escape from the reactor's outer containment vessel and the number of injuries and fatalities that might result from the accident. This study was carried out by combining all of the available information concerning the reliability of individual reactor components. It was necessary to conduct the investigation in this way because we have no experience with the actual melt down of a commercial power reactor. Several minor accidents have occurred but none has ever caused the melt down of the core or the release of any radioactivity. (The nuclear power industry has an excellent safety record, one that is better than that of any other major industry.) Only one power reactor has ever suffered an accident that led to the release of radioactivity, and this was an early British unit that did not have the double shielding wall that is in use in all U.S. commercial reactors.

The results of the study can be summarized by comparing the likelihood of a nuclear accident to the likelihoods of various natural and man-caused accidents that could have the same consequences. This comparison, taken from the report of the task force, is made in Table 13-2. The last line gives the estimates based on 100 operating reactors. If we increase this number to 1000 (a situation that will not exist until after the year 2000), we see that the probability of an accident causing 100 or more fatalities is one in 10 000 years. (Notice that this figure assumes no improvement in reactor technology during the next 25 years or so and no progress in the treatment of cancers, the main cause of fatalities in radiation exposures.) Critics of the study suggest that the task force has underestimated both the probability of a melt down and the consequences of the resulting escape of radiation. However, even if we increase these probabilities by a factor of 100, we still estimate that an accident causing 100 or more fatalities would

Table 13-2 Average Probability of Major Man-Caused and Natural Events

Type of Event	Probability of 100 or More Fatalities	Probability of 1000 or More Fatalities
Man-Caused		
Airplane crash	1 in 2 years	1 in 2000 years
Fire	1 in 7 years	1 in 200 years
Explosion	1 in 16 years	1 in 120 years
Toxic gas	1 in 100 years	1 in 1000 years
Natural		
Tornado	1 in 5 years	very small
Hurricane	1 in 5 years	1 in 25 years
Earthquake	1 in 20 years	1 in 50 years
Meteorite impact	1 in 100 000 years	1 in 1 000 000 years
Reactors		
100 plants	1 in 100 000 years	1 in 1 000 000 years

occur only once in a century. The risk of an airplane accident with the same consequences is 50 times greater.

The final threat of radioactivity release associated with nuclear power involves the manner of ultimate disposal of the radioisotopes that are formed in the fuel rods. After a fuel rod has been in a reactor for about three years (during which time about $\frac{1}{30}$ of the ^{235}U has been consumed), it begins to lose its effectiveness as an energy producer. The rod is then removed and a new rod is inserted. The spent rod is radioactively "hot"—it contains a powerful assortment of short- and long-lived radioactivity. The usual procedure is to store spent rods in a heavily shielded area for several weeks in order to allow the short-lived activities to decay. After a "cooling off" period of six weeks, the radioactivity remaining in a fuel rod is only about $\frac{1}{20}$ of that at the time of removal from the reactor.

By holding the spent fuel rods in a protected area for a relatively short period of time, most of the radioactivity will disappear. But a potentially dangerous amount of radioactivity still remains, and these radioisotopes have lifetimes ranging from years to thousands of years. Some of the more important isotopes (that is, those that are biologically important) are ^{137}Cs (half-life = 30 years), ^{90}Sr (28 y), ^{3}H (12 y), and various isotopes of plutonium (half-lives from 90 to 24 000 years).

These long-lived radioisotopes must, of course, be disposed of in some safe way. At present, radioactive waste materials are being stored in liquid form in huge million-gallon tanks that are located in concrete-shielded underground bunkers. This is only a temporary solution to the problem, however, because a final decision has not yet been made as to the method of ultimate disposal.

We must remember that concentrated radioactive materials are not only radioactively "hot" but they are also thermally hot due to the continual release of energy in the decay processes. This thermal activity is a complicating factor in arriving at a satisfactory method of disposal. One of the most promising ideas for permanent storage is to cast the radioactive materials into glassy, heat-resistant blocks and to bury these blocks in some suitable location. Such blocks should remain stable for long periods of time in spite of the continual heat load and the radiation damage produced by the radioactive decays.

What constitutes a "suitable" location for the disposal of radioactive wastes? The primary concern in selecting a storage site is to ensure that any radioactivity that might leak from the blocks will not be carried by underground water into the water supply for some nearby community. Not only must the burial site be water-free but it must also be geologically stable. We do not want any earthquakes or ground movements to disrupt or fracture the storage blocks.

There seem to be at least two different types of geological formations that satisfy these requirements—granite formations and salt deposits. Most of the Earth's regions of granite and salt have been undisturbed for many millions of years and it is supposed that they will remain inactive for many millions more. Granite has a very low porosity and so water does not flow through these formations. Salt is quite soluble in water, so the very existence of a salt deposit is evidence that the region is free of water. The main problem with using a granite depository is that extensive drilling would be necessary in order to prepare a suitable storage cavern. On the other hand, there are abandoned salt mines that could be converted to burial sites with relatively little difficulty. Detailed investigations are being made of several possible disposal sites with a view toward selecting a region with maximum stability and minimum water. A storage site with an area of only a few square miles would be sufficient to contain all of the wastes that reactors in this country will generate during the next 50 years or so.

The radioactivity produced in nuclear reactors represents a potential hazard to the public. Are we willing to face this hazard? Opinion on this issue varies widely. The extreme negative view insists that we should accept no risk of radiation exposure and therefore that the nuclear power program should be abandoned. Another group maintains that the probability of radiation accidents is far too high and that construction of new facilities should be stopped until significant improvements in reactor safety features have been made. (It should be mentioned that these attitudes are voiced in very few countries besides the United States; nuclear power programs are proceeding steadily ahead in most other countries.)

We all realize that exposure to radiation can produce biological damage, and we realize that there is always the *possibility* of a radiation accident, even though the likelihood may be very small. The question we must really ask is: Do we receive sufficient benefit from the nuclear power program for the risks we take? Most thoughtful persons who have studied the situation believe that the benefits do, in fact, far outweigh the risks. The risks involved in the production of our electricity from nuclear fission are actually extremely small in comparison to the risks we readily accept in other areas. For example, there are 50 000 deaths annually in the United States due to automobile accidents. Not even the most severe critics of nuclear power suggest that radiation accidents will cause fatalities amounting to more than a tiny fraction of this number, even when we will be deriving essentially all of our electrical power from reactors. Still, the nuclear power question is a lively issue at present and the pros and cons are being hotly debated. These debates have even led to both sides citing the same evidence in support of their views. For example, early in 1975, a technician attempted to locate an air leak at the Browns Ferry reactor in Alabama by using an open candle flame. The candle set afire some insulating material and touched off a major fire. The nuclear opponents say that the safety systems at Browns Ferry were especially weak and that the situation was dangerous. On the other hand, the nuclear advocates say the fact that there were no injuries and that a catastrophy was averted proves their point. After all, they say, even though the fire did disable 7 of the reactor's 12 safety systems, the remaining 5 were more than adequate to prevent a melt down and any release of radioactivity. It is not yet clear whether the nuclear power program in this country will proceed according to the ambitious schedule that has been laid out.

When attempting to answer the question, "Is nuclear power worth it?", it is very important to examine the role that nuclear power plays in our overall energy situation. That is, instead of trying to determine the desirability or nondesirability of nuclear power divorced from the subject of energy in general, we must also look at the alternatives. What else is possible? What else is more attractive? We will return to these questions at the end of this chapter after we have discussed the topics of breeder reactors and nuclear fusion.

Plutonium and Breeder Reactors

We have already mentioned the fact that the abundant uranium isotope, ^{238}U, does not undergo fission in the same circumstances that ^{235}U does. When a ^{235}U nucleus absorbs a slow neutron, a fission event occurs. What happens when a ^{238}U nucleus absorbs a slow neutron? The result is a simple capture that produces the isotope ^{239}U:

$$^{238}U + n \rightarrow {}^{239}U$$

The isotope ^{239}U is radioactive and undergoes β decay with a half-life of 24 min to form ^{239}Np (neptunium). This isotope is also radioactive and undergoes β decay (half-life = 2.4 days), forming ^{239}Pu (plutonium). That is,

$$^{239}U \xrightarrow{\beta \text{ decay}} {}^{239}Np \xrightarrow{\beta \text{ decay}} {}^{239}Pu$$

The end product of this sequence, plutonium-239, is again a radioisotope, but the half-life is so long (24 000 years) that appreciable quantities can be accumulated. The interesting and important feature of ^{239}Pu is that it undergoes fission upon the absorption of a slow neutron just as ^{235}U does. That is, ^{239}Pu is as good a nuclear fuel as ^{235}U. In fact, one of the two atomic weapons detonated over Japan in 1945 was constructed from ^{239}Pu, and plutonium is used in many of the weapons in the nuclear arsenals around the world today.

When ^{235}U undergoes fission, 2.5 neutrons are produced, on the average, for each fission event. In a nuclear power reactor we want to maintain a constant power output, so one of these neutrons is necessary to trigger a new fission event. If we could arrange for one additional neutron per event to be captured by a nucleus of ^{238}U, then it would be possible to produce one nucleus of ^{239}Pu for each nucleus of ^{235}U consumed. That is, just as much new fuel would be produced as is used. Not only is power generated in the process, but ^{238}U is transformed into a useful fuel at the same rate that ^{235}U disappears. A reactor that is designed to accomplish this is called a *breeder reactor*. The ratio of the amount of fuel produced to the amount consumed is called the *breeding ratio*. In a standard light-water reactor (in which the uranium fuel consists of 3 percent ^{235}U and 97 percent ^{238}U), about one plutonium nucleus is produced for every three ^{235}U nuclei consumed; that is, the breeding ratio is 0.33. A reactor is a true breeder reactor only if it produces more fissionable material than it consumes; that is, the breeding ratio must exceed one.

A breeder reactor is quite different from the light-water, slow-neutron reactors we have been discussing. First of all, in order to decrease the number of neutrons lost by capture in surrounding material and to increase the number that are captured by ^{238}U, the reactor is operated without a moderator. That is, the various events are initiated by *fast* neutrons instead of *slow* neutrons. Because it operates without a moderator, the core of a breeder reactor is much smaller than that of a conventional reactor. With the same amount of power being produced in a much smaller volume, the thermal problems are therefore much more severe in a breeder reactor. Water cannot be used to extract the heat from the core of a breeder reactor because (a) water would act also as a moderator and (b) water is not an efficient medium to use for heat transfer at the high temperatures that exist in a breeder reactor. Molten sodium is the coolant in most of the breeder reactors now operating or under study. These reactors are called *liquid metal fast breeder reactors* (LMFBRs).

Because of the high concentration of power in the core and the necessity of using a liquid metal coolant, a breeder reactor is a very complex device. Before breeder reactors can become a part of the commercial power system in the United States, a long testing program must be carried out. Other nations are ahead of the U.S. schedule. Breeder reactors have been successfully operated in France, Britain, and the Soviet Union for several years. The performance of the French reactor (Phenix), which has been operating at full power since late 1974, has so far been superior. The only significant difficulty experienced with any of these reactors has been, not with the reactors themselves, but with the systems that transfer heat from the liquid sodium to the steam generators.

The United States is now committed to the development of a commercial breeder reactor by the late 1980s, with incorporation of significant numbers into the electrical power network by the mid-1990s. At the end of this chapter we will see how breeder reactors fit into the long-range energy picture.

Nuclear Fusion

Earlier in this chapter we saw how the shape of the binding energy curve (Fig. 13-3) allows us to understand why energy is released in the fission process. Whenever there is an increase in the total binding energy of a system, there must be an accompanying release of an equivalent amount of energy. That is, whenever a process can be represented as an *upward* movement along the binding energy curve, energy is released. This is the case in the fission process: We begin with a mass number of 236 and end with two mass numbers near 100. Notice that we can also move upward along the curve by starting with nuclei that have low mass numbers. If we combine two light nuclei with mass numbers A_1 and A_2 into a single nucleus with mass number $A_1 + A_2$, the binding energy will be increased in the process and energy will be released. That is, we can release energy by fusing together two light nuclei to produce a heavier nucleus—that is the process of *fusion*. Notice that fission and fusion are exactly opposite processes, but both release energy because both involve moving upward along the binding energy curve.

When a nucleus undergoes fission, the two fragments both carry positive charges, so the repulsive electrical force that exists between them assists in pushing the fragments apart. When two nuclei attempt to fuse together there is also a repulsive electrical force between the nuclei and this force must be overcome before fusion can occur. Thus, the same electrical force that aids fission represents an obstacle that must be overcome in fusion. The only way to press two nuclei together against the repulsive electrical force is to arrange for them to collide with one another at a high speed. We know from our study of the behavior of gases that the speed of gas molecules increases as we increase the temperature. Therefore, we should be able to obtain a collision speed between nuclei sufficient to permit fusion to occur simply by increasing the temperature to the required level. This is, in fact, true. But the required temperature (in the most favorable case) is *millions* of degrees! Nuclear processes that take place only at these extremely high temperatures are called *thermonuclear reactions*.

It is clear that very special conditions are necessary for thermonuclear fusion reactions to occur. How can we produce the required temperatures? One way is to use a fission device (an atomic bomb) to generate thermonuclear temperatures in a blanket of material (usually hydrogen isotopes) that surrounds the device. The detonation of the fission device triggers the release of an even greater amount of energy by thermonuclear fusion reactions in the surrounding material. This is the principle of the *hydrogen bomb* (or *thermonuclear bomb*). The sustained and controlled release of fusion energy in the laboratory has not yet been accomplished. Scientists in all of the highly developed countries of the world are working on various ways to generate the extremely high temperatures that are required and on ways to contain the super-hot gases. However, success will not come quickly. There is almost no prospect that a prototype fusion reactor will be operating before the end of this century.

When two nuclei fuse together, the collision speed (and, hence, the temperature) that is required depends on the amount of electrical charge carried by the nuclei. If the charge is low (that is, if the nuclei have small atomic numbers), then the repulsive electrical force will be least and the *ignition temperature* will also be least. For this reason, the only practical fusion reactions appear to be those involving isotopes of hydrogen, in particular, ^{2}H and ^{3}H.

Suppose that we heat a sample of deuterium (^{2}H) to several million degrees. What will happen? When the high-speed deuterium nuclei collide with one another, two different nuclear reactions take place. In one process, nuclei of tritium (^{3}H) and normal hydrogen (^{1}H) are formed. In the other process, a nucleus of the light isotope of helium (^{3}He) and a neutron (n) are formed. Both of these reactions release several MeV of energy:

$$^2H + {}^2H \rightarrow {}^3H + {}^1H + 4.0 \text{ MeV} \tag{1}$$
$$^2H + {}^2H \rightarrow {}^3He + n + 3.3 \text{ MeV} \tag{2}$$

In these fusion-type reactions heavier nuclei are produced and energy is released. Notice that four nuclear particles are involved in each reaction and that an average of about 3.6 MeV is released in each reaction. This is almost the same amount of energy per particle that we found for the fission of uranium (0.9 MeV compared with 0.8 MeV).

The ^{3}He nuclei that are produced in the ^{2}H + ^{2}H reactions do not participate in any further activity. However, when one of the ^{3}H nuclei is struck by a deuterium nucleus, a new and even more energetic reaction occurs:

$$^2H + {}^3H \rightarrow {}^4He + n + 17.6 \text{ MeV} \tag{3}$$

That is, the tritium produced by reaction (1) above is consumed by reaction (3) with the release of additional energy.

If we add together all three of the reactions, we can express the net result of heating deuterium as

$$5\,{}^2H \rightarrow {}^4He + {}^3He + {}^1H + 2n + 24.9 \text{ MeV} \tag{4}$$

Here, we see that 10 nuclear particles are involved (5 deuterium nuclei) and that the total energy release is 24.9 MeV. The energy release per particle is approximately 2.5 MeV, or about three times greater than that for the fission of a heavy nucleus. This series of fusion reactions represents a much more efficient way to release nuclear energy than does the fission of uranium or plutonium.

In addition to increased efficiency, fusion has two other advantages over fission as a potential large-scale source of energy. First, the fuel supply (deuterium) is essentially inexhaustible. There is sufficient deuterium in the ocean waters to provide all of the world's energy requirements for millions of years. Moreover, it is not difficult to extract deuterium from sea water. Second, the end products of the fusion reaction (4) are stable isotopes, not radioisotopes as in the case of fission. (However, there will be radioactivity induced in the materials surrounding the fusion region by the capture of the neutrons released in the fusion process. Furthermore, tritium is radioactive, and even though it is largely consumed in reactions with deuterium, there will be substantial amounts present in any operating fusion reactor. Nevertheless, the radioactivity problem in a fusion reactor would be much less severe than it is in a fission reactor.)

The Outlook for Nuclear Power

Earlier in this chapter we pointed out some of the advantages that nuclear power holds over conventional methods of producing electricity. We then asked why are we not converting quickly to nuclear power, and we pointed out that one of the main reasons involves the problem of radioactivity. Now that we have discussed some of the features of fusion we seem to have the ideal solution. Fusion is an efficient way to release nuclear energy and yet it produces relatively small amounts of radioactive residue. Why should we accept any of the risks of fission power when fusion power appears so attractive?

Many scientists believe that fusion reactors will eventually supply the world with abundant and inexpensive electrical power and that this will be done safely and with the minimum effect on our environment. But we do not know when to expect the breakthrough that will lead to a practical method for tapping fusion energy, nor can we be absolutely certain that the crucial development will *ever* be made. Even the most optimistic estimates indicate that fusion reactors cannot make a substantial impact on our energy problems before the year 2020 and perhaps not until 2050. (We can make a similar comment about solar power. Although the Sun's radiations are a source of clean, cheap, and safe power, we do not yet know how to harness this energy source on a large scale. As is the case with fusion power, it will probably be well into the next century before we can expect any large contribution to our energy supply from solar power systems.)

The world population continues to grow and to strive for a higher standard of living. This involves increasing demands for energy, particularly electrical energy. How will we meet these requirements during the period before fusion power (or solar power) provides the ultimate solution?

The only chemical fuel we have in large supply is coal. We could elect to meet our rising requirements for electricity by building only coal-burning power plants and by opening new mines to provide them with fuel. It would then be unnecessary to expand the nuclear power industry. But we would pay an enormous price to accomplish this. The mining of coal not only tears up huge amounts of land, but it is also an extremely hazardous occupation. In addition to deaths and injuries caused by cave-ins, explosions, and other mine accidents, each year about 4000 miners succumb to black-lung disease. Moreover, when the coal is burned, particulate matter and noxious gases are released. The worst offender among these emissions is sulfur dioxide. Although efforts are being made to control the escape of sulfur dioxide, great quantities of this gas are still released into the atmosphere. Detailed studies have indicated that air pollution is responsible for about 50 000 excess deaths per year and that sulfur dioxide is a primary factor in these deaths. The deemphasis of nuclear power in favor of coal power would decrease the likelihood of public exposures to radiation but it would significantly increase the hazard of breathing the air.

If we must have nuclear power, can we not limit the risk by building only the well-understood standard reactors instead of developing, at great expense, new breeder reactors that are potentially more dangerous? We are pushed toward the development of breeder reactors by limitations in our uranium supplies. There has been no discovery of a significant new deposit of uranium ore in the United States in nearly 20 years, and foreign sources of this vital material are never secure. The uranium that remains in the spent fuel rods removed from reactors is not now being recovered, so each new fuel rod that is used in a reactor draws upon our ore stock. At the rate of use currently projected, our supplies of reasonable-quality uranium ores will be exhausted in 10 years or so. By reprocessing the spent fuel rods (remember, only about 3 percent of the ^{235}U is actually consumed before the rod is replaced), sufficient uranium can be recovered to extend the life of our supplies until about the year 2000. After that time we will be reduced to using low-quality ores, the processing of which is both time-consuming and expensive.

The development of breeder reactors will extend our fuel lifetime tremendously. Instead of using only the one uranium atom in 150 that is ^{235}U, we will then have the potential to use every atom. Our supply of fission fuel will therefore increase by more than a hundredfold. (In fact, one type of breeder reactor can convert thorium into nuclear fuel. There is sufficient thorium easily accessible to provide all of our electrical power needs for a hundred thousand years.)

The breeder reactor program is viewed as an interim solution to our energy problems. We need time to develop our ultimate energy sources—fusion power and solar power. Our oil and natural gas resources will not see us through. Coal, although plentiful, involves a high cost in terms of damage to the environment and to public health. Conventional reactors will consume the available supplies of uranium too rapidly. All things considered, it seems that our best chance to survive until the utopian days of cheap and abundant power arrive is to pursue the development of breeder reactors. But we must not lose sight of the fact that the ultimate goal is to draw our energy from the ocean's deuterium and the Sun's radiation. We must not suspend our efforts in these directions simply because we have a promising short-range solution.

The Risk of Illicit Nuclear Weapons

Nuclear fission power involves risks. We have already discussed several of these risks but we have not yet completed the list. What about the possibility of using nuclear materials, diverted from power reactors, for the production of nuclear weapons? The fuel that we now use in our power reactors (uranium enriched to about 3 percent ^{235}U) is not suitable for the construction of nuclear weapons. Weapons require more highly enriched uranium and it is not a simple matter to build an enrichment plant. Only a technically advanced country would be able to construct and operate such a facility. Therefore, if some non-nuclear country wishes to construct a nuclear weapon, the task will be very difficult if only natural uranium or low-enrichment reactor fuel is available. (Some research reactors require fuel enriched to 90 percent ^{235}U—this is weapons-grade uranium.)

The job of the would-be bomb makers is much easier if they have access to the spent fuel rods removed from a nuclear reactor. These rods contain plutonium that has been produced by neutron capture in ^{238}U nuclei. The techniques for the chemical separation of plutonium from uranium are well known. Such a separation procedure is made difficult and hazardous because of the fact that the materials being worked with are all highly radioactive. But a small country with a modest amount of scientific skill could carry out the separation and produce reasonably pure plutonium. About 1 g of plutonium could be extracted from each spent fuel rod, so by processing a few thousand rods, sufficient plutonium to build a nuclear weapon could be accumulated.

A country that has a research or power reactor not subject to international inspection would be able to divert sufficient plutonium for a small weapons program. In fact, this is how India acquired the plutonium for the one and only nuclear device she has exploded. Several other countries could conceivably produce weapons secretly in the same way. Some observers believe that Israel has about a dozen low-yield weapons. For a terrorist group, the problem is more difficult because the plutonium would have to be stolen, and the theft would alert the world to the possibility of a threat.

After a country or a group has illicitly acquired sufficient plutonium, there is still the problem of actually constructing a weapon that has a reasonable likelihood of exploding. Although the general design of the original plutonium bomb of 1945 has appeared in many textbooks, it is a technically demanding job to construct such a weapon. A scientifically advanced country (such as India or Israel) can support such an effort. A small group of terrorists, on the other hand, would have a difficult time. (Remember, because the theft of the plutonium was detected, they are presumably being hunted.)

Nevertheless, some experts believe there is a possibility that a dedicated, scientifically trained, and highly disciplined group of individuals could actually construct a plutonium weapon. The device would be very crude but it would probably have some explosive yield—it would not destroy Manhattan, but it might topple the World Trade Center.

To what extent has the proliferation of nuclear reactors on a worldwide scale opened the door for possible nuclear blackmail and terrorism? Some experts believe that there is a high probability that an illicit nuclear weapon will be exploded within the next decade or so. The immediate effect of such an explosion, in terms of fatalities and property damage, will not be catastrophic. But no one can predict what might be the political consequences.

It is interesting to note that the likelihood of a nuclear weapon being used for blackmail or terrorism depends very little on nuclear power activities in the United States. First of all, when U.S. reactors are sold abroad, the terms include safeguards against the diversion of plutonium. When other countries export reactors, however, there are often no such safeguards. Any country that purchases a reactor under these conditions could proceed to construct a nuclear weapon, unknown to the rest of the world. Second, we have begun to realize in this country the great importance of protecting our nuclear materials. Within the last few years, significant improvements have been made in the security systems associated with the processing, transportation, and use of nuclear materials. Any terrorist group intent on stealing sufficient plutonium to construct a bomb would probably find the job easier and the risks smaller in some other country.

Our conclusions regarding the possible illicit use of nuclear materials in weapons for blackmail or terrorism are distressing. There appears to be some reasonable likelihood that we will again see a nuclear weapon exploded during this century. And it appears that we in the United States have very little control over the situation. All we can do is to maintain high standards of security over our own nuclear materials, to insist on adequate safeguards when exporting reactors, and to encourage other exporters to institute controls over the possible diversion of nuclear materials.

Is nuclear power worth it? We have identified several areas in which the nuclear power program involves risks to the general public. The most important hazards are those associated with nuclear melt downs, the release of radioactivity from waste storage sites, and the construction of nuclear weapons from diverted plutonium. Most critics of nuclear power have emphasized the first of these hazards. However, it is probably the last—the threat of international blackmail with nuclear weapons—that is the most serious hazard resulting from the spread of nuclear power. (We have not discussed the hazards involved in the maintaining by the large powers of huge arsenals of nuclear weapons because these risks are independent of the nuclear power program.)

In spite of the various risks associated with the use of nuclear power, there appears to be no alternative that is more attractive, at least none that will be available for 50 to 75 years. If we can prevent the use of nuclear weapons by international blackmailers and terrorists, the risks associated with the normal peaceful uses of nuclear energy do not appear to be excessive. (But, remember, this view is not supported by everyone; there is a substantial opposing sentiment.) For the risks that we do take, the benefits are considerable: no air pollution, no degrading of the landscape, and an assured fuel supply that is not at the mercy of foreign governments.

Questions and Exercises

1. Suppose that there is a nucleus consisting of N neutrons and Z protons. The mass of this nucleus is $N \times$ (neutron mass) $+ Z \times$ (proton mass). Is the nucleus stable? Explain carefully.

2. Would energy be released by the fission of a nucleus with $A = 60$ into two fragments with equal masses? (Refer to Fig. 13-3.)

3. In designing the control system of a reactor, what quantities could be monitored to determine whether the control rods should be inserted or withdrawn in order to maintain the neutron multiplication factor at exactly one?

4. Suppose that ^{235}U produces, on the average, 3 neutrons per fission event (instead of 2.5). Would a light-water reactor then require fuel with a greater or a lesser enrichment of ^{235}U?

5. The jacket that encases the uranium in a fuel rod is made from a special alloy. What properties must this material have?

6. The fraction of electrical power produced by reactors in the Chicago area is about 30 percent and in New England it is about 20 percent. In the Pacific Northwest and in the Texas-Oklahoma-Louisiana region, the fraction is considerably less. Why is this so?
7. An important consideration in determining whether to expand the nuclear power industry is to evaluate the benefit versus the risk. How do you judge the benefit versus risk when getting a chest X ray?
8. What are some of the potentially hazardous features of breeder reactors that are not found in light-water reactors?
9. What are some of the advantages of fusion power compared to fission power?

Additional Details for Further Study

Nuclear Reactions in the Sun

The energy that we receive from the Sun in the form of radiation is actually fusion energy. The source of the Sun's energy is deep within its core where both the temperature and the pressure are extremely high. In fact, at the center of the Sun, the temperature is approximately 10 million degrees and the density is more than 100 times that of water! Under these extreme conditions, thermonuclear fusion reactions readily take place, thereby supplying the Sun with the energy that it radiates into space.

Most of the material in the Sun's core is hydrogen (^{1}H). Because the temperature and pressure are so high, a fusion reaction takes place that has never been observed under the laboratory conditions possible on Earth. This reaction is the fusing together of two hydrogen nuclei (protons) to produce a deuterium nucleus:

$$^1H + {}^1H \rightarrow {}^2H + e^+$$

where e^+ represents a *positron*, a particle that is identical to an ordinary electron except that it carries a positive charge (see Chapter 9). The deuterium produced in this process reacts with hydrogen according to

$$^2H + {}^1H \rightarrow {}^3He + \gamma$$

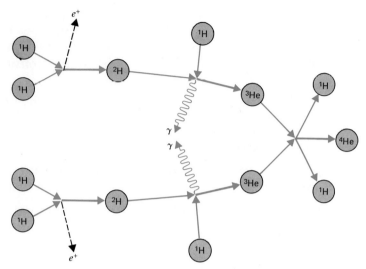

Figure 13-10 *Schematic representation of the sequence of thermonuclear fusion reactions (the* proton-proton *chain of reactions) that provide the Sun's energy. Six hydrogen nuclei (protons) are involved in the chain of reactions but two are returned to the medium in the final reaction. The net result of the sequence is 4 $^1H \to {}^4He$.*

where γ represents the gamma radiation emitted in this capture process. The next step involves a pair of ^{3}He nuclei, each produced by the reaction above:

^{3}He + ^{3}He → ^{4}He + ^{1}H + ^{1}H

This sequence of reactions (called the *proton-proton* or *p-p* chain of reactions) is shown schematically in Fig. 13-10. Because *two* ^{3}He nuclei are involved in the final reaction in the chain, the two preceding reactions that lead to ^{3}He are shown separately. This gives an accurate overall picture of the nuclei that participate. Notice that there are *six* different hydrogen nuclei that enter the reactions (indicated by the *inward* arrows). In addition there are *two* hydrogen nuclei that are emitted in the final reaction. Thus, the net result of the five reactions taken together is the conversion of four hydrogen nuclei into one helium nucleus:

4 ^{1}H → ^{4}He + energy

The energy released in this fusion process is approximately 26 MeV or 6.5 MeV per particle involved—a very high efficiency for producing nuclear energy.

In the Sun, the *proton-proton* reactions convert hydrogen at a rate of about 600 billion kilograms per second (6×10^{11} kg/s), which corresponds to a power level of 400 billion billion MW (4×10^{20} MW)! Although energy is produced in the Sun at a tremendous rate, the mass decrease that accompanies this energy production is only a tiny fraction of the Sun's mass. Thus, the Sun will continue to generate and radiate energy at the current rate for another few billion years without appreciable decrease in mass.

Additional Exercises

1. The conditions of temperature and pressure in the Sun's core are much higher than those that can be produced in the laboratory. Explain why this accounts for the fact that fusion reactions take place readily in the center of the Sun but cannot be observed on Earth.
2. The disc of the Earth intercepts approximately 4×10^{-10} of the energy radiated by the Sun. This solar power incident on the Earth is equivalent to the output of how many 1000-MW power plants? [Ans. 160 million.]

Index

Absolute temperature scale, 158
Absolute zero, 158
Acceleration, 10, 40, 55
   centripetal, 57, 73
   due to gravity, 52, 58
Adams, John Counch, 62
Adiabatic process, 165, 173
Alpha particles, 301, 375
Alternating current (AC), 140
Ampere (unit), 121
Anderson, Carl D., 285
Angstrom (unit), 304
Antiparticle, 286
Aperture, 233
Apogee, 65
Atomic bomb, 411
Atomic spectra, 303
Atomic structure, 300, 334

Balmer, Johann, 305
Balmer series, 305
Battery, 114
Becquerel, Henri, 301, 374
Bel (unit), 203
Bell, Alexander Graham, 203
Beta rays, 375
Biological effects of radiation, 392
Bohr, Niels, 303, 306
Boyle, Robert, 159
Boyle's law, 160
Breeder reactors, 102, 426

de Broglie, Louis Victor, 311
de Broglie wavelength, 312, 328

Calorie (unit), 108, 167
Cameras, 232
Cathode-ray tube, 201
Celsius temperature scale, 157
Chadwick, James, 376
Chain reaction, 410
Charles, Jacques, 161
Charles-Gay-Lussac law, 162
Chemical compounds, 339
Chemical reactions, 42, 87, 113, 352, 387
Coherence, 315
Color, 240, 249, 303
Compton, Arthur H., 385
Compton effect, 385
Copernicus, Nicolaus, 52
Coulomb (unit), 121
Coulomb's law, 145
Covalent binding, 342
Critical mass, 411
Crystals, 341, 349
Curie, Marie and Pierre, 374
Current, electric, 121
   alternating (AC), 140
   direct (DC), 140

Decibel (unit), 204
Diffraction, 243
Diode, 360

Direct current (DC), 140
Doppler, Christian Johann, 206
Doppler effect, 206
Dose, 393

Efficiency, 90
Einstein, Albert, 262, 268, 286, 298
Electrical charges, 114
Electrical conductors, 113, 121
Electrical insulators, 113
Electrical resistance, 123, 363
Electric current, 121
Electric field, 118
Electric generators, 142
Electric motors, 138
Electric power, 125, 146
Electrolyte, 114
Electromagnet, 134
Electromagnetic induction, 142
Electrons, 113, 331
   charge of, 121
   wave properties of, 313
Electronvolt (unit), 369
Ellipses, 62
Energy, 82
   chemical, 88
   conservation of, 84
   electrical, 87
   of fuels, 92
   geothermal, 103
   gravitational, 86
   kinetic, 84, 108
   mass, 284, 406
   nuclear, 88, 101, 285, 418
   photon, 299, 307, 370
   potential, 82
   resources, 96
   solar, 103
   sources of, 95
   thermal, 85, 151, 164
   usage of, 93
Ether, 261
Exclusion principle, 336, 364

Fahrenheit temperature scale, 157
Faraday, Michael, 143
Field, electric, 118
   forces due to, 134
   magnetic, 128
Film, 247
Fission, 102, 285, 408
f-Number (of lens), 232
Force, 23, 40, 55
   electric, 118, 135
   gravitational, 58, 74
   lines of, 118
   magnetic, 135
   nuclear, 88
Frequency, 187
   characteristic, 213
   fundamental, 198
   harmonic, 198
Frisch, Otto, 408
Fusion, 102, 428, 437

Galileo, 58
Gamma rays, 368, 375
Gargarin, Yuri, 37
Gas law, 163, 176
Gay-Lussac, Joseph Louis, 161
Goddard, Robert H., 29, 32, 37
Gravitation, 52, 58, 74
   force constant, 74

Hahn, Otto, 408
Half-life, 380
Halley, Edmund, 64
Halley's comet, 64
Harmonics, 198
Heat, 163
Heat of fusion, 167
Heat pump, 174
Heat of vaporization, 168
Heisenberg, Werner, 331
Hertz, Heinrich, 187
Hertz (unit), 187
Holography, 326
Horsepower (unit), 108
Hydrogen atom, 305
Hydrogen bomb, 429
Hydrogen bonds, 348

Ideal gas law, 163, 176
Image formation, 227
Index of refraction, 224, 303

Induction, electromagnetic, 142
Integrated circuits, 361
Interference, 197, 245, 312
Ion, 30, 340
Ionic binding, 340
Ionization, 383
Isothermal process, 166, 172
Isotopes, 376

Joule (unit), 107

Kepler, Johannes, 52, 62
Kinetic energy, 84, 108

Lasers, 316
Length contraction, 272
Lenses, 224, 252
Leverrier, Urbain, 62
Light, 59, 218, 263, 296, 303
Loudness, 203

Mach number, 194
Magnetic field, 128
    of Earth, 129
    produced by electric currents, 131
Magnetic poles, 128
Magnets, 127
Mass, 7, 76, 280
    and energy, 284
    increase with speed, 280, 292
Meitner, Lise, 408
Meltdown, 421
Metallic binding, 346
Metric system, 4
Microscope, 230
Mirrors, 219, 255
Moderator, 413
Molecules, 42, 87, 158, 167, 339
Momentum, 23, 42, 46
    conservation of, 28, 46, 292
Muons, 276
Musical sounds, 199

Neutron, 376, 410
Neutron activation analysis, 391
Newton, Isaac, 26, 58
Newton (unit), 48
Newton's law of gravitation, 61, 74

Newton's law of motion, 47
Noise, 205
Nuclear accidents, 423
Nuclear binding energy, 406
Nuclear bomb, 411, 429
Nuclear energy, 101, 285, 418, 431
Nuclear fission, 102, 285, 408, 437
Nuclear fusion, 102, 428
Nuclear masses, 404
Nuclear reactors, 101, 404, 412
Nuclear safety, 420
Nuclear weapons, 433
Nuclei, 302, 376, 404

Ohm, Georg Simon, 123
Ohm (unit), 124
Ohm's law, 123
Orbits, parking, 66
    of planets, 62
    of space craft, 65
    synchronous, 67, 75
Oscilloscope, 201
Overtones, 198

Pascal, Blaise, 153
Pascal's law, 153
Pauli, Wolfgang, 335
Pauli principle, 336
Perigee, 65
Period, 186
Periodic table of elements, 335
Photoelectric effect, 296
Photons, 298, 307
Planck, Max, 299
Planck's constant, 299, 331
Planetary orbits, 62
Plastics, 350
Plutonium, 426, 434
Polar binding, 347
Polymers, 352
Positron, 286
Potential difference, 117
Potential energy, 82
    electrical, 87
    gravitational, 86
Power, 90, 108
    electric, 125, 146
Powers-of-10, 15

Pressure, 152
    atmospheric, 153
Prism, 240
Probability, 331, 344
Proton-proton reaction chain, 438

Quanta, 298
Quantum number, 308
Quantum theory, 299, 329

Rad (unit), 394
Radiation, 368
Radiation applications, 389
Radiation damage, 395
Radiation dose, 393
Radiation effects, 383
Radiation exposure, 395, 420
Radioactivity, 301, 374, 400
Radioisotopes, 377, 424
Reflection, 219
Refraction, 220
Refrigeration cycle, 169
Relativity, 260
    general theory, 286
Rem (unit), 395
Resistance, 123, 363
Resonance, 213
Rocket propulsion, 22, 29, 42
Rutherford, Ernest, 301, 408

Satellites, 65
    synchronous, 67, 75
Semiconductors, 356
Shawlow, Arthur, 320
Shock waves, 208
Sonic boom, 209
Sound, 192
Space flight, 35, 65
Specific heat, 178
Spectra, 303
Speed, 8
    of camera lens, 232
    of film, 249
    of light, 263
    of sound, 194
    of waves, 187
States of matter, 166
Stimulated emission, 314

Strassman, Fritz, 408
Sublimation, 168
Superconductors, 363
Superposition, 199

Telescope, 231, 255
Temperature, 156
    scales for, 157
Thermal pollution, 418
Thermodynamic laws, 150
Thermonuclear reactions, 429
Time dilation, 269, 290
Time travel, 276
Townes, Charles, 320
Transformers, 147
Transistor, 361
Twin paradox, 279

Uncertainty principle, 330
Uranium, 285, 374, 378, 400, 407, 410, 412, 426, 434

Vectors, 12, 17
Velocity, 13, 266
Volt (unit), 117
Voltage, 116

Watt (unit), 91
Wavelength, 185, 311
Waves, 182
    compressional, 188
    electromagnetic, 221
    electron, 313
    infrasonic, 193
    light, 220, 298
    longitudinal, 188
    matter, 312
    shock, 208
    sound, 192
    standing, 196
    transverse, 188
    traveling, 184
    ultrasonic, 193
    water, 211
Weight, 7, 76
Work, 80

X rays, 368